Prospects and Challenges of Community-Based Tourism and Changing Demographics

Ishmael Mensah
University of Cape Coast, Ghana

Ewoenam Afenyo-Agbe
University of Cape Coast, Ghana

A volume in the Advances in Hospitality, Tourism, and the Services Industry (AHTSI) Book Series

IGI Global
PUBLISHER of TIMELY KNOWLEDGE

Published in the United States of America by
IGI Global
Business Science Reference (an imprint of IGI Global)
701 E. Chocolate Avenue
Hershey PA, USA 17033
Tel: 717-533-8845
Fax: 717-533-8661
E-mail: cust@igi-global.com
Web site: http://www.igi-global.com

Library of Congress Cataloging-in-Publication Data

Names: Mensah, Ishmael, editor. | Afenyo-Agbe, Ewoenam, 1985- editor.
Title: Prospects and challenges of community-based tourism and changing demographics / Ishmael Mensah, and Ewoenam Afenyo-Agbe, Editor.
Description: Hershey, PA : Business Science Reference (an imprint of IGI Global) [2021] | Includes bibliographical references and index. | Summary: "This book provide theoretical and empirical insights into contemporary issues relating to community-based tourism (CBT) and highlights critical lessons and trends in CBT from both established and new community-based tourism initiatives to inform the design, management, marketing and sustainability of community-based tourism projects"-- Provided by publisher.
Identifiers: LCCN 2021018255 (print) | LCCN 2021018256 (ebook) | ISBN 9781799873358 (hardcover) | ISBN 9781799873365 (paperback) | ISBN 9781799873372 (ebook)
Subjects: LCSH: Community-based tourism. | Community development. | Sustainable tourism.
Classification: LCC G156.5.C65 P76 2021 (print) | LCC G156.5.C65 (ebook) | DDC 910.68--dc23
LC record available at https://lccn.loc.gov/2021018255
LC ebook record available at https://lccn.loc.gov/2021018256

This book is published in the IGI Global book series Advances in Hospitality, Tourism, and the Services Industry (AHTSI) (ISSN: 2475-6547; eISSN: 2475-6555)

British Cataloguing in Publication Data
A Cataloguing in Publication record for this book is available from the British Library.

Advances in Hospitality, Tourism, and the Services Industry (AHTSI) Book Series

ISSN:2475-6547
EISSN:2475-6555

MISSION

Globally, the hospitality, travel, tourism, and services industries generate a significant percentage of revenue and represent a large portion of the business world. Even in tough economic times, these industries thrive as individuals continue to spend on leisure and recreation activities as well as services.

The Advances in Hospitality, Tourism, and the Services Industry (AHTSI) book series offers diverse publications relating to the management, promotion, and profitability of the leisure, recreation, and services industries. Highlighting current research pertaining to various topics within the realm of hospitality, travel, tourism, and services management, the titles found within the AHTSI book series are pertinent to the research and professional needs of managers, business practitioners, researchers, and upper-level students studying in the field.

COVERAGE

- Leisure & Business Travel
- Cruise Marketing and Sales
- Hotel Management
- Health and Wellness Tourism
- Casino Management
- Destination Marketing and Management
- Travel Agency Management
- Service Management
- Tourism and the Environment
- Service Design

IGI Global is currently accepting manuscripts for publication within this series. To submit a proposal for a volume in this series, please contact our Acquisition Editors at Acquisitions@igi-global.com or visit: http://www.igi-global.com/publish/.

The Advances in Hospitality, Tourism, and the Services Industry (AHTSI) Book Series (ISSN 2475-6547) is published by IGI Global, 701 E. Chocolate Avenue, Hershey, PA 17033-1240, USA, www.igi-global.com. This series is composed of titles available for purchase individually; each title is edited to be contextually exclusive from any other title within the series. For pricing and ordering information please visit http://www.igi-global.com/book-series/advances-hospitality-tourism-services-industry/121014. Postmaster: Send all address changes to above address.

Titles in this Series

For a list of additional titles in this series, please visit:
www.igi-global.com/book-series/advances-hospitality-tourism-services-industry/121014

Rebuilding and Restructuring the Tourism Industry Infusion of Happiness and Quality f Life
André Riani Costa Perinotto (Universidade Federal do Delta do Parnaíba, Brazil) Verônica Feder Mayer (Federal Fluminense University, Brazil) and Jakson Renner Rodrigues Soares (Universidade da Coruña, Spain & Universidade Estadual do Ceará, Brazil)
Business Science Reference • © 2021 • 330pp • H/C (ISBN: 9781799872399) • US $195.00

Handbook of Research on the Impacts and Implications of COVID-19 on the Tourism Industry
Mahmut Demir (Isparta University of Applied Sciences, Turkey) Ali Dalgıç (Isparta University of Applied Sciences, Turkey) and Fatma Doğanay Ergen (Isparta University of Applied Sciences, Turkey)
Business Science Reference • © 2021 • 906pp • H/C (ISBN: 9781799882312) • US $415.00

Socio-Economic Effects and Recovery Efforts for the Rental Industry Post-COVID-19 Strategies
Maximiliano Korstanje (University of Palermo, Argentina)
Business Science Reference • © 2021 • 281pp • H/C (ISBN: 9781799872870) • US $195.00

Risk, Crisis, and Disaster Management in Small and Medium-Sized Tourism Enterprises
Diego R. Toubes (University of Vigo, Spain) and Noelia Araújo-Vila (University of Vigo, Spain)
Business Science Reference • © 2021 • 308pp • H/C (ISBN: 9781799869962) • US $195.00

Resiliency Models and Addressing Future Risks for Family Firms in the Tourism Industry
Anita Zehrer (MCI The Entrepreneurial School, Austria) Gundula Glowka (MCI The Entrepreneurial School, Austria) Katrin Magdalena Schwaiger (MCI The Entrepreneurial School, Austria) and Victoria Ranacher-Lackner (MCI The Entrepreneurial School, Austria)
Business Science Reference • © 2021 • 361pp • H/C (ISBN: 9781799873525) • US $195.00

Strategic International Restaurant Development From Concept to Production
Angelo A. Camillo (Sonoma State University, USA)
Business Science Reference • © 2021 • 497pp • H/C (ISBN: 9781799843429) • US $195.00

IGI Global
PUBLISHER of TIMELY KNOWLEDGE

701 East Chocolate Avenue, Hershey, PA 17033, USA
Tel: 717-533-8845 x100 • Fax: 717-533-8661
E-Mail: cust@igi-global.com • www.igi-global.com

Table of Contents

Detailed Table of Contents

Though CBT is an attractive model for the attainment of both community development and environmental conservation, its implementation is fraught with challenges. This chapter examines the principles, forms, barriers, and challenges to CBT. The chapter further proposes a model for the achievement of sustainable CBT and draws implications from the barriers and benefits for the management of CBT. The chapter begins with an explanation of the community concept which is defined in the context of community-based tourism. It continues with a review of some models of community participation. The chapter further assesses the rationale for community participation and examines the forms of CBT. Though community participation remains a good model for tourism development at the community level, there are several limitations that make the benefits unachievable, especially in developing countries. The chapter ends by proposing a model of sustainable CBT by drawing implications for the management of CBT projects.

Preface

Community-based tourism (CBT) has been widely acclaimed as a model of tourism that is suitable for communities in developing countries. This is because it helps to improve rural economies, improve the livelihoods of community members, empower local communities, alleviate poverty and enhance environmental conservation. Unlike conventional mass tourism, CBT is owned and operated by local communities. This has resulted in a proliferation of CBT projects around the world over the years. Though CBT remains an attractive model of tourism development, the actual implementation is fraught with challenges. There are many operational, structural and cultural barriers that hinder the operations of CBT projects and prevent local communities from reaping the full benefits of tourism. These barriers include centralization of decision-making processes, external control of tourism development, lack of community involvement in decision-making, and a lack of local capacity to manage tourism projects.

Additionally, contemporary trends and forces are shaping CBT. Factors such as technological advances, changing demographics and the COVID-19 pandemic have implications for the operation and sustainability of CBT initiatives. In the face of these forces and trends, coupled with the challenges associated with the operationalization of CBT, governments, destination marketing organizations and managers of CBT projects must adopt appropriate policies and programmes that promote partnerships and collaborations, empowerment of local communities, adoption of technology and building of resilience.

Though issues relating to CBT have become topical in recent times, the experiences in different parts of the world have been different. The fortunes, challenges and impacts of CBT vary across different contexts. This book addresses the issues related to CBT from different perspectives. Cases are drawn from different countries including Ghana, Nigeria, Tanzania, Thailand, Pakistan, and Uganda. The book provides theoretical and empirical insights into the prospects, challenges and trends associated with CBT from a multidisciplinary global perspective. By highlighting the critical issues and trends in CBT from different contexts, this book seeks to generate further debates and shape the design, operations, management, and marketing of community-based tourism projects to ensure their sustainability. Lastly, the book provides students, academics and practitioners with a comprehensive knowledge of CBT.

ORGANIZATION OF BOOK

Introduction

Community-based tourism (CBT) remains an attractive model for improving the socio-economic conditions of communities through tourism. It is also a tool for fostering environmental conservation. However, studies have shown that while many communities have derived some socio-economic benefits from CBT, there is also evidence to show that the actual implementation of this model of tourism development is fraught with challenges. The anticipated benefits do not seem to reach ordinary community members. This is essentially as a result of certain operational, structural and cultural barriers that hinder genuine participation by community members. In this chapter, the community-based tourism and community development nexus is examined based on evidence from developing countries. The chapter further examines the implications of changing demographics and the COVID-19 pandemic for community-based tourism development and provides a broad outlook of the structure of the book.

Chapter 1

Though CBT is an attractive model for the attainment of both community development and environmental conservation, its implementation is fraught with challenges. This chapter examines the principles, forms, barriers and challenges to CBT. The paper further proposes a model for the achievement of sustainable CBT and draws implications from the barriers and benefits for the management of CBT. The chapter begins with an explanation of the community concept which is defined in the context of community based tourism. It continues with a review of some models of community participation. The chapter further assesses the rationale for community participation and examines the forms of CBT. Though community participation remains a good model for tourism development at the community level, there are several limitations that make the benefits unachievable, especially in developing countries. The chapter ends by proposing a model of sustainable CBT by drawing implications for the management of CBT projects.

Chapter 2

Many Nigerians are unemployed and live in extreme poverty. The Nigerian government has therefore identified tourism as one of the promising sectors in its efforts to diversify the economy from the oil sector. CBT which is a kind of tourism that encourages environmental, social, cultural and economic sustainability in communities could be a good model for addressing the socio-economic challenges confronting Nigeria. Based on literature derived from JSTOR, Google scholar and Emerald, this chapter draws implications from the inclusive pro-poor tourism planning in Uganda for CBT development in Nigeria.

Chapter 3

Uganda is among the most bio-diverse countries and a competitive wildlife-based tourism destination in the world. Community-based tourism approach has been adopted in the country's conservation areas as a strategy to ensure that local communities benefit and support wildlife conservation. This chapter analyses local communities' perceptions of conservation and the benefits they get from tourism in Queen Elizabeth Conservation Area. The study reveals that local communities were concerned about loss of protected resources and support their conservation irrespective of the benefits they get from tourism in the conservation area. There is the need to design conservation programmes that focus on local community-conservation-benefits nexus which take into consideration the perceived conservation values, strategies for benefit sharing and incorporation of indigenous knowledge systems.

Chapter 4

Participatory development has become an integral strategy for achieving sustainable tourism in view of the varying impacts brought by mass tourism. Community participation ensures local control of the pace of tourism development and the creation of a more personalised tourism product. However, some studies have stated tension between governments and locals, which has hindered tourism development. This study, thus, sought to explore the extent to which communities are involved in decision-making and how useful the outcome has been. To achieve the study objectives, two tourism sites in Ghana implemented based on participatory principles, namely, Kakum National Park, and Afadjato Mountain were compared. Results of the study suggest there is a fair degree of community involvement at the Gbledi Traditional Area (Afadjato) leading to satisfaction (smiling-faces), than Abrafo-Odumase (Kakum) which exhibited dissatisfaction (frowning-faces).The policy and research implications relate mainly to the challenge of achieving sustainable tourism development at the community level.

Chapter 5

Conventional tourism and its attendant challenges for nature and communities have necessitated the need for tourism to be more sustainable, equitable, and responsible, hence the introduction of community-based tourism (CBT). CBT was developed as a model and a viable instrument for poverty reduction, offering opportunities for conservation and rural economic development. However, despite the potential of CBT to generate welfare for communities, many initiatives have failed to deliver on their promises. Since tourism and justice issues cannot be separated from each other, concerns over the discussion of justice related issues have emerged as a community concept which requires critical investigation. This chapter presents a theoretical exploration of how justice is conceptualized, with an emphasis on distributive justice in the context of CBT, and how it is shaping the production of CBT. It also explores some critical approaches to tourism studies and practice, and how justice for local communities is conceptualized in CBT contexts.

Chapter 6

Cultural tourism is an increasingly visible trend in the tourism industry since last few years. The current chapter is one of the first academic studies to specifically analyze the possible impacts of cultural tourism on the local culture itself and the wellbeing of the host community while highlighting cultural conservation as a mediator. The study utilizes primary data collected from rural areas in the Himalayan Gilgit Baltistan region in Northern Pakistan. This region has experienced a significant rise in cultural tourism due to China-Pakistan Economic Corridor (CPEC). As a result, there has been a significant and visible development of touristic facilities in the region. Hence, there is a visible shift from earlier tourism, which was mostly linked to mountaineering adventure, to more relatively mass cultural tourism. The results suggest that cultural conservation serves as a partial mediator and that cultural tourism tends to positively and significantly influence the host community's wellbeing, and cultural conservation partially mediates this relationship.

Chapter 7

English has not only become a lingua franca for global communication, but it also carries a symbolic power that influences local people and stakeholders in community-based tourism (CBT) in Thailand. Thai villagers encounter communicative challenges due to their limited English proficiency. Thai-English speaking tour guides are valued, as they provide translation services on behalf of CBT members whose interactions with foreign tourists are minimal and marginalized. Losses in translation regarding local wisdom and heritage are inevitably unavoidable. This chapter addresses a conceptual model of crowdsourcing translation for CBT. The study hopes that the sociolinguistic and cultural gaps can be bridged for the enhancement of other CBT communities. Crowdsourcing in the form of collaborative translation and volunteering translators' efforts can potentially provide an alternative lens for possible translation scenarios. Tourism operators and stakeholders can apply these practices to assist the local people to promote lesser-known tourist attractions in remote areas in Thailand.

Chapter 8

The approaches to natural resources management have evolved. Disparities in their adoption are likely to produce a long-lasting negative impact on the resources and the livelihood security of the community depending on them. The use of geoinformation by the local community is a critical measure to the sustainability of its resources. Nonetheless, the application of geoinformation technologies to the Community-Based Natural Resources for the tourism industry is highly unknown. This chapter reviewed the application of geoinformation technology to the management of community-based natural resources in the Pangani District of northern Tanzania. It considers how geoinformation technology is used in the management of tourism activities for community development. Specifically, the paper discusses community developments resulting from that as well as challenges associated with the use of Geographical Information Systems and Remote Sensing technologies. The chapter concludes with key recommendations for improving those challenges.

Conclusion

At the community level, community-based tourism has been projected as a pathway for promoting more inclusive and sustainable tourism industry. In practice, there are mixed perspectives regarding whether community-based tourism is mere rhetoric or a plausible reality. This book contributes to the ongoing debate by providing both theoretical and empirical evidence from multidisciplinary perspectives. This concluding chapter provides a reflection on the main issues discussed in this book. It further examines the future outlook for CBT and the implications for sustainable CBT development.

Ishmael Mensah
University of Cape Coast, Ghana

Ewoenam Afenyo-Agbe
University of Cape Coast, Ghana

Acknowledgment

We wish to express our sincerest appreciation to some individuals and organizations without whose support this book would not have seen the light of day. First, we wish to thank our mentors and senior colleagues Prof. Laud A. Dei, Prof. Kofi Awusabo-Asare, Prof. Albert Abane, Prof. Francis E. Amuquandoh, Prof. Alex Boakye Asiedu and Prof. Oheneba Akyeampong for setting the pace in research on community-based tourism in Ghana and serving as sources of inspiration. We also wish to acknowledge all the authors for believing in this project and giving up valuable time to contribute to this book. We are indebted to all the anonymous reviewers for their valuable comments which helped in shaping the manuscripts. Our special thanks also go to the team at IGI Global, especially Eric Whalen and Katie McLoughlin, Assistant Development Editor for Book Development for their immense assistance and professionalism. Finally, we acknowledge Ms Eunice Sakyi Kyeremah for editorial assistance. You all have our heartfelt gratitude.

Introduction

Community-Based Tourism and Community Development: The Prospects, Challenges and Trends

There has been a growing interest among academics and practitioners regarding community-based tourism (CBT) since it emerged as an alternative to conventional tourism. This interest has been particularly significant in developing countries where natural and cultural resources for CBT abound. Community-based tourism has been presented as a form of tourism where the local community has substantial control over, and involvement in its development and management, and a major proportion of the benefits remain within the community' (WWF, 2001). However, for local communities to derive the desired outcomes from tourism development, their participation in tourism development is imperative (Blackstock, 2005). CBT involves a bottom-up participatory approach to tourism development. It is a form of tourism where decision-making and control of the pace of tourism development are vested in local communities. CBT also serves as a tool for the conservation of the environment, as the revenue derived from tourism provides an incentive for the conservation of natural and cultural resources by local communities. Thus, there is the recognition that CBT can create better linkages between conservation, economic development and sustainable development (Stronza & Godillo, 2008; West, 2006). In order to use tourism to stimulate growth, communities should ensure that it supports local economic development, promotes equitable participation by community members, is ecologically sustainable, reduces negative environmental impacts, conserves the culture of the community and educates visitors about culture and nature (Asker et al, 2010). However, the benefits of CBT do not only accrue to the local community, but also tourists who seek individualistic, participatory, sustainable, and distinctive spiritual and cultural features of the community (López-Guzmán, Borges, & Castillo-Canalejo, 2011).

Despite the importance of CBT for community development, it has been observed that policies and practices adopted by governments have concentrated power and resource control in the state. This erodes traditional tenure and resource management systems and creates tensions between the two systems (Kasanga & Kotey, 2001; Kuusaana & Gerber, 2015). Though CBT appears to be an attractive model for local community development, it appears the actual implementation is fraught with some challenges. This is further compounded by changing demographics, technological advancements and crises such as the COVID-19 pandemic that are driving change within the tourism industry Thus for the full benefits of CBT to be derived, CBT projects must take into consideration these factors. This chapter examines the prospects and challenges of CBT as a tool for community development. It further outlines and sets the tone for the subsequent chapters in the book.

THE PROSPECTS OF COMMUNITY-BASED TOURISM AS A TOOL FOR COMMUNITY DEVELOPMENT

Generally, CBT is perceived as a viable tool for community development by linking income generation, social inclusion, gender equity, and environmental sustainability (Burgos et al., 2017). However, the success of the CBT model hinges on the active and genuine participation of community members in the tourism development process. It is only through the active participation of communities that the benefits derived from tourism will trickle down directly to community members and accelerate the socio-economic development process. The prospects of CBT lies in its potential to create jobs and provide income-earning opportunities for local communities as well as generate revenue for infrastructural development in particular and socio-economic development as a whole. CBT provides direct employment opportunities like tour guiding, wildlife rangering, resource management, cultural performances, catering and housekeeping as well as indirect employment opportunities in areas such as agriculture, transportation and trade. These are jobs that do not usually require specialised training and therefore suit community members in especially rural communities in developing countries. For instance, at the Khama Rhino Sanctuary Trust (KRST), a community-based wildlife project near Serowe, Botswana, the community makes money from activities such as guided game drives, rhino and giraffe trails, birdwatching, nature walks and the provision of accommodation (Stone, & Stone, 2011).Also, Li (2006) found that in the Jiuzhaigou Biosphere Reserve which is a UNESCO World Heritage site situated in a valley in the Sichuan province of China, almost the entire community residents within the reserve had received economic benefits from tourism in a variety of ways. These included direct employment in the tourism industry, small business ownership or employment in related jobs. According to Li (2006), these benefits were a result of the replacement of traditional means of subsistence farming and hunting with tourism.

The revenue and income derived from CBT could be used to enhance local economic development through linkages with the local economy, support for local families and expenditure on local products and services (Lapeyre, 2010). In effect, CBT helps to inject more money into local economies and to ignite the multiplier effect of such revenues and incomes. It is in this regard that Nelson (2004) concludes that community-based ecotourism has benefited Ololosokwan, a village in Tanzania as both villagers and the village council have had more money, and this has created new development options.

Moreover, CBT provides much needed income for the development of infrastructure. According to Brunt and Courtney (1999), the development of CBT increases the number of facilities, roads, parks, and recreational and cultural attractions, which help to improve residents' quality of life and preserve their culture. In Sirigu, a village in the Upper East Region of Ghana, for instance, the community has received some benefits since the establishment of a community-based ecotourism project. Community members have benefited from a total of 10 toilet facilities and five boreholes from tourism proceeds while a community library and a school block were also donated by tourists (Manu & Kuuder, 2012). In a case study of the Chobe Enclave Conservation Trust (CECT) in Botswana, Stone (2015) reported that in 2008 the community purchased a total of five tractors with trailers and ploughing equipment to boost agriculture and two general shops to provide service to the community. Also, in 2009, the community decided to fund three mechanized corn grinding mills, a cement brick moulding project and five large size tents with chairs to be used during wedding and funeral ceremonies. In the same vein, Mensah (2017) found that one area that the Mesomagor community in Ghana had seen some tangible improvement since the advent of CBT development in the community was infrastructure, including a school, teachers' accommodation block, roads, a guest house, and a health post. These modest infrastructural projects do not only enhance the wellbeing of communities but also help alleviate poverty.

Several studies have shown that wide-scale community participation in community-based organisations has a significant effect on poverty alleviation (Yang & Hung, 2014). CBT which is based on the principle of community participation is therefore seen as a strategy for poverty alleviation in developing countries (Spenceley, 2008; Baktygulov & Raeva, 2010). CBT is usually people-centred and helps in improving community livelihoods by providing alternative sources of income. Given this, China, the most populous country in the world, has adopted CBT initiatives based on a model of tourist villages in the countryside, to help secure and improve people's livelihoods (Yang, 2015).

CHALLENGES OF COMMUNITY-BASED TOURISM AS A TOOL FOR COMMUNITY DEVELOPMENT

Many researchers have noted that community participation in sustainable tourism development in developing countries is fraught with many challenges (Marzuki, Hay, & James, 2012; Saufi, O'Brien, & Wilkins, 2014). Hence, the expectation that local communities will benefit from CBT has not always been the case. In many protected areas in developing countries, lack of local participation is a common flaw in ecotourism projects (Bruyere, Beh, & Lelengula, 2009; Lai & Nepal, 2006; Xu et al., 2009). Several studies indicate that, despite host residents' interest in participating in the tourism development process, they have simply not been given adequate opportunities or lack the knowledge and understanding to meaningfully do so (Saufi, O'Brien, & Wilkins, 2014)

A study by Bello, Lovelock and Carr (2017) in Malawi revealed some significant limiting factors affecting community participation in ecotourism, including apathy, inadequate financial resources, inadequate information, low education levels, unfair distribution of benefits, inadequate trained human resources, a centralised public administrative system, lack of coordination and human-wildlife conflicts. In a related study by Chili and Ngxongo (2017) at the Didima Resort in Bergville, South Africa, the Umhlwazini community highlighted a lack of information and awareness, corruption, monopolising of government resources, and a lack of dynamic leadership as some of the most significant challenges confronting them. Cole (2006) in a study on sustainable tourism of a community in Eastern Indonesia found that the local people did not only lack information about how to participate but also knowledge about development matters that are required for meaningful participation. In a related study on the district of Shiraz in Iran, Aref (2011) found that centralization of governance, was the major issue as community members were hardly involved in decision-making on tourism. There was also lack of resources, knowledge, skills and education and a lack of a sense of ownership.

Regarding the Chobe Enclave Conservation Trust (CECT) in Botswana, Stone (2015) demonstrated that the adoption of CBT under the principles of community-based natural resource management (CBNRM) may not always bring the desired outcomes since there are operational, structural and cultural limits that impede community participation. The inability of CBT to deliver on the proposed benefits is mainly a result of lack of community participation. Tasci et al. (2013) argued that many CBT initiatives had failed as a result of the lack of community involvement and participation, lack of a sense of communal ownership of the initiative and a lack of local financial resources. Also, communities do not benefit from CBT because their needs are not factored into such projects as lamented by Mbaiwa (2005) in a study on the Okavango Delta in Botswana. According to him, the tourism industry in the

Okavango Delta did not consider the socio-cultural, economic and environmental needs of the host economy. Instead, it was characterised by the marginalisation of local companies and investors, leakages and repatriation of tourism revenue from Botswana to developed countries, failure to employ tourism to promote rural development and poverty alleviation, and, the failure to consider local environmental regulations in the conservation of the Okavango Delta as a natural ecosystem.

It appears the socio-political structure of local communities in developing countries engender elite control and foreign ownership of the tourism industry. Kontogeorgopoulos (2005), argues that due to the cultural, social and political circumstances of southern Thailand, tourism would not thrive without foreign ownership. The reality is that many CBT projects require significant investment in infrastructure development which has resulted in high dependency (Stronza & Gordillo, 2008) as communities do not have the financial resources to undertake such projects. Community members have at best a limited role in tourism planning as the main actors are external agents, tourism businesses and government agencies (Moscardo (2011). In a study by Chili and Ngxongo (2017) in Didima Resort and Umhlwazini Community in the KwaZulu-Natal province of South Africa, virtually all the respondents agreed that the community was not involved in the tourism decision-making process. This is against the background that many CBT projects in developing countries claim to involve community members in the tourism development process. However, in reality, there is no genuine participation by community members or participation is characterized by inducements or some degree of tokenism. This is what Tosun (2006) describes as induced participation which according to him, is typical of the situation in most developing countries. Here, host communities simply endorse decisions made by external agencies and government institutions regarding tourism development rather than being allowed to make their own decisions. In reality, ownership and control of CBT projects are often vested in local elites and external agencies. According to Marzuki et al. (2012), in most developing countries, tourism activities are usually dominated by very few elites in the communities, resulting in a lack of meaningful communal ownership. Marzuki (2008) also asserts that the highly monopolised municipal and government processes are operational barriers that reduce the involvement of the local community to merely informing them on future projects and the exchange of information demanded by the community.

Goodwin (2006) reports on an evaluation of CBT projects around the world which was carried out by the International Centre for Responsible Tourism at Leeds Metropolitan University, UK. The report shows that CBT projects failed because of the following;

- A lack of understanding of the need for local people to engage in commercial activities;
- A lack of engagement with the private sector enterprises including travel agents, tour operators and hoteliers;
- Poor location of CBT projects which does not allow poor people in communities to benefit from the tourists;
- Inadequacy of appropriate tourism facilities for generating income; and
- Overreliance on tourist revenue for conservation at the expense of local communities' benefit.

CHANGING DEMOGRAPHICS AND COVID-19

Changing Demographics

One of the most enduring developments in any human society is demographic change. One industry that has been impacted significantly by such demographic changes, is the tourism industry. The revolution in human reproduction in contemporary times has implications for travel, tourism and hospitality (Webster, 2019). Much of the demographic changes have occurred in developed countries which serve as source markets for developing countries and CBT destinations. According to a report by the German Federal Ministry of Economics and Technology [FMET] (2009), the major socio-demographic changes that may impact the development pattern of leisure and tourism are higher life expectancy, healthier ageing population, smaller family size, higher percentage of the population with high level of education and population migration. These socio-demographic changes affect people's participation in tourism and their travel behaviour specifically the decision to participate in tourist activities, the type of tourist who will travel to particular destinations, the type of destination chosen, the length of stay at the destination and how much to spend (see Glover & Prideaux, 2009).

A recent report on demographic change and tourism by the World Tourism Organization & European Travel Commission (2010) has indicated that the demographic changes in modern societies have created new challenges for the development of tourism due to their implications on tourism participation behaviour. However, these demographic changes could also present opportunities to destinations. For instance, it has been observed that tourist behaviour has begun to change towards a desire for authentic local culture. Therefore, they do not only intend to travel to new destinations, but also to explore the culture, ethics, ancient indigenous remains, and local history of tourist destinations (Lo & Janta, 2020). This is an opportunity for CBT development as communities could benefit from packaging and promoting their unique and authentic cultures that will appeal to tourists.

A significant demographic change is the growing influence of millennials who now account for more than one-third of the US labour force (Desilver, 2019). Millennials who were born between 1981 and 1996 are avid users of the Internet, mobile devices, and social media. As such they are sometimes referred to as digital natives. Thus, to gain a fair share of this market, destinations, including rural communities must employ ICT in the marketing, management and delivery of CBT products. Technology is one of the forces that is undeniably driving change within the tourism industry. CBT destinations cannot be insulated from this wave of change. CBT destinations should employ ICT for activities such as resource management, customer database management, marketing communications and enhancing the entire visitor experience.

Covid-19

A major crisis that is shaping all aspects of life is the COVID-19 pandemic. Tourism is one of the industries that have been most adversely impacted by the pandemic(Chang et al., 2020). According to the United Nations Conference on Trade and Development [UNCTAD] (2021), due to COVID-19, inbound tourist arrivals declined by 74%, representing 1 billion trips, between January and December 2020 and this resulted in a drop in global tourism receipts by $1 trillion. COVID-19 has also affected conservation and the livelihoods of communities due to a general decline in travel to developing countries where these CBT projects are found.

COVID-19 has therefore become a major impediment to conservation which is a *sin qua non* for CBT development. Social distance restrictions, lockdowns and the general loss of livelihoods will drive community members to engage in poaching and other activities in order to survive. Also, declines in tourism revenues and incomes are a disincentive for conservation. In some communities, people who work directly in the tourism industry have either been rendered unemployed or have seen some drastic cuts in their incomes. Therefore, they have had to resort to actions like encroachments, poaching and the consumption of bushmeat, and trophy hunting in order for them to survive.

Studies have shown that in India, Kenya, South Africa, and Uganda the poaching of wildlife has more than doubled during lockdowns (Athumani, 2020; Badola, 2020).

According to Lindsey et al. (2020), COVID-19 can increase food insecurity and poverty. This is because it threatens conservation, leading to the loss of tourism and trophy hunting revenue. This can intensify the risks associated with conservation and lead to land conversions and loss of stewardship in wildlife conservation.

A drop in tourism revenue occasioned by a general decline in tourist arrivals during the COVID-19 pandemic period has seriously impacted the livelihoods of, especially communities that depend on tourism. In Botswana, for instance, COVID-19 has had an impact on community development through abrupt losses of employment and income (Stone, Stone, Mogomotsi & Mogomotsi, 2021). Thus, the current unprecedented COVID-19 pandemic calls for a review of how conservation in rural areas can remain viable despite such disruptions, and how communities can continue to derive benefits from conservation activities (Kaelo et al. 2020).

STRUCTURE OF THE BOOK

The book is composed of eight chapters. In the Introduction chapter, Ishmael Mensah and Ewoenam Afenyo-Agbe set the tone for the book by introducing readers to the central theme of the book i.e., "community-based tourism and community development: the prospects, challenges and trends". Ewoenam Afenyo-Agbe and Ishmael Mensah further examine the theoretical and empirical basis of the book in "Principles, benefits and barriers to community-based tourism: Implications for management" in Chapter 1. This is followed by the implications of inclusive pro-poor tourism planning in Uganda for strategic planning for CBT development in Nigeria by Olukemi Deborah Fagbolu in Chapter 2. Jim Ayorekire, Francis Mugizi, Joseph Obua and Grace Ampaire then provide a case study of community-based tourism and local people's perceptions of conservation from the Queen Elizabeth Conservation Area, Uganda in Chapter 3. In chapter 4, Ama Birch-Freeman, Ishmael Mensah and Kwabena Barima Antwi compare two scenarios of impacts of CBT on communities, Abrafo-Odumase and Gbledi Traditional Area in Ghana. The issue of distributive justice in CBT is discussed by Victor M. Agbo in chapter 5. This is followed by an analysis of the links between cultural tourism and the wellbeing of local citizens and the mediating role of cultural conservation in Northern Pakistan by Saranjam Baig, Arifa Shabbnum, Ahmad Arslan in Chapter 6.

In Chapter 7, Narongdej Phanthaphoommee and Singhanat Nomnian, examine the application of crowdsourcing translation for CBT Empowerment in Thailand. Norbert John Ngowi then examines the application of Geoinformation Technology to the management of community-based natural resources for tourism development in Northern Tanzania in Chapter 8. In the Conclusion chapter, Ishmael Mensah reflects on the issues raised in the various chapters and provides some suggestions for the sustainability of CBT projects in contemporary times.

CONCLUSION

There is ample evidence to show that CBT has a lot of prospects especially for rural communities in developing countries. The injection of tourism revenue and income into the communities helps to generate employment for local people, improve infrastructure and enhance the livelihoods of the people. However, studies have also shown that many communities have not derived the full benefits of CBT due to operational, structural and cultural barriers to their participation in the tourism development process. Surmounting these barriers requires stakeholder collaboration in tourism development at the local level. For communities to reap the full benefits of CBT, their active participation in decision-making, planning and implementation is crucial. It is also important for CBT projects to take into consideration certain forces driving change within the industry Factors such as changing demographics, pandemics and technological changes have implications for CBT development. Throughout this book, these issues will be examined from different perspectives and in different contexts.

REFERENCES

Aref, F. (2011). Barriers to community capacity building for tourism development in communities in Shiraz, Iran. *Journal of Sustainable Tourism*, *19*(3), 347–359. doi:10.1080/09669582.2010.517314

Asker, S., Boronyak, L., Carrard, N., & Paddon, P. (2010). *Effective Community Based Tourism: Best Practice Manual*. Sustainable Tourism Cooperative Research Center.

Athumani, H. (2020, July 7). *Wildlife poaching doubles in Uganda during COVID-19 lockdown*. Voice of America. https://www.voanews.com/covid-19-pandemic/wildlifepoaching-doubles-uganda-during-covid-19-lockdown

Badola, S. (2020). *Indian wildlife amidst the COVID-19 crisis: An analysis of status of poaching and illegal wildlife trade*. TRAFFIC-WWF. Available at:https://www.traffic.org/site/assets/files/12885/wildlife-amidst-covid-19-india-web.pdf

Baktygulov, S., & Raeva, R. (2010). *Creating Value for All: Community-Based Tourism*. United Nations Development Programme.

Bello, F. G., Lovelock, B., & Carr, N. (2017). Constraints of community participation in protected area-based tourism planning: The case of Malawi. *Journal of Ecotourism*, *16*(2), 131–151. doi:10.1080/14724049.2016.1251444

Blackstock, K. (2005). Acritical look at community based tourism. *Community Development Journal: An International Forum*, *40*(1), 39–49. doi:10.1093/cdj/bsi005

Brunt, P., & Courtney, P. (1999). Host perceptions of sociocultural impacts. *Annals of Tourism Research*, *26*(3), 493–515. doi:10.1016/S0160-7383(99)00003-1

Bruyere, B. L., Beh, A. W., & Lelengula, G. (2009). Differences in perceptions of communication, tourism benefits, and management issues in a protected area of rural Kenya. *Environmental Management*, *1*(43), 49–59. doi:10.100700267-008-9190-7 PMID:18777191

Burgos, A., & Mertens, F. (2017). Participatory management of community-based tourism: A network perspective. *Community Development (Columbus, Ohio)*, *48*(4), 546–565. doi:10.1080/15575330.2017.1344996

Chang, C. L., McAleer, M., & Ramos, V. (2020). A charter for sustainable tourism after COVID-19. *Sustainability*, *12*(9), 3674. doi:10.3390u12093671

Chili, N. S., & Ngxongo, N. A. (2017). Challenges to active community involvement in tourism development at Didima Resort–a case study of Umhlwazini community in Bergville. *African Journal of Hospitality, Tourism and Leisure*, *6*(2), 1–15.

Cole, S. (2006). Information and empowerment: The keys to achieving sustainable tourism. *Journal of Sustainable Tourism*, *14*(6), 629–644. doi:10.2167/jost607.0

Desilver, D. (2019, August 29). *10 facts about American workers*. Pew Research Center.

Federal Ministry of Economics and Technology. (2009). *The Impact Of Demographic Change On Tourism And Conclusions For Tourism Policy*. Federal Ministry of Economics and Technology.

Glover, P., & Prideaux, B. (2009). Implications of population ageing for the development of tourism products and destinations. *Journal of Vacation Marketing, 15*, 25-37.

Goodwin, H. (2006). *Community-based tourism failing to deliver? ID 21 Insights, (62).*

Kaelo, D., Sopia, D., Bell, D., Diggle, R., & Nelson, F. (2020). *From crisis to solutions for communities and African conservation* (commentary). https://news.mongabay.com/by/dickson-kaelo-daniel-sopia-damian-bell-richard-diggle-andfred-nelson/

Kasanga, K., & Kotey, N. A. (2001). Land management in Ghana: Building on tradition and modernity. *The Journal of the Bertrand Russell*, 1–42. Retrieved from https://www.eldis.org/vfile/upload/1/document/0708/DOC5021.pdf

Kontogeorgopoulos, N. (2005). Community-based ecotourism in Phuket and Ao Phangnga, Thailand: Partial victories and bittersweet remedies. *Journal of Sustainable Tourism*, *13*(1), 4–23. doi:10.1080/17501220508668470

Kuusaana, E., & Gerber, N. (2015). Institutional synergies in customary land markets – selected case studies of large-scale land acquisitions (LSLAs) in Ghana. *Land (Basel)*, *4*(3), 842–868. doi:10.3390/land4030842

Lai, P.-O., & Nepal, S. K. (2006). Local perspectives of ecotourism development in Tawushan nature reserve, Taiwan. *Tourism Management*, *27*(6), 1117–1129. doi:10.1016/j.tourman.2005.11.010

Lapeyre, R. (2010). Community-based tourism as a sustainable solution to maximise impacts locally? The Tsiseb Conservancy case, Namibia. *Development Southern Africa*, *27*(5), 757–772. doi:10.1080/0376835X.2010.522837

Lindsey, P., Allan, J., Brehony, P., Dickman, A., Robson, A., Begg, C., Bhammar, H., Blanken, L., Breuer, T., Fitzgerald, K., Flyman, M., Gandiwa, P., Giva, N., Kaelo, D., Nampindo, S., Nyambe, N., Steiner, K., Parker, A., Roe, D., ... Tyrrell, P. (2020). Conserving Africa's wildlife and wildlands through the COVID-19 crisis and beyond. *Nature Ecology & Evolution*, *4*(10), 1300–1310. doi:10.103841559-020-1275-6 PMID:32728187

Lo, Y. C., & Janta, P. (2020). Resident's perspective on developing community-based tourism–a qualitative study of Muen Ngoen Kong Community, Chiang Mai, Thailand. *Frontiers in Psychology*, *11*, 1493. doi:10.3389/fpsyg.2020.01493 PMID:32848976

López-Guzmán, T., Borges, O., & Castillo-Canalejo, A. M. (2011). Community-Based Tourism in Cape Verde- A Case Study. *Tourism Management*, *17*, 35–44.

Manu, I., & Kuuder, C. J. W. (2012). Community-based ecotourism and livelihood enhancement in Sirigu, Ghana. *International Journal of Humanities and Social Science, 2*(18), Marzuki, A. (2008). Decision making and community participation: A case study of the tourism industry in Langkawi. *Turizam: Znanstveno-stručničasopis, 56*(3), 227-241.

Marzuki, A., Hay, I., & James, J. (2012). Public participation shortcomings in tourism planning: The case of the Langkawi Islands, Malaysia. *Journal of Sustainable Tourism*, *20*(4), 585–602. doi:10.1080/09669582.2011.638384

Marzuki, A., Hay, I., & James, J. (2012). Public participation shortcomings in tourism planning: The case of the Langkawi Islands, Malaysia. *Journal of Sustainable Tourism*, *20*(4), 585–602. doi:10.1080/09669582.2011.638384

Mbaiwa, J. E. (2005). The Problems and Prospects of Sustainable Tourism Development in the Okavango Delta, Botswana. *Journal of Sustainable Tourism*, *13*(3), 203–227. doi:10.1080/01434630508668554

Mensah, I. (2017). Benefits and challenges of community based ecotourism in park-fringe communities, the case of *Mesomagor* of Kakum National Park, Ghana. *Tourism Review International*, *21*(1), 81–98. doi:10.3727/154427217X14866652018947

Moscardo, G. (2011). Exploring social representations of tourism planning: Issues for governance. *Journal of Sustainable Tourism*, *19*(4-5), 423–436. doi:10.1080/09669582.2011.558625

Nelson, F. (2004). *The Evolution and Impacts of Community-Based Ecotourism in Northern Tanzania* (No. 131). London: International Institute for Environment and Development.

Saufi, A., O'Brien, D., & Wilkins, H. (2014). Inhibitors to host community participation in sustainable tourism development in developing countries. *Journal of Sustainable Tourism*, *22*(5), 801–820. doi:10.1080/09669582.2013.861468

Spenceley, A. (2008). Local impacts of community-based tourism in Southern Africa. In A. Spenceley (Ed.), *Responsible Tourism. Critical Issues for Conservation and Development* (pp. 285–304). Earthscan.

Stone, L. S., & Stone, T. M. (2011). Community-based tourism enterprises: Challenges and prospects for community participation; Khama Rhino Sanctuary Trust, Botswana. *Journal of Sustainable Tourism*, *19*(1), 97–114. doi:10.1080/09669582.2010.508527

Stone, L. S., Stone, M., Mogomotsi, P., & Mogomotsi, G. (2021). The Impacts of Covid-19 on nature-based tourism in Botswana: Implications for community development. *Tourism Review International*, *25*(2-3), 263–278. doi:10.3727/154427221X16098837279958

Stone, M. T. (2015). Community empowerment through community-based tourism: the case of Chobe Enclave Conservation Trust in Botswana. In *Institutional arrangements for conservation, development and tourism in Eastern and Southern Africa* (pp. 81–100). Springer. doi:10.1007/978-94-017-9529-6_5

Stronza, A., & Gordillo, J. (2008). Community views of ecotourism. *Annals of Tourism Research*, *35*(2), 448–468. doi:10.1016/j.annals.2008.01.002

Tasci, A. D., Semrad, K. J., & Yilmaz, S. S. (2013). *Community Based Tourism Finding the Equilibrium in the COMCEC Context Setting the Pathway for the Future*. COMCEC Coordination Office.

Tosun, C. (2006). Expected nature of community participation in tourism development. *Tourism Management, 27*(3), 493–504. doi:10.1016/j.tourman.2004.12.004

United Nations Conference on Trade and Development. (2021). *COVID-19 and Tourism An Update: Assessing the economic consequences.* Retrieved 5th August, 2021 from https://unctad.org/system/files/official-document/ditcinf2021d3_en_0.pdf

Webster, C. (2019). Halfway there: The transition from 1968 to 2068 in tourism and hospitality. *Zeitschrift für Tourismuswissenschaft, 11*(1), 5–23. doi:10.1515/tw-2019-0002

West, P. (2006). *Conservation is our governance now: The politics of ecology in Papua New Guinea.* Duke University Press.

World Tourism Organization & European Travel Commission. (2010). *Demographic Change and Tourism.* World Tourism Organization & European Travel Commission.

World Wildlife Fund. (2001). *Guidelines For Community-Based Ecotourism Development.* WWF International.

Xu, J., Lu, Y., Chen, L., & Liu, Y. (2009). Contribution of tourism development to protected area management: Local stakeholder perspectives. *International Journal of Sustainable Development and World Ecology, 1*(1), 30–36. doi:10.1080/13504500902757189

Yang, L. (2015). *Tourism Development and Poverty Alleviation.* Available online: http://scholarworks.umass.edu/cgi/viewcontent.cgi?article=1144&context=ttra

Yang, X., & Hung, K. (2014). Poverty alleviation via tourism cooperatives in China: The story of Yuhu. *International Journal of Contemporary Hospitality Management, 26*(6), 879–906. doi:10.1108/IJCHM-02-2013-0085

Chapter 1
Principles, Benefits, and Barriers to Community-Based Tourism:
Implications for Management

Ewoenam Afenyo-Agbe
https://orcid.org/0000-0001-6470-6844
University of Cape Coast, Ghana

Ishmael Mensah
https://orcid.org/0000-0002-8068-5446
University of Cape Coast, Ghana

ABSTRACT

Though CBT is an attractive model for the attainment of both community development and environmental conservation, its implementation is fraught with challenges. This chapter examines the principles, forms, barriers, and challenges to CBT. The chapter further proposes a model for the achievement of sustainable CBT and draws implications from the barriers and benefits for the management of CBT. The chapter begins with an explanation of the community concept which is defined in the context of community-based tourism. It continues with a review of some models of community participation. The chapter further assesses the rationale for community participation and examines the forms of CBT. Though community participation remains a good model for tourism development at the community level, there are several limitations that make the benefits unachievable, especially in developing countries. The chapter ends by proposing a model of sustainable CBT by drawing implications for the management of CBT projects.

DOI: 10.4018/978-1-7998-7335-8.ch001

INTRODUCTION

The Community-based approach to tourism development has gained currency, especially in the developing world. This is because it is considered a viable tourism development option for the attainment of both socio-economic development and environmental conservation (Lee & Jan 2019). Unlike other conventional models of tourism development, it places the power of control over tourism development in the hands of local people thereby ensuring that the level of tourism development reflects the aspirations of local people (Murphy, 1985). However, it is important to note that the level of community involvement in tourism development tends to vary from non-involvement to citizen control. Despite the attractiveness of community-based tourism (CBT) as a viable model for tourism development at the community level, its practical implementation is often fraught with challenges which Tosun (2000) broadly classifies into operational barriers, structural barriers and cultural barriers. Generally, the inadequate capacity of local people in terms of knowledge, skills, funding and other resources to undertake tourism development is seen as the major barrier to CBT development. In spite of these challenges, CBT has proven to be more beneficial to several communities in the developing world. However, for the CBT projects to be truly participatory and beneficial to local communities, policymakers and managers must institute appropriate mechanisms for ensuring fairness and transparency as well as building resilience to ensure sustainable tourism development. This chapter begins with a discussion on the community-based approach to tourism development, including the community concept, community-based tourism, community participation, the rationale for community participation and forms of community participation. This is followed by the barriers to community participation and the benefits of CBT. The chapter ends with some implications for the management of CBT projects.

THE COMMUNITY-BASED APPROACH TO TOURISM DEVELOPMENT

The Community Concept

The community concept has been defined, severally, usually in terms of space, social organization, as well as people and social interactions (Harwood, 2010). In terms of space, a community is defined as "a group of people, often living in the same geographical area, who identify themselves as belonging to the same group" (Sproule, 1996, p. 236). A community as a social organization comprises a social system made up of groups and institutions. When a community is defined in terms of people and their social interactions, their common identity, shared set of beliefs and practices are highlighted. James (2006) has categorized communities based on three basic standpoints, namely, place, identity-based and organizationally-based. According to him, "communities of place" ranges from the local neighbourhood, village, town, city, region up to the planet as a whole; "identity-based communities" groups individuals based on their ethnicity, religion or culture and "organizationally-based communities" are made up of different kinds of informal or formal associations and networks.

However, it is important to note that a group of people merely occupying a geographical space does not constitute a community in the context of Community-Based Tourism (CBT). There is a need for cohesion and shared aspirations among people in a community if their common problems are to be resolved for their progress and well-being. Though people belonging to a community have shared values and culture, there is hardly any homogeneous community. Indeed, all communities are heterogeneous, and community members could be stratified based on their gender, ages, levels of education, religion, marital status, income and social status. In the context of CBT, a community is defined as comprising of a group of people with shared values, living in a defined geographical area, who interact to achieve shared goals.

Community-Based Tourism

Brohman (1996) offers a very comprehensive definition of CBT: describing it as development that seeks to strengthen institutions designed to enhance local participation and promote the economic, social and cultural well-being of the popular majority. It also seeks to strike a balanced and harmonious approach to development that would stress considerations such as the compatibility of various forms of development with other components of the local economy; the quality of development, both culturally and environmentally, and the divergent needs, interests and potentials of the community and its inhabitants.

There are however contestations regarding whether CBT is a type of tourism or an approach to tourism development. The chapter aligns with the latter because the main focus of CBT is the promotion of community participation in the tourism development process which could be applied to any form of tourism. When CBT is perceived as an approach or a framework, it can guide the development of other forms of tourism. The existence of concepts and related practices such as Community-based Ecotourism, Community-Based Heritage Tourism, Community-Based Cultural Tourism is an attestation of this position (see Jones, 2005; Kontogeorgopoulos, 2005; Salazar, 2012; Schmidt, 2017).

Regardless of the debates on what CBT is, there are common issues that cut across the various definitions; community ownership, participation, empowerment, development and sustainability. These reflect the goals, aspirations and principles of this approach to tourism development. As suggested by Suansri (2003) and ASEAN (2016), the development of any form of tourism with the CBT approach should be guided by the following principles:

- Recognition, support and promotion of community ownership of tourism;
- Establishment of partnerships with relevant stakeholders;
- Promotion of community pride and cross-cultural learning;
- Improvement in the quality of life of community members;
- Ensuring environmental and cultural sustainability;
- Improvement in social well-being and maintenance of human dignity,
- Inclusion of a fair and transparent benefit-sharing mechanism,
- Enhancement of linkages to local and regional economies,
- Improvement in the quality of visitor experiences by strengthening meaningful host and guest interaction; and
- Attainment of financial self-sufficiency of community-based initiatives.

Community Participation

Participation connotes inclusiveness. It is the involvement of all identifiable stakeholders in policy formation, planning and development of programmes and activities that would help change their communities. Arnstein (1969) states that the purpose of participation is the redistribution of power, to enable society to fairly redistribute the benefits and costs of development. According to Cornell (1997:250) participation is "not only about achieving efficient and equitable distribution of material resources: it is also about the sharing of knowledge and the transformation of the process of learning itself in the service of people's self-development". With community participation, citizens are given the opportunity to voluntarily and democratically partake in development processes. Community participation is characterized by a bottom-up approach to development as opposed to imposition from central authorities. The call for community participation is based on the assumption that participation lessens opposition to development, minimises negative impacts and revitalises economies (Hardy et al., 2002). However, in reality, there are different degrees of community participation depending on the extent to which communities are involved in developmental processes. This is explained in Table 1 (Arnstein, 1969; Pretty, 1995; Tosun, 2006). Based on these models, community participation usually follows a spectrum, ranging from passive to active community participation and community ownership. Thus, community participation could evolve from a tokenistic and therapeutic manipulation where community participation is a facade to a situation where there is real empowerment and where resource control and decision making are transferred to the local community. Arnstein (1969) has conceptualized the degree of community participation in the form of a ladder or typology of citizen participation. The ladder comprises eight levels and is classified into three categories relative to the authenticity of citizen participation. The lowest end of the ladder represents non-participation, the middle of the ladder indicates degrees of citizen tokenism and the highest end refers to degrees of citizen power.

Pretty (1995) on the other hand proposed seven levels of participation, namely manipulative participation, passive participation, participation by consultation, participation for material incentives, functional participation, interactive participation and self-mobilization. Pretty (1995) refers to "manipulative participation" as a form of participation when actual power lies within groups beyond the local community. The opposite of this is "interactive participation" and "self-mobilisation" when local people actively get involved in decision-making.

In the context of community participation in tourism development Tosun's (1999) also classifies community participation into three namely; spontaneous participation, coercive participation and induced participation. Spontaneous participation, which is the highest level of community participation is the ideal form of participation that can generate trust, ownership and social capital among the residents.

The Rationale for Community Participation in Tourism Development

Several arguments have been made in favour of community participation in CBT. These include enhancing the benefits derived from tourism by local people; readjusting the balance of power for greater local control over tourism development; empowerment of local people; enhancing the tourist experience and image of the destination; resolving potential conflicts in tourism development and ultimately ensuring sustainable tourism development.

Ensuring Community Support for CBT to Enhance the Tourist Experience

The participation of local people in the tourism planning and development process is essential for enhancing the quality of the tourist product and the quality of experience gained by tourists. This is because, with CBT, the community constitutes the resource for tourism. The land, natural, historical heritages and cultural resources are all part of the local community. Thus, the use of these resources for tourism affects the local people. Given this, their input is necessary for delivering quality

Table 1. Typologies of community participation

<table>
<tr><th colspan="2">Arnstein's Typologies of Community Participation</th><th>Pretty's Typologies of Community Participation</th><th>Tosun Typologies of Community Participation in Tourism</th></tr>
<tr><td rowspan="3">Degrees of citizen power</td><td>Citizen control
Residents are in full charge of policy and managerial aspects and can negotiate the conditions under which outsiders may change them.</td><td>Self-mobilization
People participate by taking initiatives independently of external institutions to change systems.</td><td rowspan="3">Spontaneous participation
Bottom-up; active par.; direct participation; par. in decision making, authentic participation; self-planning</td></tr>
<tr><td>Delegated power
Officials give up at least some degree of control, management, decision-making authority, or funding to citizens. Locals now have significant control and can ensure that plans suit them.</td><td rowspan="2">Interactive
People participate in joint analysis, development of action plans and formation or strengthening of local institutions. Participation is seen as a right, not just the means to achieve project goals.</td></tr>
<tr><td>Partnership
Officials allow citizens to negotiate better deals, veto decisions, share funding, or put forward requests that are at least partially fulfilled. However, such power is not voluntarily shared by public institutions, but rather taken by the citizens through actions such as protests.</td></tr>
<tr><td rowspan="3">Degree of citizen tokenism</td><td>Placation
Local people are granted a limited degree of influence in the decision process and could advise or offer alternative plans. However, authority representatives hold the right to change or reject their proposals.</td><td>Functional
People may participate by forming groups to meet objectives related to the project which have already been determined by external agents. Participation is seen by external agencies as a means to achieve project goals, especially reduced costs.</td><td rowspan="3">Induced participation
Top-down; passive; formal; mostly indirect; degree of tokenism, manipulation; pseudo-participation; participation in the implementation and sharing benefits; choice between proposed alternatives and feedback.</td></tr>
<tr><td>Consulting
Participants are asked to provide feedback but there is no guarantee that their opinion will be considered or accepted. Such participation remains just a window-dressing ritual.</td><td>Material incentives
People participate by contributing resources, for example, labour, in return for food, cash or other material incentives.</td></tr>
<tr><td>Informing
There is a one-way flow of information from officials to citizens who simply accept the information with no channel provided for feedback and no power for negotiation.</td><td>Consultation
People participate by being consulted or by answering questions. External agents define problems.</td></tr>
<tr><td rowspan="2">Non-participation</td><td>Therapy
Community participation initiatives are staged to appease the anger or frustration of local people (cure) but not to deal with the real problem or consider their perspectives</td><td>Passive
People participate by being told what has been decided or has already happened.</td><td rowspan="2">Coercive participation
Top-down, passive; mostly indirect, formal; participation in implementation, but not necessarily sharing benefits; choice between proposed limited alternatives or no choice; paternalism, non-participation, high degree of tokenism and manipulation.</td></tr>
<tr><td>Manipulation
Participation is merely the creation of community committees or associations that are used by those in power to "demonstrate" the use of citizen participation.</td><td>Manipulative
Participation is simply a pretence, with 'people's representatives on official boards, but who are un-elected and have no power.</td></tr>
</table>

Source: Arnstein (1969); Pretty (1995); Tosun (2006)

experiences to tourists. Failure to involve the local community in the tourism planning and development process could result in a backlash as they could develop a hostile attitude towards the tourists (Pearce, 1994). On the other hand, if the local people are happy as a result of their participation, they would be more hospitable towards tourists and this will create a positive image of the destination which in turn generates positive word-of-mouth (Zamani-Farahani & Musa, 2012).

Empowerment of Local People

One of the most vigorous arguments in favour of community participation in CBT is the fact that unlike other conventional forms of tourism, it empowers local people by readjusting the balance of power and reasserting local community views against those of the developers or the local government authority (Tosun, 2000). Local people are given greater ownership and authority to decide how tourism should be developed in their communities. Through this, local people can play a more active role in shaping tourism development in their communities. For a CBT project, the local people are empowered to decide on the kind of tourism facilities and wildlife conservation programmes that should be developed in their communities and how the costs and benefits of the tourism projects should be shared among the various stakeholders (Akama, 1996).

Resolving Conflicts

In most communities, the use of shared resources like forests and heritage sites for tourism development could result in conflicts between community members and tourism developers. This is because there could be conflicting uses of these resources. However, the participation of community members in tourism development usually helps to diffuse tensions and settle conflicts that may arise. This is because when local communities are involved in CBT development, they tend to have a sense of ownership of the plans and projects and would want the project to succeed. Community participation usually assists in reducing any possible conflicts because the views of all the stakeholders including community members regarding resource use are understood and respected.

Improving Benefits of Tourism to Local People

Community participation in CBT development ensures that more benefits from tourism development accrue to community members (Bello et al., 2018). It helps to balance the needs of community members with that of other stakeholders. This ensures that the aspirations of communities are included in tourism plans and programmes so that they can derive the right economic, environmental and socio-cultural returns from tourism development. Community participation also provides entrepreneurial and employment opportunities for local residents and help improve the local infrastructure.

Ensuring Sustainable Tourism Development

Community participation in CBT ensures the sustainability of the tourism industry. Community participation engenders a sense of ownership of tourism resources by community members. This sense of ownership can contribute to the sustainable use of the resources by local residents in a destination (Mbaiwa, 2007). Also, sustainable tourism is likely to be achieved through community participation in the tourism planning process because the inputs of local residents are factored into decisions on how tourism should be developed and this makes the planning process more effective, equitable and legitimate (Buanes et al., 2005; Simpson, 2008). Involving local residents in various stages of tourism development including informing them about relevant aspects of projects, incorporating their views and ensuring that they gain a fair share of benefits helps to achieve sustainable tourism development.

Forms of Community-Based Tourism

Community-Based Ecotourism

Generally, ecotourism has been defined as environmentally responsible travel and visitation to relatively undisturbed natural areas, to enjoy and appreciate nature (and any accompanying cultural feature—both past and present) that promotes conservation, has low visitor impact, and provides for beneficially active socio-economic involvement of local populations (Ceballos-Lascurain, 1996). CBE involves travel to natural areas that help to conserve the environment and sustain the well-being of local people (TIES 2007). These features are in tune with the fundamental principles of community-based tourism. CBE is essentially ecotourism that is managed and controlled by the host community. It usually involves local residents engaging in conservation practices, development and marketing of natural resources for tourism. Through this, environmental awareness is enhanced among residents and as well as improvement in their socio-economic well-being. CBE is premised on the fact that the natural environment must pay for itself by generating economic benefits for the local community and that the economic benefits derived from the environment through CBE would promote pro-environmental attitudes and behaviours among local residents (Kiss, 2004). CBE tends to be small-scale, community-led, local development-oriented, ecologically sustainable and socially equitable.

Community-Based Cultural Tourism

Cultural tourism involves all kinds of travel, whereby travellers learn about the history and heritage of others or about their contemporary way of life or thinking (McIntosh & Goeldner, 1986). The cultural tourist, by travelling to cultural attractions seeks new information and experiences to satisfy their cultural needs (Richards, 1996). Such experiences include music, dances, folklore, handicrafts, rituals, food, hospitality, festivals and an insight into the way of life of a community. Community-based cultural tourism is a form of CBT in which members of a host community are directly involved in designing, developing and marketing their cultural resources to tourists, as well as organising tours to cultural sites and showing tourists aspects of their lives. The distinction between CBE and community-based cultural tourism is that with the latter, the main attraction is the culture of the community. However, community-based cultural tourism is based on the same principles of CBT and CBE. Hence, community-based cultural tourism could be used as a tool to tackle poverty and empower the poor and the marginalized (Manyara & Jones, 2007; Moswete & Lacey, 2015). In rural communities in Africa especially, community-based cultural tourism has been used to promote women empowerment such as in Kenya where at the Umoja Uaso Women's Group's cultural village Samburu women have developed their own enterprises, thereby gaining not only economic power but substantial local political influence as well. Unlike cultural tourism in an urban context, community-based cultural tourism is usually on a small scale and does not require the development of museums, theatres and other superstructure, it simply involves repackaging what the community has accumulated or even what is actually taking place, instead of developing other tourist products.

Community-Based Natural Resource Management

A Community-based Natural Resource Management (CBNRM) project can be defined as a project whereby a community organize itself to derive benefits from the use and conservation of local natural resources. The communities tend to receive exclusive rights and responsibilities for the management of the natural resource from the government (Arntzen et al., 2003, 12). It is usually undertaken where local communities have land and/or resource user rights and make their own decisions regarding the land and resources. It is a bottom-up approach to natural resource conservation and management, unlike the traditional top-down conservation strategies which adopted the 'fences and fines' approach resulting in low levels of local benefits and community participation in conservation. CBNRM adopts a participatory approach to natural resource management by actively involving community members in the management of their natural resources. The rationale behind CBNRM is that governments cannot successfully and efficiently protect natural resources outside protected areas. Therefore, allowing the community to manage their own natural resources is a better development and conservation strategy (Arntzen et al., 2003). Placing the management of communal natural resources in the hands of community members increases environmental awareness, promotes conservation, enhances the economic well-being of communities and ultimately ensures sustainability. To ensure the sustainability of natural resources, under CBNRM, tourism is one of the few economic activities that are permitted and it generates financial benefits for the local communities and enhances the conservation of natural resources.

CBNRM could also takes the form of protected area outreach and collaborative management. With *protected area outreach*, a government agency such as the Wildlife Service reaches out to the communities fringing protected areas and provide some benefits such as infrastructural development to help offset the costs of denying the community the use of natural resources in the protected area and to ensure that the local people do not undermine the integrity of the protected area. A typical example is the Integrated Community Development Projects (ICDP) of the Ngorongoro Conservation Area in Tanzania. *Collaborative management* on the other hand seeks to involve local stakeholders in the management of natural resources through some form of negotiated agreements. Usually, the arrangements are between the government agency in charge of the protected area and residents of the communities around the protected area. An example is the Community Resource Management Areas (CREMAs) in Ghana. It is an arrangement between the Wildlife Division of the Ghana Forestry Commission and some forest-fringe communities. Through this arrangement, the communities through their elected representatives, together with the Wildlife Division and District Assembly, formulate the CREMA constitution, bylaws and natural resource management plan for each area. The local people are therefore involved in determining how the natural resources should be protected, managed and conserved. This has helped to foster the conservation of natural resources through a reduction in illegal activities (Murray et al., 2019).

BARRIERS TO COMMUNITY PARTICIPATION

Though community-based tourism has been considered as a viable model for achieving environmental conservation and economic development, studies have shown that several limitations make the benefits unachievable, especially in developing countries. The limitations to community participation in tourism development can be broadly classified into three namely, operational barriers, structural barriers and cultural barriers (Tosun, 2000).

Operational Barriers

Operational barriers result from the way institutions and stakeholders are coordinated and how tourism development is communicated to local residents. Among these barriers are the centralization of public administration, lack of coordination among agencies and lack of information to community members.

Centralization of Public Administration of Tourism

It is usually the case that in most developing countries, administrative and planning powers are concentrated under a central authority (Tosun & Timothy, 2001) and the centralised public administration system is characterized by bureaucracy. This usually results in delays in decision-making and renders community members powerless, unable to make decisions on their own. According to Marzuki (2008), these highly monopolised municipal and government processes reduces the involvement of the local community to mere recipients of information on projects. For instance, under the Community Resource Management Area (CREMA) in Ghana, links to non-local state actors have been described as ineffective because of inadequate fiscal decentralization, weak/absent lower-level governance structures and inattention to conservation and development as a distinct dual project (Murray et al., 2019).

Lack of Co-ordination

The lack of coordination is often due to the fragmented nature of the tourism industry. Typically, there are NGOs and donors, government agencies, landowners, local businesses, community groups and tourism operators. The activities of all these stakeholders have to be coordinated to ensure that the project objectives are achieved. However, in most cases, this is not the case. In a study on community-based tourism development at Gajah Mungkur Wonogiri in Indonesia, Jannah et al. (2018) identified the lack of coordination between government and community as one of the main challenges with the CBT project. Lack of coordination of the various stakeholders could compromise the quality of the tourism product and reduce the effectiveness of the participatory tourism development approach. Even within the community, there are usually diverse interest groups who have numerous, differing objectives, interests, roles, powers and competencies which tend to hinder coordination.

Lack of Information

According to Sharma (2004), most barriers to tourism development result from limited access to information, lack of awareness and insufficient community funding. For community members to fully participate in the tourism development process, the necessary information on the project must be disseminated to them so that they will know their roles and what to expect. However, in most developing countries where there are CBT projects, most of the local community members are not well-informed about tourism development in their areas to elicit their needed participation (Tosun, 2000). Information on tourism is usually insufficient and is disseminated to the public in incomprehensible forms (Hall, 2008; Marzuki et al., 2012). In a study on sustainable community-based tourism in Indonesia, Cole (2006) found that the local people lacked understanding and knowledge of development matters that are required for their meaningful participation in tourism.

Structural Barriers

Structural barriers relate to the capacity of the host community. The features of the community and its membership makes it difficult for the people to take full advantage of the opportunities of CBT and to reach their full potentials. These include the attitude of professionals, lack of expertise, lack of trained human resources and lack of financial resources.

Attitudes of Professionals

In most cases, CBT projects are spearheaded by technocrats who may be employees of NGOs and governmental institutions such as the National Tourism Authority. These are professionals who have the technical know-how and engage in activities such as drafting plans and programmes. There is however the tendency for them to look down on community members and disregard their suggestions. Such attitudes do not foster community participation. According to Gray (1985), governments and other key stakeholders do not always regard local residents as equal partners and they tend to keep power in their own hands.

Lack of Expertise

Also, in most developing countries, there is a lack of qualified personnel in community planning as the available planners are only conversant with the traditional planning techniques and have little experience of how to incorporate citizen participation in their planning (Timothy, 1999; Tosun, 2000). Planners are usually concerned about outputs rather than objectives, therefore objectives relating to sustainability and community empowerment remain largely unattainable. Employing the bottom-up approach in tourism planning requires specialist skills and knowledge which wildlife and tourism officials may lack.

Elite Domination

Elites are influential individuals such as traditional rulers and wealthy individuals who play an important role in decision-making on community-based tourism projects. Such people are usually appointed or self-appointed onto tourism development committees to serve as representatives of the wider community. However, their opinions may not truly reflect the broader community as they pursue their interest rather than the collective well-being of the community. According to Scheyvens (2002, p. 9), "elites often dominate community-based development efforts and monopolize the benefits of tourism". In most communities, tourism activities are largely dominated by few elites in communities and this results in the alienation of the majority of community members and a lack of meaningful ownership of the initiatives by community members (Marzuki et al., 2012). Therefore, the concerns and aspirations of community members may not be factored into the CBT project. A major reason for the failure of Administrative Management Design for Game Management Areas (ADMADE) in Zambia, was placing management in the hands of traditional authorities rather than the local people (Virtanen, 2003). Also, the lucrative agritourism industry in the Philippines has been noted to be dominated by persons who have strong political links or belonged to wealthy families. The properties of these individuals also receive more attention and publicity as compared to the properties owned by ordinary farmers who have no such links. Montefrio and Sin (2019) argue that such a situation may only promote the dominance of privileged persons and entities in the country's agritourism industry to the detriment of the local farmers.

Lack of Trained Human Resources

The lack of qualified human resources in a local community limits the community's participation in tourism development. A meaningful way of involving community members in tourism development is by offering them jobs in the tourism industry. However, this cannot be achieved in a situation where they lack the requisite qualifications, knowledge and skills. This leads to a situation whereby outsiders and foreigners occupy the most attractive jobs whiles local people are engaged in janitorial jobs. Though community involvement in tourism development is a laudable idea, the communities often do not have the experience and the trained human resource to handle the projects professionally (Tosun, 2000; Baum, 2018). Moreover, the fact that local residents, involved in tourism development do not possess the requisite competencies can make the management of the complex organizational processes which tourism planning entails very difficult (Tosun, 2000; Hung et al., 2011).

Relatively High Cost of Community Participation and Lack of Financial Resources

The community participation approach to tourism planning is considered an expensive process in terms of time, money and skills required (Hall, 2008; Tosun, 2000). The reality is that most of the host communities in developing countries do not have the financial resources to cater for the high transaction costs involved and the requisite business skills to manage the project (Sebele, 2010; Lundberg & Fredman, 2012). Usually, the communities lack the needed financial resources to make the necessary investments in the project (Tosun, 2000). Hence most CBT projects are not able to flourish in the long term. According to Steven and Jennifer (2002), the inadequacy of community resources for tourism development hinders the community's ability to effectively influence and develop tourism policy. Communities usually become overdependent on external donors and NGOs for funding and by so doing become subject to the dictates of the donors which may not coincide with that of community members. The challenge often is that where there is high dependence, immediately the donor support ends, the tourism project ceases to strive. Goodwin (2006) lamented that donor dependency is common in CBT since nine of the 15 CBT projects identified in his research were still dependent upon or seeking donor funding. Also, overreliance on external donors could compromise the economic viability of such projects. The Kharma Rhino Sanctuary Trust (KRST) in Botswana for instance is said to be too financially dependent on external donors including the African Development Foundation, Environmental Heritage Foundation of Botswana, European Development Fund, EU and Global Environmental Facility (implemented by the United Nations Development Programme).

Cultural Barriers

Cultural barriers relate to the way of life of community members, including practices that tend to limit their participation in tourism development. They include low competency levels of host community members in dealing with the impacts of tourism development. Specifically, insufficient levels of formalised education in a host community may result in a lack of knowledge about tourism, which may lead to limited access to tourism benefits.

Limited Capacity of Poor People

The lack of community capacity and limited understanding of tourism impacts have been recognized as barriers to effective tourism development in third world countries (Moscardo, 2008). It is generally acknowledged that the ultimate barrier to community participation is the inadequate capacity of local people in terms of expertise, understanding and awareness of the local community (Cole, 2006; Tosun, 2000). The majority of people living in especially rural communities in developing countries are usually preoccupied with meeting their basic necessities of life and may not be concerned about communal issues. This is coupled with the fact that they often do not even know how to initiate and manage the community participation process (Joppe, 1996). Also, the patriarchal nature of most communities in African societies often leads to the marginalization of women. Studies on CBT projects in countries like Namibia and Zimbabwe have shown that women have been marginalized (Kellert et al., 2000; Sullivan, 2003).

Apathy and Low Level of Awareness in the Local Community

For community participation in tourism development to be effective, community members must be motivated, accept and be willing to take part in the process. In a study on a rural fishing community in Kenya for instance, Bodin and Crona (2008) found that community members who had less or no trust in the management of the CBNRM schemes felt disenfranchised and were less likely to participate in any community-based initiative. Thus, low levels of interest is a disincentive for active participation by community members. Apathy is usually the result of a lack of trust in officials and the project, unfair distribution of benefits and lack of a sense of community ownership. Lachapelle et al. (2003, p. 187) observe that public apathy 'may be derived from a sense of cynicism, disenchantment, or lack of trust, leading to the belief that agencies don't acknowledge, hear, or honour what the public say'. Also, failure to distribute the benefits of CBT equally to all beneficiaries results in conflicts in the community and this reduces community interest in the management of CBT projects. Evidence provided by Sandbrook and Adam (2012) in Uganda pointed to the fact that the well-educated, rich, the youth and those who lived closer to the project benefited more from the Bwindi Impenetrable National Park. The fallout of such unequal distributions is the withdrawal of support for the project by aggrieved parties and this can also weaken the trust and cohesion in local communities (Jones, 2005) and affect the sustainability of CBT initiatives. Unequal distribution of benefits could result in conflicts as some parties will feel they have been short-changed. A case in point is the community-based coral reef management programme in Gili Indah, Indonesia. Because there were different stakeholders with different interests, conflicts arose especially between the local fishermen who felt they were not benefiting and the tourism entrepreneurs whose businesses were dependent on the coral reefs (Satria et al, 2006).

Lack of awareness about the potential and current costs and benefits of tourism development is also a barrier to community participation. To gain the full commitment and participation of community members, they must be apprised of full information about the tourism project. This will not only instil confidence in them but also help manage their expectations. To Simmons (1994), the lack of knowledge of tourism among local people requires considerable efforts to persuade them to participate in the tourism development process.

BENEFITS OF COMMUNITY-BASED TOURISM

Proponents of the CBT model believe that this approach to tourism development is more beneficial to host communities. Possible benefits that could accrue from CBT development include employment and income, infrastructural development, empowerment of host communities and environmental conservation.

Employment and Income

Host communities are often lured into tourism development because of the economic benefits they could derive from it. There are several instances where employment opportunities have been created for local residents through CBT initiatives. Three forms of employments are often created by CBT namely direct, indirect and induced employment. The first form is local residents getting direct employment in the CBT project as tour guides, cooks, receptionists, handicrafts among others. The second type takes an indirect form when local residents get the opportunity to supply materials, farm produce or render services towards the CBT initiative. The final form is induced employment where the money that is earned from tourism employment is used to create new jobs. Correspondently, four types of income (Ashley, 2000) could also accrue to the local residents and the community as a whole. These are wages from direct tourism employment, income from indirect tourism employment, profit from individually owned tourism enterprises and income from community-based tourism enterprises.

Butler (2017) in his study of CBT in Dullstroom, South Africa noted that the local tourism industry had generated several employment opportunities which offered consistent income for the locals. Similarly, Abukhalifeh & Wondirad (2019) reported that in Pulau Redang, Malaysia, the production and sale of handicrafts, traditional games, provision of traditional cuisines and performances of traditional dances were the main tourism-related income-earning activities for the local residents. Earlier, Stronza (2007) in their study on the Posada Amazonas eco-lodge in Peru reported that in 2004 and 2005 $ 182,583 and $ 208,328 respectively were made as profit. Out of this, 60 per cent went directly to the community for development while the rest was evenly distributed among the families in the community by their leaders. As of 2006, the average household income had increased by 25% of the dividends from tourism.

Infrastructural Development

The idea of CBT is partly premised on the belief that tourism can stimulate community development. One visible form this has taken is infrastructural development. Sometimes, the needed infrastructure is provided by an external body to promote tourism activities in the community. There have also been cases where local communities depended on profits from the CBT initiatives to address their infrastructural needs. In Tafi Atome, Ghana, Afenyo & Amuquandoh (2014) noted that profit from the CBT initiative in the community had become an alternative source of funds for community projects. These profits were used to fund the construction of two public toilets, a health post, housing quarters for nurses, kindergarten school block for the community. Besides, many electricity transmission poles were purchased to facilitate the electrification of the community. Selected communities engaged in CBT in Nicaragua also experienced an improvement in their communal infrastructures such as water supply, cleaning of public spaces, creation of paths and gardens. This was made possible through the collective agreement of the communities to plough back a component of revenue generated from the CBT activities into these developments (Zapata, Hall, Lindo & Vanderschaeghe, 2011).

Empowerment

One important component of CBT is the empowerment of local communities. Empowerment in tourism has been described as the enablement of individuals or communities to have more control and influence decisions that affect their lives, making sure that these decisions inure to their benefits (Murphy, 2014). Scheyvens (2009, p. 464) defines empowerment as 'the activation of the confidence and capabilities of previously disadvantage or disenfranchised individuals for groups so that they can exert greater control over their lives, mobilise resources to meet their needs and work to achieve social justice'. Empowerment could be economic, social, political or psychological. CBT's potential to bring out individual and community empowerment has been highlighted in the extant literature. Cases have been cited where CBT initiatives have contributed to expanding the assets and capabilities of local residents to participate in and influence the tourism development process to their benefit. For instance, Moswete and Lacey (2015) in their study of women empowerment through cultural tourism in Botswana found that women who were either single parents or breadwinners could now earn money from tourism to cater for their household needs. There have been instances where women engaged in tourism got the confidence to return to school to learn how to read and write because of the exposure and confidence built as a result of their tourism-related interactions (Scheyvens, 1999). Butler (2017) also reported that tourism activities in Dullstroom, South Africa, provided financial stability to locals engaged in it. As a result, they were able to afford the fees of local schools for their children. In the Manicaland Province of Zimbabwe, local residents were pleased that the local culture and indigenous knowledge systems were incorporated into the tourism projects within their localities. This boosted their self-esteem, pride and support for the projects. In addition, the local communities were able to embark on developmental projects such as the construction of early childhood classrooms funded by tourism revenue (Gohori & van der Merwe, 2021). The formal and informal education and training of local residents have also contributed to the expansion of their knowledge and skillset (Butler, 2017; Gohori & van der Merwe, 2021).

Environmental Conservation

Environmental conservation is very fundamental to CBT because in most cases, the main attraction is a natural resource. Tourism has always provided local communities with an economic reason for the conservation of wildlife and natural resources. As a result, local communities have become more environmentally conscious and are more committed to protecting their environment. However, it has been observed that local communities are more committed to protecting the environment when they are well compensated and their livelihoods are not compromised. For instance, villages that were part of the CAMPFIRE project in Zambia were taught how to use village mapping wildlife population census and monitoring to ensure a wildlife population that could deal with wildlife conservation. In return, income from the CAMPFIRE programme was channelled into community development projects (Dixey, 2008). Another example is the Tafi Atome community in Ghana. The community indicated their commitment to protecting their monkey sanctuary by developing a set of bye-laws to regulate residents and tourists' behaviour (Afenyo & Amuquandoh, 2014). This was because they could identify with the benefits they were getting from the tourism project. In Cambodia, Ota, Lonn and Mizoue (2021) noted that community-based ecotourism reduced deforestation in the CBET area by about 2% and further promoted forest recovery by about 0.5% as compared to other places which were not involved in CBT.

IMPLICATIONS FOR THE MANAGEMENT OF COMMUNITY-BASED TOURISM PROJECTS

It has been emphasized throughout the above discussions that CBT has an important role to play in achieving community development and environmental conservation. Based on the emergent lessons from the implementation of the CBT approach worldwide, it is suggested that for sustainable CBT to be achieved, management of the process will require the active participation of all stakeholders including government, community organizations and landowners (Figure 1). Stakeholder collaboration will only be effective if it is organized on the aegis of a local CBT development committee with representations from all stakeholders. Such a structure will foster greater community involvement and empowerment in a situation where there is transparency, accountability and fairness in the sharing of benefits. Community involvement and empowerment are necessary for the attainment of environmental conservation and community development. For community members to fully participate in tourism development, they need to be equipped with the relevant resources, information, knowledge and skills especially in planning, marketing and finance. The challenges and benefits associated with CBT coupled with some contemporary issues such as the COVID-19 pandemic provide some implications for the management of CBT projects.

First, given the limited capacity of community members to successfully manage CBT projects, there is the need for a legal local governance structure in place to have oversight responsibility for the management of CBT projects. This could be in the form of a local tourism management board or committee. This institution should have a rotational tenure and be made up of representatives of all identifiable sub-groups in the community. The board must be transparent in its operations and must develop and enforce bye-laws with the active participation of community members.

Secondly, periodic communal meetings can become an avenue for improving the democratization process in local communities. It will also help to ensure greater transparency and accountability to stakeholders. These communal meetings will provide a platform for local residents to elicit information, express their views and concerns and possibly have them addressed. It will also enable the local tourism management board to interact with the community they are representing and to regularly update them on the project.

Also, the fairness of benefits-sharing schemes has always been a bone of contention among stakeholders. While individual benefits are good, communal benefits are better. If benefits are to be far-reaching, they should be communal. Thus, emphasis should be placed on community projects or the provision of social amenities which have unrestricted access. This is the surest way of ensuring that the benefits of CBT trickle down to even the most marginalized groups in the community. With this, all local residents can identify with the benefits of CBT initiatives.

Furthermore, the role of the private sector in fostering sustainable community-based tourism cannot be underestimated. This explains why private-community partnership and joint venture arrangements have become very popular among CBT initiatives. Usually, the private sector partner is expected to bring on board capital and business know-how whiles the community partner usually brings land, labour and local knowledge. In reality, the government or an external donor is usually a third partner, providing basic infrastructure, finance, capacity building and sometimes mediating negotiations between private and community partners. Joint ventures can help address the knowledge, management and experience deficiencies of communities involved in tourism management. NGO's can play a valuable role in training, education and capacity building, as well as directing environmental conservation and benefits from tourism. Nevertheless, dependency on such external assistance has to be watched and weighed carefully.

Finally, COVID-19 and other crises in the past have accentuated the need for the tourism industry to build resilience. The tourism industry is highly susceptible to threats and shocks e.g. health scares, natural disasters, terrorism, economic recession, climate change, seasonality (Baker & Coulter, 2007; Baum & Hagen, 1999; Moreno & Becken, 2009; Nyaupane & Chhetri, 2009). However, the industry has demonstrated resilience and the ability to recover from the effect of these threats in the past. However, the unprecedented effect of the COVID-19 pandemic on the tourism industry has brought to the fore the industry's preparedness towards threats of such scale and severity. It has been recognized that many CBT project managers do not have the skills and resources to implement plans or take steps to safeguard their initiatives against risks or threats (Irvine & Anderson (2004). It is important that all CBT projects or initiatives mandatorily have a risk preparedness plan in place. This calls for the education and training of the managers to ensure the effective implementation of such plans. Post Covid-19, the emphasis has to be on supporting CBT projects or initiatives to be more proactive to risks and threats rather than being reactionary.

CONCLUSION

The community-based approach to tourism development is imperative if the benefits of tourism development are to trickle down to community members. This is because unlike other conventional models of tourism development, the CBT model places power in the hands of community members, involving them at the various stages of tourism development. They are therefore able to pursue tourism development that truly reflects their aspirations. CBT, therefore, helps to address the developmental challenges of communities by creating employment opportunities and providing income for infrastructural projects. The appeal of CBT is that it is a two-edged sword for achieving both socio-economic development and environmental conservation, which is a win-win situation. Thus, community participation remains a good model for tourism development at the community level. However, because the process is often fraught with challenges, there is a need for all stakeholders to collaborate to ensure sustainable tourism development.

REFERENCES

Abukhalifeh, A. N., & Wondirad, A. (2019). Contributions of community-based tourism to the socio-economic well-being of local communities: The case of Pulau Redang Island, Malaysia. *International Journal of Tourism Sciences*, *19*(2), 80–97. doi:10.1080/15980634.2019.1621521

Afenyo, E. A., & Amuquandoh, F. E. (2014). Who benefits from community-based ecotourism development? Insights from Tafi Atome, Ghana. *Tourism Planning & Development*, *11*(2), 179–190. doi:10.1080/21568316.2013.864994

Akama, J. S. (1996). Western environmental values and nature-based tourism in Kenya. *Tourism Management*, *17*(8), 567–574. doi:10.1016/S0261-5177(96)00077-5

Arnstein, S. R. (1969). A ladder of citizen participation. *Journal of the American Institute of Planners*, *35*(4), 216–224. doi:10.1080/01944366908977225

Arntzen, J. W., Molokomme, D. L., Terry, E. M., Moleele, N., Tshosa, O., & Mazambani, D. (2003). *Final report of the review of community-based natural resource review in Botswana: Report no. 1*. Gaborone: Centre for Applied Research for the National CBNRM Forum.

ASEAN. (2016). *ASEAN community based tourism standard*. Jakarta: Association of Southeast Asian Nations. Retrieved from https://www.asean.org/storage/2012/05/ASEAN-Community-Based-Tourism-Standard.pdf

Ashley, C. (2000). *The Impacts of Tourism on Rural Livelihoods: Namibia's Experience*. Overseas Development Institute.

Baker, K., & Coulter, A. (2007). Terrorism and tourism: The vulnerability of beach vendors' livelihoods in Bali. *Journal of Sustainable Tourism*, *15*(3), 249–266. doi:10.2167/jost643.0

Baum, T. (2018). Sustainable human resource management as a driver in tourism policy and planning: A serious sin of omission? *Journal of Sustainable Tourism, 26*(6), 873–889. doi:10.1080/09669582.2017.1423318

Baum, T., & Hagen, L. (1999). Responses to seasonality: The experiences of peripheral destinations. *International Journal of Tourism Research, 1*(5), 299–312. doi:10.1002/(SICI)1522-1970(199909/10)1:5<299::AID-JTR198>3.0.CO;2-L

Beeton, S. (2006). *Community development through tourism*. Landlinks Press. doi:10.1071/9780643093881

Bello, F. G., Lovelock, B., & Carr, N. (2018). Enhancing community participation in tourism planning associated with protected areas in developing countries: Lessons from Malawi. *Tourism and Hospitality Research, 18*(3), 309–320. doi:10.1177/1467358416647763

Bodin, O., & Crona, B. I. (2008). Management of natural resources at the community level: Exploring the role of social capital and leadership in a rural fishing community. *World Development, 36*(12), 2763–2779. doi:10.1016/j.worlddev.2007.12.002

Brohman, J. (1996). *Popular Development Rethinking the Theory and Practice of Development*. Blackwell Publishers.

Buanes, A., Jentoft, S., Maurstad, A., Søreng, S. U., & Karlsen, G. R. (2005). Stakeholder participation in Norwegian coastal zone planning. *Ocean and Coastal Management, 48*(9), 658–669. doi:10.1016/j.ocecoaman.2005.05.005

Butler, G. (2017). Fostering community empowerment and capacity building through tourism: Perspectives from Dullstroom, South Africa. *Journal of Tourism and Cultural Change, 15*(3), 199–212. doi:10.1080/14766825.2015.1133631

Ceballos-Lascurain, H. (1996). *Tourism, ecotourism, and protected areas*. IUCN. doi:10.2305/IUCN.CH.1996.7.en

Cole, S. (2006). Information and empowerment: The keys to achieving sustainable tourism. *Journal of Sustainable Tourism, 14*(6), 629–644. doi:10.2167/jost607.0

Cornell, D. (1997). Participatory development: An approach sensitive to class and gender. *Development in Practice, 7*(3), 248–259. doi:10.1080/09614529754486 PMID:12348314

Dixey, L. (2008). The unsustainability of community tourism donor projects: Lessons from Zambia. In A. Spenceley (Ed.), *Responsible Tourism: Critical Issues for Conservation and Development* (pp. 323–340). Earthscan.

Gohori, O., & van der Merwe, P. (2021). Tourism and Community Empowerment: The Perspectives of Local People in Manicaland Province, Zimbabwe. *Tourism Planning & Development*, 1–19. doi:10.1080/21568316.2021.1873838

Goodwin, H. (2006). Community-based Tourism: Failing to Deliver? Id1 Insights 62. Brighton: Institute of Development Studies.

Gray, B. (1985). Conditions facilitating interorganizational domain. *Human Relations*, *38*(10), 911–936. doi:10.1177/001872678503801001

Hall, C. M. (2008). *Tourism Planning: Policies, Processes and Relationships*. Pearson Education.

Hardy, A., Beeton, R., & Pearson, L. (2002). Sustainable tourism:An overview of the concept and its position in relation to conceptualisations of tourism. *Journal of Sustainable Tourism*, *10*(6), 475–496. doi:10.1080/09669580208667183

Harrison, D., & Schipani, S. (2007). Lao tourism and poverty alleviation: Community-based tourism and the private sector. *Current Issues in Tourism*, *10*(2-3), 194–230. doi:10.2167/cit310.0

Harwood, S. (2010). Planning for community based tourism in a remote location. *Sustainability*, *2*(7), 1909–1923. doi:10.3390u2071909

Hung, K., Sirakaya-Turk, E., & Ingram, L. J. (2011). Testing the efficacy of an integrative model for community participation. *Journal of Travel Research*, *50*(3), 276–288. doi:10.1177/0047287510362781

Irvine, W., & Anderson, A. R. (2004). Small tourist firms in rural areas: Agility, vulnerability and survival in the face of crisis. *International Journal of Entrepreneurial Behaviour & Research*, *10*(4), 229–246. doi:10.1108/13552550410544204

James, P. (2006). *Globalism, nationalism, tribalism: Bringing theory back in towards a theory of abstract community*. Sage Publications.

Jannah, E. U., Muryani, C., & Rindarjono, M. G. (2018). Community-Based Tourism Development at Gajah Mungkur Wonogiri Tourist Attraction. *IOP Conference Series. Earth and Environmental Science*, *145*, 798–807. doi:10.1088/1755-1315/145/1/012017

Jones, S. (2005). Community-based ecotourism: The significance of social capital. *Annals of Tourism Research*, *32*(2), 303–324. doi:10.1016/j.annals.2004.06.007

Joppe, M. (1996). Sustainable community tourism development revisited. *Tourism Management*, *17*(7), 475–479. doi:10.1016/S0261-5177(96)00065-9

Kellert, S. R., Mehta, J. N., Ebbin, S. A., & Lichtenfeld, L. L. (2000). Community natural resource management: Promise, rhetoric and reality. *Society & Natural Resources, 13*(8), 705–713. doi:10.1080/089419200750035575

Kiss, A. (2004). Is community-based ecotourism a good use of biodiversity conservation funds? *Trends in Ecology & Evolution, 19*(5), 232–237. doi:10.1016/j.tree.2004.03.010 PMID:16701261

Kontogeorgopoulous, N. (2005). Community-based ecotourism in Phuket and Ao Phangnga, Thailand: Partial victories and bittersweet remedies. *Journal of Sustainable Tourism, 13*(1), 4–23. doi:10.1080/17501220508668470

Lachapelle, P. R., McCool, S. F., & Patterson, M. E. (2003). Barriers to effective natural resource planning in a 'Messy' world. *Society & Natural Resources, 16*(6), 473–490. doi:10.1080/08941920309151

Lee, T. H., & Jan, F. H. (2019). Can community-based tourism contribute to sustainable development? Evidence from residents' perceptions of the sustainability. *Tourism Management, 70*, 368–380. doi:10.1016/j.tourman.2018.09.003

Lundberg, C., & Fredman, P. (2012). Success factors and constraints among nature-based tourism entrepreneurs. *Current Issues in Tourism, 15*(7), 649–671. doi:10.1080/13683500.2011.630458

Lyttleton, C., & Allcock, A. (2002). *Tourism as a tool for development: UNESCO-Lao National Tourism Authority Nam Ha Ecotourism Project*. External Review July 6–18, 2002. Vientiane: NTAL and UNESCO.

Manyara, G., & Jones, E. (2007). Community-based tourism enterprises development in Kenya: An exploration of their potential as avenues of poverty reduction. *Journal of Sustainable Tourism, 15*(6), 628–644. doi:10.2167/jost723.0

Marzuki, A. (2008). Decision making and community participation: A case study of the tourism industry in Langkawi. *Turizam: međunarodni znanstveno-stručni časopis, 56*(3), 227-241.

Marzuki, A., Hay, I., & James, J. (2012). Public participation shortcomings in tourism planning: The case of the Langkawi Islands, Malaysia. *Journal of Sustainable Tourism, 20*(4), 585–602. doi:10.1080/09669582.2011.638384

Mbaiwa, J. E. (2007). Local community attitudes towards wildlife conservation and community based natural resources management in Ngamiland District, Botswana. In B. Schuster & O. T. Thakadu (Eds.), *Natural Resources Management and People* (pp. 19–26). IUCN CBNRM Support Programme.

McIntosh, R. W., & Goeldner, C. R. (1986). *Tourism: Principles, practices and philosophies* (5th ed.). John Wiley & Sons.

Menkiti, I. A. (1984). Person and community in African traditional thought. *African Philosophy: An Introduction*, *3*, 171–182.

Montefrio, M. J. F., & Sin, H. L. (2019). Elite governance of agritourism in the Philippines. *Journal of Sustainable Tourism*, *27*(9), 1338–1354. doi:10.1080/09669582.2019.1621327

Moreno, A., & Becken, S. (2009). A climate change vulnerability assessment methodology for coastal tourism. *Journal of Sustainable Tourism*, *17*(4), 473–488. doi:10.1080/09669580802651681

Moscardo, G. (Ed.). (2008). Building community capacity for tourism development. CAB International.

Moswete, N., & Lacey, G. (2015). "Women cannot lead": Empowering women through cultural tourism in Botswana. *Journal of Sustainable Tourism*, *23*(4), 600–617. doi:10.1080/09669582.2014.986488

Murphy, J. W. (2014). *Community-based interventions: Philosophy and action.* Springer. doi:10.1007/978-1-4899-8020-5

Murphy, P. E. (1985). *Tourism: A community approach.* Methuen.

Murray, G., Agyare, A., Dearden, P., & Rollins, R. (2019). Devolution, coordination, and community-based natural resource management in Ghana's community resource management areas. *African Geographical Review*, *38*(4), 296–309. doi:10.1080/19376812.2018.1426022

Nyaupane, G. P., & Chhetri, N. (2009). Vulnerability to climate change of nature-based tourism in the Nepalese Himalayas. *Tourism Geographies*, *11*(1), 95–119. doi:10.1080/14616680802643359

Ota, T., Lonn, P., & Mizoue, N. (2021). Contribution of Community-Based Ecotourism to Forest Conservation and Local Livelihoods. In T. Yahara (Ed.), *Decision Science for Future Earth*. Springer. doi:10.1007/978-981-15-8632-3_9

Pearce, P. L. (1994). Tourist-resident impact: Examples and emerging solutions. In W. F. Theobald (Ed.), *Global Tourism: The Next Decade* (pp. 103–123). Butterworth-Heinemann.

Pretty, J. (1995). The many interpretations of participation. *Focus (San Francisco, Calif.)*, *16*(4), 4–5.

Richards, G., & Richards, G. B. (Eds.). (1996). *Cultural tourism in Europe*. Cab International.

Ross, S., & Wall, G. (1999). Evaluating ecotourism: The case of North Sulawesi, Indonesia. *Tourism Management, 20*(6), 673–682. doi:10.1016/S0261-5177(99)00040-0

Salazar, N. B. (2012). Community-based cultural tourism: Issues, threats and opportunities. *Journal of Sustainable Tourism, 20*(1), 9–22. doi:10.1080/09669582.2011.596279

Satria, A., Matsuda, Y., & Sano, M. (2006). Questioning community based coral reef management systems: Case study of Awig-Awig in Gili Indah, Indonesia. *Environment, Development and Sustainability, 8*(1), 99–118. doi:10.100710668-005-0909-9

Scheyvens, J. (2009). Empowerment. In R. Kitchin & N. Thrift (Eds.), *International Encyclopaedia of Human Geography* (pp. 464–470). Elsevier. doi:10.1016/B978-008044910-4.00092-4

Scheyvens, R. (1999). *The potential for ecotourism to facilitate the empowerment of local communities in Southern Africa: A summary report using selected case studies*. Massey University.

Scheyvens, R. (2002). *Tourism for development: Empowering communities*. Prentice Hall.

Schmidt, P. R. (2017). *Community-based heritage in Africa: Unveiling local research and development initiatives*. Routledge. doi:10.4324/9781315270050

Sebele, L. S. (2010). Community-based tourism ventures, benefits and challenges: Khama rhino sanctuary trust, central district, Botswana. *Tourism Management, 31*(1), 136–146. doi:10.1016/j.tourman.2009.01.005

Sharma, K. K. (2004). *Tourism and socio-cultural development*. Sarup and Sons.

Simmons, D. G. (1994). Community participation in tourism planning. *Tourism Management, 15*(2), 98–108. doi:10.1016/0261-5177(94)90003-5

Simpson, M. C. (2008). Community benefit tourism initiatives—A conceptual oxymoron? *Tourism Management, 29*(1), 1–18. doi:10.1016/j.tourman.2007.06.005

Sproule, K. W. (1996). Community-based ecotourism development: Identifying partners in the process. In E. Malek-Zadeh (Eds.), The Ecotourism Equation: Measuring the Impacts (pp. 233–250). New Haven, CT: Yale University.

Steven, D., & Jennifer, T. (2002). *Challenges and barriers to community participation in policy development.* Retrieved from http://www.ruralnovascotia.ca/documents/policy/challenges %20and%20barriers.pdf

Stronza, A. (2007). The economic promise of ecotourism for conservation. *Journal of Ecotourism, 6*(3), 170–190. doi:10.2167/joe177.0

Suansri, P. (2003). *Community based tourism handbook.* Responsible Ecological Social Tour-REST. Retrieved from https://www.mekongtourism.org/wp-content/uploads/REST-CBT-Handbook-2003.pdf

Sullivan, S. (2003). Protest, conflict and litigation: dissent or libel in resistance to a conservancy in north-west Namibia. In E. Berglund & D. Anderson (Eds.), *Ethnographies of Conservation: Environmentalism and the Distribution of Privilege* (pp. 69–86). Bergahn Press.

The Mountain Institute. (2000). *Community based tourism for conservation and development: A resource kit.* Retrieved from http://www.mountain.org

TIES (The International Ecotourism Society). (n.d.). *Definition and principles: Ecotourism.* http://www.ecotourism.org/webmodules/webarticlesnet/templates/eco_template.aspx?articleid=95&zoneid=2

Timothy, D. J. (1999). Participatory planning a view of tourism in Indonesia. *Annals of Tourism Research, 26*(2), 371–391. doi:10.1016/S0160-7383(98)00104-2

Tosun, C. (1999). Towards a typology of community participation in the tourism development process. *Anatolia, 10*(2), 113–134. doi:10.1080/13032917.1999.9686975

Tosun, C. (2000). Limits to community participation in the tourism development process in developing countries. *Tourism Management, 21*(6), 613–633. doi:10.1016/S0261-5177(00)00009-1

Tosun, C. (2006). Expected nature of community participation in tourism development. *Tourism Management, 27*(3), 493–504. doi:10.1016/j.tourman.2004.12.004

Tosun, C., & Timothy, D. J. (2001). Shortcomings in planning approaches to tourism development in developing countries: The case of Turkey. *International Journal of Contemporary Hospitality Management, 13*(7), 352–359. doi:10.1108/09596110110403910

Virtanen, P. (2003). Local management of global values: Community-based wildlife management in Zimbabwe and Zambia. *Society & Natural Resources, 16*(3), 179–190. doi:10.1080/08941920309164

Wallace, G. N., & Pierce, S. M. (1996). An evaluation of ecotourism in Amazonas, Brazil. *Annals of Tourism Research, 23*(4), 843–873. doi:10.1016/0160-7383(96)00009-6

Wearing, S. L., Wearing, M., & McDonald, M. (2010). Understanding local power and interactional processes in sustainable tourism: Exploring village-tour operator relations on the Kokoda Track, Papua New Guinea. *Journal of Sustainable Tourism, 18*(1), 61–76. doi:10.1080/09669580903071995

Zamani-Farahani, H., & Musa, G. (2012). The relationship between Islamic religiosity and residents' perceptions of socio-cultural impacts of tourism in Iran: Case studies of Sare'in and Masooleh. *Tourism Management, 33*(4), 802–814. doi:10.1016/j.tourman.2011.09.003

Zapata, M. J., Hall, C. M., Lindo, P., & Vanderschaeghe, M. (2011). Can community-based tourism contribute to development and poverty alleviation? Lessons from Nicaragua. *Current Issues in Tourism, 14*(8), 725–749. doi:10.1080/13683500.2011.559200

APPENDIX

Figure 1. A model for sustainable CBT

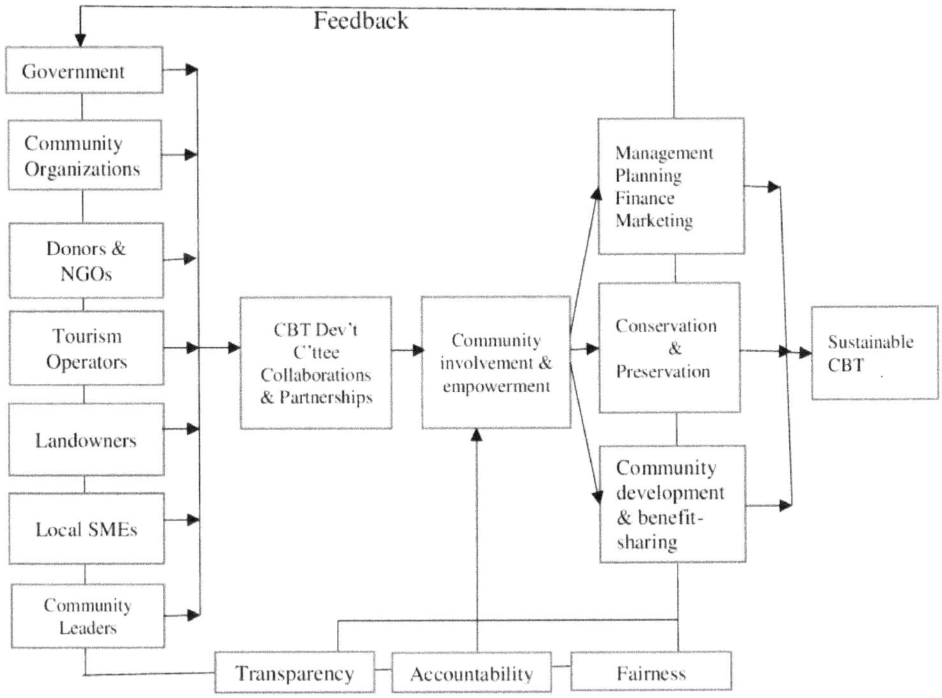

Chapter 2
Implications of the Inclusive Pro-Poor Tourism Planning in Uganda for Strategic Planning for Community-Based Tourism Development in Nigeria

Olukemi Deborah Fagbolu
National Open University of Nigeria, Nigeria

ABSTRACT

Many Nigerians are unemployed and live in extreme poverty. The Nigerian government has therefore identified tourism as one of the promising sectors in its efforts to diversify the economy from the oil sector. CBT which is a kind of tourism that encourages environmental, social, cultural, and economic sustainability in communities could be a good model for addressing the socio-economic challenges confronting Nigeria. Based on literature derived from JSTOR, Google Scholar, and Emerald, this chapter draws implications from the inclusive pro-poor tourism planning in Uganda for CBT development in Nigeria.

INTRODUCTION

Poverty is multi-faceted; hence, the millennium development goals (MDGs) focus on reducing the numbers of people living in abject poverty, that is, those living on less than 1USD per day. Many Nigerians are faced with poverty although, poverty reduction remains one of the Development Goals (Bassey & Egon, 2016; Fagbolu,

DOI: 10.4018/978-1-7998-7335-8.ch002

2021; UNWTO, 2002; World Bank, 2000; Yang, & Hung, 2014). Disharmony, therefore, exists between the current situation in Nigeria and one of the MDGs, i.e., poverty reduction. Nonetheless, the Nigerian government has been making several concerted efforts towards diversifying her mono oil economic-based sector to multi-economic sectors including tourism.

Globally, tourism comprises the activities of persons travelling outside their usual environment for not more than one calendar year and whose main purpose of travel is other than the exercise of an activity remunerated from the place(s) visited (Giampiccoli & Mtapuri, 2020). Tourism as the science, art, and business of attracting and transporting visitors, fosters accommodating them and catering for their needs and wants throughout their stay. The industry has developed into a great industry worldwide comprising of various businesses in different sectors (Esu, 2015; World Tourism Organization (UNWTO), 2016).

Research agenda on tourism vary from one destination to another. Strategic planning research tends to focus on developing different forms of tourism and related issues depending on situations and needs (Burton & Steane, 2004; David, 2011; Getz, 2008; Gossling, 2009; Jennings, 2010; Ritchie, 2004; Veal, 2011). Nickols (2016, p.7) refers to strategy as "a complex web of thoughts, ideas, insights, experiences, goals, expertise, memories, perceptions, and expectations that provides general guidance for specific actions in pursuit of particular ends". In a way, it describes, where we are, where we are going and, how to get there. It embraces the overall mission, vision and objectives of an organization. Strategy provides direction, takes cognizance of resources and opportunities, controls resistance and barriers in order to reach the affected.

As a result, strategic planning connotes a specific vision, mission, competencies and an articulated set of individuals with discrete actions capable of supporting strategic goals. Hence, the idea is choosing options that are unique and also deciding what not to do. Meanwhile, strategic planning is usually geared towards proffering solutions to problems in a process that engages different stakeholders. However, in recent years there have been calls for greater community-based tourism (CBT) development due to the emergence of alternative forms of tourism.

CBT fits into the MDGs as a form of tourism by which hosts extend an invitation to tourists to visit their communities to experience their local tourism products. CBT has become a pertinent and critical element with considerable potentials for improving the well-being of local communities and for poverty reduction. CBT can boost employment and incomes of local communities and has been found to be a good approach for developing sustainable tourism and local communities (Comerio & Strozzi, 2019; Mtapuri & Giampiccoli, 2016; Sène-Harper & Séye, 2019).

This chapter specifically focuses on strategic planning for CBT development in Nigeria. The chapter reviews literature retrieved from Emerald, JSTOR and Google scholar databases. The chapter in a logical manner presents the current status and challenges of CBT in Nigeria. Furthermore, it provides the case of strategic planning for inclusive pro-poor tourism in Uganda and the implications for strategic planning for CBT development in the Nigerian context.

COMMUNITY-BASED TOURISM (CBT)

CBT is a kind of tourism that considers environmental, social, and cultural sustainability. It is usually owned and operated by residents to ensure that guests, experience the local and community way of life (Kwangesh, 2014). CBT supply comprises natural and cultural resources, its demand could be individuals or groups of domestic or foreign tourists while suppliers include the locals, tour operators and travel agents and other tourism service providers. The strategic positioning of CBT could help local communities in developing countries to satisfy customer needs (Burgos & Mertens, 2017). It is also capable of providing tangible and intangible products for tourists in various operating segments locally. The products include transportation, recreation, cultural activities, shopping, accommodation, local food centres and other tourism-related services in a hospitable environment (Aref, Gill & Aref, 2010; Asuk, Ebu & 1Ifebueme, 2018).

The number of travels in industrial-based countries is increasing while this trend is forcing the tourist market to become a CBT demand market instead of a supply market (Schiffman & Kanuk, 2000; Seaton & Bennett, 1996). In local communities, visitors can enjoy available local tourism resources in a friendly atmosphere. CBT activities include village trekking, village biking, fieldwork and boating, and it affects the local agricultural sector as well as small scale businesses such as local drinks processing and local crafting. CBT tends to encourage employment and revenue generation for locals thereby resulting in poverty alleviation. CBT provides opportunities for locals to be gainfully employed due to its labour inclusive nature as well as the potential to build on the tourism assets of rural communities with few other growth options (Comerio & Strozzi, 2019; Mayaka, Croy & Cox, 2019).

Tourists' demand authentic local cultural practices, heritage, ecological and natural resources that offer a rich and great and tourism experience. Every CBT experience is unique and immensely depends on the destination and the needs of the host communities. It involves a real insight into local lives, ensuring that the tourism experience is of significance to local people.

The benefits of CBT could be broadly classified into economic, social, and environmental. CBT results in community empowerment, cultural preservation and environmental conservation (Adetola, 2015; Kwangseh, 2014). CBT ensures that local people possess full ownership and management of tourism in the communities, in order to allow the benefits of tourism to stay within the local environments (Comerio & Strozzi, 2019; Mayaka, Croy & Cox, 2019; Mtapuri & Giampiccoli, 2016; Sène-Harper & Séye, 2019). CBT can as well make significant contributions to development, particularly in the areas of agricultural transformation, community development, infrastructural development, as well as social and economic empowerment. It can enhance the social wellbeing of communities and change both the environment and the lives of the members of those communities (Briones, Yusay & Valdez, 2017).

CURRENT STATUS AND CHALLENGES FACING COMMUNITY-BASED TOURISM (CBT) DEVELOPMENT IN NIGERIA

Africa is known to be the world's second-most populous and second-largest continent. The continent of Africa covers about 20.4 percent of the total land area representing six percent of the Earth's total surface area. Geographically, it is sub-divided into west, central, east, south and north with countries like Ghana, Cote d'Ivoire, Togo, Mali and Nigeria in the West of the continent.

Nigeria with a total population of about 200 million people is the most populous country in Africa. The country has 36 states and a Federal Capital Territory. The country is endowed with cultural, natural and eco-tourism resources located across six geo-political zones in the country. Harnessing the resources could yield numerous benefits (Ahmadu, 2011; Esu, 2015; Igbojekwe, 2011). Nigeria is clamouring for economic diversification from its mono-oil sector to other sectors including tourism which has the potential to alleviate poverty (Asuk, Ebu & Ifebueme, 2018).

A study by Blackstock (2005) revealed that CBT tends to take a functional approach to community involvement treating the host community as a similar entity that neglects the structural limitations to local control of the industry. CBT can be crucial to the development and provision of local tourism products in the rural communities by the locals in different geo-political zones of Nigeria.

Development is complex and people view development more as a political term. As a political term and process, it connotes who has the power to get something done for whom (Snyder et al., 1991). However, it could be defined as stimulating social change to enable people to attain their human potential. It is a process and not an outcome. As a dynamic concept, it entails positive change, from one state to another; growth, improvement and so on. Development brings about the transformation or conversion of resources into other outputs for improvement in a system. Efforts at developing the CBT potentials of some states like Anambra, Bayelsa, Cross-river, Enugu and Lagos, Ondo are on course (Emeka, Ejim & Amaka, 2015; Esu, 2015; Eze-Uzomaka, 2006). Esu (2015) proposes a tourism sector development conceptual model that portrays tourism entrepreneurship as the potential driver of transforming the tourism industry in Nigeria as shown in Figure 1.

Figure 1. Conceptual model for tourism industry and destination development in Nigeria
Source: Esu, B. B. (2015, p. 572).

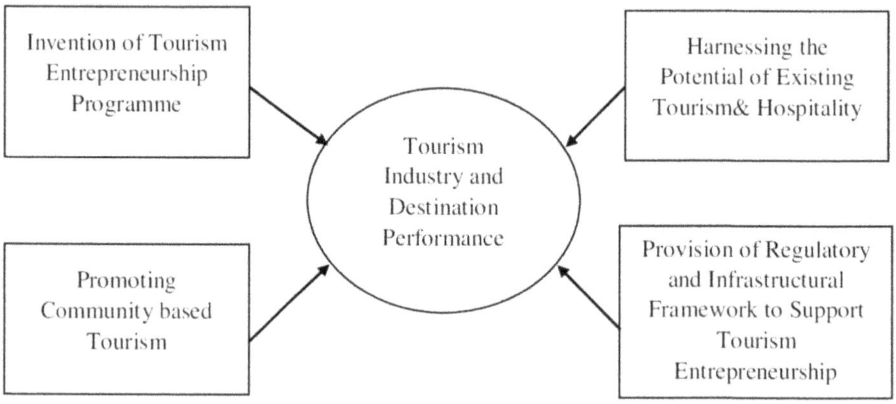

The study highlights specific entrepreneurial tourism operations that are necessary to optimize the country's tourism goals while the model presents CBT promotion in Nigeria as one of the measures for transforming the tourism industry and destination performance.

CBT development in Nigeria will likely create a more sustainable tourism market and serve as a poverty reduction tool due to its potential to foster growth, that enables the locals to actively participate in and significantly benefit from social and economic activities (Kruczek, Kruczek & Szromek, 2018; Terencia, 2018). This can produce expected development and growth in the local communities thereby reducing rural-urban migration and alleviating poverty among the locals in the country.

Efforts at CBT development is still low compared to conventional tourism promotion in Nigeria. The government has been making several efforts towards developing the tourism industry in order to address some of the problems in the sector. The Federal Government set up Nigeria

Tourism Master Plan Committee to look at the tourism-related issues in the country. The government promulgated her first National Tourism Policy during the National Conference on Tourism Development held in Lagos between 10th and 12th July 1990 which was aimed at achieving certain objectives. The main objectives of the National Tourism Policy are:

Table 1. Findings of the Nigeria Tourism Development master plan committee on human resources capabilities

The human resources capabilities of the tourism and hospitality sectors are lagging far behind in terms of quality, standards, and skills delivery. There is a lack of balance between management and supervisory training and craft skills training. The numbers receiving craft skills training need to be greatly increased.
The National Institute for Hospitality and Tourism (NIHOTOUR) is grossly overstaffed and ineffective in its present role as a provider of educational services. The institute has no credible on-site practice facilities nor has it any experience of delivering skilled personnel to the workforce.
The various education providers do not specify skills or program outcomes. There is no way of ensuring consistent delivery of quality outcomes nor is there any way for auditing the quality of outcomes of the various training and education institutions. The conditions in the training, institutes are very poor, dirty, ill-equipped, badly maintained and are under-funded. The management and staff morale is poor and at a low level. A number of institutions train their course attendees in theoretical aspects but have no on-site practice training facilities.
There is a significant shortage of suitably qualified instructors, teachers and lecturers Across the teaching faculty. They also lack industry experience and an international aspect. The consequences are that the teaching methods and course content are out of date.

Source: Tourism Development International (2006, p.7).

1. To protect and promote Nigeria's cultural heritage as a resource for homegrown socio-economic development.
2. To encourage community and public partnerships in tourism development.
3. To generate foreign exchange, enhance income redistribution, alleviate poverty, and create employment.
4. To promote Nigeria as a desirable tourism destination within the context of Africa's cultural renaissance.
5. To promote geopolitical integration, healthy international cooperation and understanding
6. To ensure environmental sustainability in the development of tourism resources. (Bassey & Egon, 2016; Tourism Development International, 2006).

However, not all tourism stakeholders were involved in the formulation of the policy. The policy is based on the New Partnership for Africa's Development (NEPAD) baseline study on tourism and inspired by the pronouncement and determination of the Secretary-General of the WTO to develop and market Nigeria as a tourism destination. The findings of the Nigeria Tourism Development Master Plan Committee during the planning process are presented in Table 1.

The policy "aimed at developing a competitive sustainable tourism within the confines of the World Tourism Market and open up Nigeria as a major tourist destination in Africa through the promotion of domestic tourism" (Bassey & Egon, 2016, p. 26). However, implementation of the policy is fraught with some problems across federal, state, local government area and community levels as portrayed in Table 2.

Table 2. Nigeria tourism development master plan implementation problems

Tier of government	Implementation problems
Federal Level	Low and inadequate budgetary allocation to the tourism sector in the annual national appropriation bill (check %) Bureaucratic bottlenecks at the apex tourism body of the country: Federal Ministry of Culture, Tourism and National Orientation and the Nigeria Tourism Development Corporation Poor working relationship between the Federal Ministry of Culture, Tourism and National Orientation and the Nigeria Tourism Development Corporation (NTDC) in the formulation of strategic direction for the achievement of the national tourism objectives. Conflicts in the roles of Federal Ministry of Culture, Tourism and National Orientation, NTDC and National Institute for Hospitality and Tourism (NIHOTOUR) in the implementation of national tourism policies as it concerns manpower development The fuzziness of the NTDC Law in the distribution of responsibilities between NTDC and State Tourism Organizations (STOs). A good example is seen in the licensing of hospitality facilities in Nigeria; the case of Lagos State and the Federal Government of Nigeria.
State Level	The failure of the State Governments to domesticate the content of the National Tourism Master Plan. The inefficiency of most State Tourism Boards and the attendant inability of the State Tourism Organizations (STOs) to effectively develop tourism at the state level
Local Level	The failure of Local Government Council Areas (LGAs) to see tourism development as part of their legislated functions, thereby leading to the absence of tourism plans to guide tourism businesses at that level, despite the fact that the NDTC Law provides for it. The unwillingness of the local governments to drive the development of tourism in the rural areas.
Community Level	The failure to participate in tourism development as a result of ignorance and lack of empowerment

Source: Esu, B. B. (2015, P. 571).

Furthermore, the government established the Nigeria Tourism Development Corporation (NTDC) through Decree 81 of 1992 and in 2005 formulated the Nigeria Tourism Development Master Plan to boost the industry in Nigeria (Bassey & Egon, 2016; Esu, 2015; Fallon & Fagbolu, 2021; Tourism Development International, 2006). However, the slow pace of implementation of the policy remains a matter of serious concern to every tourism stakeholder in the country. The efforts have not really achieved much due to the fact that the nation is still battling with how to harness tourism resources into tourist products and make them attractive enough to draw visitors.

CBT development in the country is faced with some challenges which include the roles of CBT within the overall socio-economic development of the destination region, the type of destination that will most effectively fulfil the desired roles, sources of finance for the tourism sector, terms, nature and direction of product development (Amalu & Ajake, 2018; Amalu, et al., 2017; Amalu, et al., 2019; Amerta, 2017). Other issues facing CBT in the country include inadequate transportation access and infrastructure, lack of regulation of service providers, poor environmental practices, poor industry image, lack of credibility, poor industry-community relationships, inadequate human resources and labour supply, technology, and international travel restrictions. The challenges extend to insurgency, kidnapping, abduction, crisis, disaster, stakeholders' mismanagement, lack of local development, weak power relations and competition. Others are cultural/religious differences, unsustainable practices and inadequate strategic planning for CBT (Dimoska & Trimcev, 2012; Menbere & Menbere, 2017; Mensah, 2017).

These issues demand adequate strategic planning for CBT in order to create employment, generate revenue and reduce poverty among local communities. The planning can promote a more equitable and sustainable CBT development in Nigeria. However, it involves various levels of planning which include CBT products and development. The argument on tourism policy is either based on competitiveness or sustainability of destinations (Goeldner & Ritchie, 2009; Guo, Jiang, & Li, 2019). Proper processes observed during CBT policy formulation are critical to strategic planning for CBT development.

CBT is becoming a significant and even essential part of the local communities in African countries like Cameroun, Southern Africa, and Uganda. However, this chapter considers strategic tourism planning for inclusive pro-poor tourism program in Uganda as a good model for strategic planning for CBT development in Nigeria (Pillay & Rogerson, 2013; Theerapappisit, 2009; Tucker, 2011).

THE CASE OF STRATEGIC PLANNING FOR INCLUSIVE PRO-POOR TOURISM IN UGANDA

As part of the planning process, a survey was conducted within three months i.e. between August and October 2011 by a team of 13 international and national tourism experts and consultants in collaboration with the Ministry of Tourism, Wildlife and Heritage, the Uganda Tourism Board (UTB), and the Uganda Export Promotion Board (UEPB). UNDP funded the study, which made it possible for the researchers to hire 10 additional national consultants to survey selected exit points, thereby providing the team with a clear understanding of both the Tourism Value Chain in Uganda and the existing opportunities for vulnerable populations to benefit more from a rapid increase in international arrivals. Qualitative and quantitative methods were employed for data collection. The approach was in different stages.

First, a team of three consultants (national and international) worked under the supervision of the ITC Sector Competitiveness Trade Promotion Adviser and with the support of the Uganda Export Promotion Board and the Uganda Tourism Board. Throughout the research, the team consulted widely with stakeholders in several tourist destinations. Second, discussions were held with government departments, namely: The Uganda Tourism Board (UTB), The Uganda Wildlife Authority (UWA) and the Uganda Export Promotion Board (UEPB). Third, a stakeholders' workshop was organized in Kampala to present both the study objectives and itinerary to tourism professionals and government officials. Heads of the Association of Uganda Tour Operators (AUTO), Uganda Hotels Owners Association (UHOA), Department of Industrial Art and Design of Makerere University, Uganda Tourism Board and National Crafts and Arts Association of Uganda (NACAU), the Uganda Community Tourism Association (UCOTA) and the representatives of the USAID/STAR project attended that meeting.

Furthermore, the team met with local associations and communities namely; the Boomu women's group, the Kyaninga community, the Albert Nile Conservation and Tourism Association, and the Mubako local women Craft Association, Others were the Nyabwina women groups, the Kibale Association for Rural and Environmental Development (KAFRED), the Njara Muslim Women Group, the Kikongoro Women Community Tourism Group, the Katwe community, the Buhoma community rest camp, the Bunyoni community, and the Exposure Africa Project.

In order to collect raw data, the consultants conducted face-to-face interviews with representatives of tourism businesses such as hotels, safari lodges, resorts, tour operators, handicraft vendors, wholesalers, and restaurant managers. During the study duplication of works was avoided by extending the opportunity study to a full value chain analysis of the tourism sector. Interviews and meetings were conducted with focus groups; 28 hotel managers, 117 staff of hotels, 21 restaurant owners, 15 tour operators who specialized in tours for Western tourists, 29 handicraft producers and wholesalers, five food producers, 11 community projects, and three operators of NGOs. Three hundred and thirty-five international tourists completed a survey within five days. Data collected related to their behaviour, satisfaction and average daily spending while visiting Uganda. The interviews took place at the main exit points of Uganda (Entebbe, Busia, Mutukula, Katuna and Malaba) in order to cover all different categories of tourists visiting the country.

Inclusive tourism development referred to as CBT in this context serves as means of providing alternative or a new avenue of livelihood for the poor, alleviate poverty by linking agriculture, creative arts, and cultural performances to the tourism industry (Pillay & Rogerson, 2013). The government of Uganda urged the International

Trade Centre's (ITC) to develop an approach to tourism that is able to strengthen the capacities of local people in a competitive and sustainable manner and generate employment opportunities for poor and marginalized communities (Thomas, Barya, & Katongole, 2011). The aim of the strategic planning project on pro-poor inclusive tourism for Uganda project was to identify suitable project interventions through which the ITC expertise could impact tourism for the poor (Edgell & Swanson, 2013) within three to five years. The development and implementation of the project commenced in 2011 and will end in 2014.

ITC is the technical cooperation agency of the United Nations (UN) and the World Trade Organization (WTO). Its main activities include design of projects aimed at developing export strategies, increasing the competitiveness of sectors and enterprises, production and dissemination of trade intelligence, improving the capacity of trade support institutions as well strengthening business and trade policies.

The country's population according to the World Bank in 2011 was about 34.51 million. The country has had a checkered history from the period before 1894, through British rule between 1894 and 1962, to the early independence era of 1962 - 1971, the reign of Idi Amin between 1971-1979, and from 1986 until date. The nation witnessed British Colonial rule beginning from the late 1800s with the formulation of administrative law. However, Uganda fought and gained independence from British Colonial Rule on October 9, 1962. The country has more than 113 districts comprising 24 districts in the eastern region, 21 districts in the northern region, 19 districts in the western region and 16 districts in the central region. Official languages spoken by the people are Swahili and English but a language in the south called Luganda is mainly spoken alongside other languages, while the currency is known as Ugandan Shilling. The country's climate is not uniform because of modifications by altitude.

The Republic of Uganda popularly called the "Pearl of Africa" is geographically located in Eastern Africa, occupying 241,550.7 square kilometres of land. The name Uganda is taken from the Buganda Kingdom while the capital is called Kampala. Uganda shares borders with South Sudan in the North, Kenya in the East, Tanzania and Rwanda in the South and the Democratic Republic of Congo in the West.

The 1950s was the beginning of commercial tourism with notable growth of about 34% between 1962 and 1971. As of that period, the tourism industry was the third-largest source of foreign exchange revenue for Uganda apart from coffee and cotton. However, political instability between 1971 and 1986 resulted in wildlife poaching, destruction of tourism infrastructure, superstructure and facilities. Consequently, the earnings from the tourism industry declined by 13.8percent annually. Most tourists usually visit the Northern part of the country especially Lake Victoria National Park. Only 1,000 tourists visited the country in 1980 as compared to 85,000 in 1969.

The tourism industry was revived in 1987 following a peaceful political atmosphere in the country. A Tourism Master Plan was developed in 1993 with a forecast of an annual growth rate of 13 percent but this rose to 53 percent in 1996. In 1995, the tourism earnings were $90 million compared to $0.125 million in 1980. In the last few decades, the industry has been witnessing few setbacks consequent to the Great Lakes Region's political instability. A study by Thomas, Barya & Katongole (2011) shows that tourism is becoming a fast-growing segment of the Ugandan economy with an annual growth rate of 21 percent over 1992 through 2000. In 2010, 946,000 tourists visited Uganda while tourist spending as of 2011 was $662 million.

The Uganda Wildlife Authority (UWA) in conjunction with other stakeholders are charged with the conservation, sustainability and development of tourism economic resources. Also, the Uganda Tourist Board was established in 1994 to promote and popularize Uganda as a viable holiday destination both locally and internationally in order to increase the contribution of tourism to GDP, improve Uganda's competitiveness as an international tourism destination, and increase its share in the African and world tourism market. The government of Uganda indicates that actions should be taken to improve the quality of public infrastructure in the National Development Plan (2010-2014).

Additionally, the World Bank has recently funded four separate sustainable tourism development plans for four districts in Western Uganda. The government has positively articulated its intent to prioritize and strongly support the tourism sector, through resource allocation and strategic prioritization have not favoured the sector. Even though tourism is yet to be fully developed as expected, it has been contributing significantly to the country's revenue. Tourism is also causing conflicts between the various poor local residents in the destination areas. This condition led to the development of inclusive tourism in order to provide new or alternative sources of income for the locals.

Basic features of tourism in Uganda comprise large and small lakes such as Lake Victoria, Lake George and others. The source of the Nile was found in Uganda by British explorers in the 1860s. Uganda has abundant wildlife and game in about 60 protected areas, nine national parks, eco-tourism sites, and other nature reserves, as well as religious and cultural attractions such as dance ceremonies and lifestyles (Ashley, Roe & Goodwin, 2001).

The policy on pro-poor inclusive tourism is also aimed at developing and promoting eco-tourism, community development and sustainable conservation. Other key goals highlighted include revenue generation through the promotion of longer lengths-of-stay with higher daily expenditure and wider distribution of revenue earnings through large-scale participation of communities and districts in tourism; channelling tourism revenue towards the conservation of natural resources to generate local interest in environmental protection, and emphasizing tourism development based on Uganda's rich culture and history and the diverse traditions and hospitality of its people to forge a Ugandan identity and encourage cultural pride.

Results of the study show that tourists visit the country for family visitation, study, internship or volunteer work, round trip, game viewing, visit local communities, bird watching, spiritual holiday, cultural holiday, honeymoon, adventure, festivals, eco-tourism and spa or wellness. The tourists came from America, Kenya, Europe, United Kingdom, Tanzania, Rwanda, Asia Pacific, Congo, Sudan, Middle East and Other Africa countries.

Policies that affect tourism development in Uganda, include the Uganda Wildlife Policy (2014), the Uganda Wildlife Act (2019) and Wildlife Conservation by the Uganda Wildlife Authority (2011). However, it was realized that the increasing number of visitors to Uganda annually and the key tourism sites for pro-poor inclusive tourism could have a strong impact on poverty reduction in the country (Marx, 2011; Mitchell & Ashley, 2010). Uganda's poorest people include millions of subsistence farmers living in remote, scattered areas throughout the country.

As a result, the government has acknowledged that some indirect effects of tourism such as buying local products, souvenirs, supplies and services by tourists could create direct and indirect employment for the poor. The poorest regions are the north and northeast of the country, where outbreaks of civil strife and cattle stealing have disrupted the lives of smallholder farmers, agricultural production and other economic activities. Four kinds of inclusive tourism products were identified in the country. First, ethical fashion products. Second, lifestyle products such as homeware and street fashion. The fourth kind is visual arts usually sold in hotels, resorts and galleries and lastly, music performance at tourist locations and records sold to tourists. Some of the locally made items produced from available raw materials in the country are basketry, mats, ceramics, beads, pottery, hand textiles, woven products, toys, jewellery, bags, ornaments, leather products, batiks, and wood crafts. The crafts sector is culturally practised by different generations and dominated by the youth, seeking additional income but there is the need for more sustainable incomes (Lansing & De Vries, 2007; Islam, 2015).

This inclusive tourism strategic planning project carried out was based on a request from the Ugandan Government to ITC. The government sought to develop competitive micro establishments geared towards providing employment opportunities and sources of income for the marginalized and poor communities. The three stages in the strategic planning process are comprised of the following:

Diagnosis of Current Situation and Context: This stage involved the mapping of the tourism economy of selected destinations through the active participation of the poor to enable them to understand the financial inflows into the current tourism sector.

Project Opportunities, Prioritization, and Feasibility: This process enabled a systematic approach to develop a 'long list' of project options and guided movement towards a 'short list' of high priority interventions for achieving poverty reduction.

Table 3. Stakeholders covered in the planning process

Location	Stakeholders
Kampala	hotels, tour companies, airport and bus terminals together with craft markets, including Exposure Africa, the National Theatre Market, Banana Boat and Ndere Cultural Centre
kafu local community	Market vendors and their leaders
Packwach,	Anyayo Hotel, Global Village Hotel, Albertine Conservation and Tourism Association, Packwach Rural Environmental Agro-Forestry, Pacer Village Black Smiths Community, Bero Pamungu Community Based Organization, BeroPamungu Drama and Farmers Group
Murchison Falls National Park	Boomu women's group community tourism, Red Chili Hideaway Camp, Para Lodge, Mubako Community Integrated Association, Uganda Wildlife Authority Office.
Fort Portal and Rwenzori Area	Kyaninga Lodge, Kyaninga Community Group, Nyabwina Women and Cultural Group, Nubian-Njara Muslim Women Catering and Hiring Services Group, Whispering Gardens Camping Centre, Kbarole Tourism Association and Kabarole Tours and Travel.
Queen Elizabeth National Park	Kikorongo Women Community; Simba Safari Lodge, Katwe Salt Lake Community, Katwe Tourism Information Centre, Katunguru Highway Crafts Centre, Kyambura Community Lodge and Kingfisher Lodge.
Bwindi National Park	Buhooma Community Camp Ground, Batwa Cultural Centre Crafts Shop and Booking Office, Batwa Development Program and Volcanoes Safari Lodge
Kabale-Lake Bunyonyi Area	Nyambugo Women Group, Bufumuka Village Community and Birds Nest Tourist Hotel
Governmental and non-governmental agencies	Uganda Tourism Board, Uganda Export Promotion Board, USAID, Sustainable Tourism in the Albertine Rift (STAR), Uganda Community Tourism Association (UCOTA) and Ministry of Trade (MOT).

Source: Thomas, F., Barya, G., & Katongole, C. (2011, pp.100-101).

Project Planning: At this phase, the focus was on packaging proposed interventions into a project document for future interventions.

In line with the purpose of the study, the strategic planning on inclusive pro-poor tourism identified the tourism-related problem during the study and provided possible interventions to help address the problems within the International Trade Center's (ITC) logical framework. ITC broadened the scope of the study which initially only focused on enabling artisans that work as micro-producers along with Ugandan artists to reach tourists. It involved two other UN agencies (UNDP and UNCTAD) that were in favour of the sustainable development of the tourism sector in Uganda.

The consultants engaged stakeholders across the country in general and the tourism industry in particular. Nine areas were covered during the study as presented in Table 3 below.

The areas of expertise included tourism marketing, specific service provisions such as computer and information technology, and tourism and hospitality management. They linked agriculture, artisanal product and artistic services to tourism by enhancing local communities' development and environmental management. The general analysis focused on the tourism competitiveness, including the competitiveness of the crafts sectors, before looking at the policy objectives and strategies to manage the development of the tourism sector. The main issues classified within three broad categories in the planning process were: communication and marketing, products and services competitiveness and tourism management.

Also, a number of issues were identified among various stakeholders including the following:

- **Central and local governments**: ineffective tourism management: lack of competitiveness (costing, pricing, standardization), the ineffectiveness of tourism marketing and communication, poor mismanagement of tourism funds (park entrance fees), poor infrastructure in some areas (roads, power, telecommunications, access to water). Weaknesses in the decentralization process: management of tourism is at the central government level and is not assimilated within local government plans, district commercial offices are not empowered, and there is a lack of effective awareness of and sensitization of people in general issues pertaining to tourism.
- **Private sector**: Lack of competitiveness of middle-range/low-end supply (low cost/value products), lack of support for the establishment of linkages with local communities, low remuneration or absence of contracts, absence of regional leadership, recruitment of non-professional and unqualified personnel and weak, disjointed and uncoordinated opportunistic approach to business.
- **Local communities**: Low benefits derived from tourism by the poor, lack of skills and capacities, lack of knowledge and appreciation of tourism development, and exploitation by the private sector.
- **NGOs and CBOs**: Lack of capacities and technical skills for product diversification and innovation, lack of funds, lack of coordination among tourism actors to create strong linkages among CBOs, communities and other tourism sectors.
- **Tourists**: Lack of information about community-based activities, product origin and poverty reduction opportunities.

Based on the identified challenges, a number of interventions embarked upon which included the following:

- Enhancement of skills of producers, as well as instituting production and sales mechanisms by developing and diversifying products from raw materials to packaging (including pricing and costing).
- Upgrading of the quality and production of (selected) commercially viable products by organizing training on sustainability and business development for women and other groups in the communities.
- Providing community assistance towards registration of businesses and fundraising as well as setting up (spaces/areas) to sell food and souvenirs.
- Strengthening backward linkages through capacity building to increase production and respond to the needs of sellers.
- Developing a corporate and business pro-poor product catalogue for the business tourism sector in Kampala.
- Supporting a national competition on crafts, visual arts, entertainment and cuisine (Ketwu Festival).
- Ensuring that the private sector and local producers in the area enter into contractual agreements.

The study showed that consultants collaborated with the government of Uganda in an effort to improve the economic empowerment of poor locals in the country. The consultants were able to approach UNDP and secure financial assistance for the project and this helped in accomplishing the project. The government of Uganda engaged foreign consultants towards the success of the project and that act helped them to complete the task within three (3) months as scheduled.

An improvement in the benefits of CBT to local communities also means a proper engagement of both local communities and tourism businesses. This enabled the communities to support and create branded crafts and enhanced the production of diversified and specialized products such as agricultural products (Fariborz, Sarjit & Farshid, 2010; Ijasan & Izobo, 2012).

At the end of the study, the consultants made certain recommendations to the government in order to help solve the challenges facing the development of inclusive tourism. Recommended actions are a crucial part of the strategic planning process. They, therefore, recommended actions that could be taken which include "design support and brand creation to enhance the production of diversified/specialized products, business development and other types of sales opportunities.

IMPLICATIONS OF THE UGANDA INCLUSIVE PRO-POOR TOURISM STRATEGIC PLANNING FOR CBT DEVELOPMENT IN NIGERIA

The process and outcomes of strategic planning on pro-poor inclusive tourism for Uganda is a good example of strategic planning that can be adopted and applied in future tourism planning initiatives in any community due to the impacts that it has generated.

The inclusive PPT planning in Uganda conducted in 2011 has been making significant impacts on employment generation, supply of goods and services in tourism enterprises by the poor and direct sales of goods and services to visitors by the poor (informal economy). Others are running of tourism enterprises by the poor (SMMEs or CB enterprises); taxes or levies on tourism revenues or profits with proceeds benefiting the poor and donations in cash and kind by tourists and enterprises in ways that benefit the locals. It has also resulted in infrastructural investment that can also provide better means of livelihood to the poor and communities in Uganda (Adiyia & Vanneste, 2018; MacKAY & Campbell, 2012).

The lessons learnt from the Ugandan initiative above implies that for Nigeria to develop CBT, tourism planners, government and other stakeholders should see themselves as partners in strategic planning for CBT development (Brunet & Aubry, 2016; Di Minin, Leader-Williams & Bradshaw, 2016; Ullah & Kim, 2020). Undoubtedly, CBT will impact the locals in the communities in Nigeria through employment generation. However, CBT development requires strategic planning to create a unique and valuable position for CBT destinations in Nigeria. Figure 2 depicts a proposed strategic planning process for CBT development in Nigeria.

There is a need for strong leadership in the tourism sector by tourism authorities including the government. Essentially, the Ministry of Tourism can liaise with the development agencies in identifying the priorities for the tourism sector. Thus, CBT strategic planners ought to be committed, well-informed, willing to delegate and empower others (Tanjua, 2003). First, there should be a formulation, implementation and review of unique policies and strategies in CBT by the destination management team. Next, a certain level of both physical facilities and service quality within the CBT destination areas should be maintained. Lastly, there should be consultations and collaboration with other related bodies to ensure that the relevant aspects of the destination conform to both national and international standards. Having a National CBT policy as an extension of National Tourism policy is imperative to CBT development in Nigeria.

The destination management organization (DMO) needs to collaborate with stakeholders. The Ministry of Tourism can take a lead role in sustainable CBT development in Nigeria (Christie, 2002; Dieke, 2003; Fagbolu, 2021).

Figure 2. Proposed strategic planning process for CBT development in Nigeria
Source: *Author's Construct*

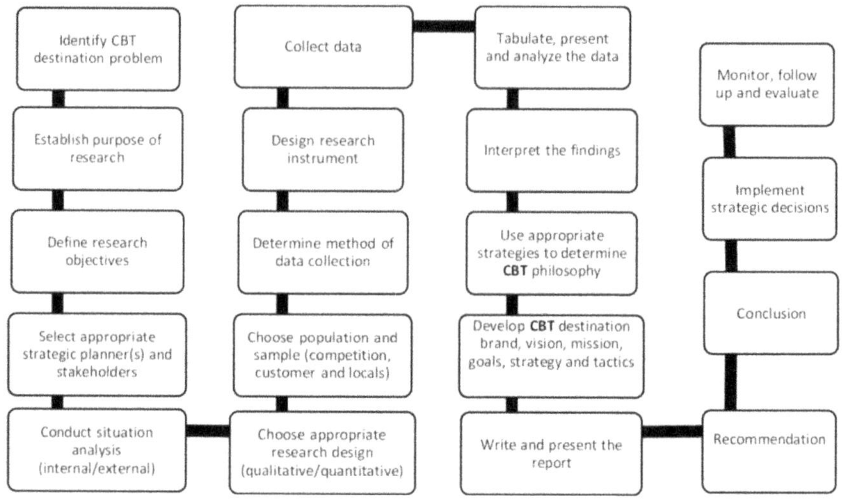

The basic elements of the policy structure include CBT philosophy, vision, objectives and constraints on supply with demand strategies, operational and tactical supply and demand programs and macro-environmental structure (Edgell & Swanson, 2013).

CBT policy formulation can be in four stages, namely, definitional, analytical, operational and implementation stages. At the definitional stage, it is important to define the CBT destination system, identify CBT philosophy, and develop a vision and goals. Situation analysis is conducted during the analytical stage that involves analysis of the micro and macro environment and review of competitive and sustainable development. The operational stage focuses on identifying strategic conclusions for supply and demand together with recommendations on CBT policy.

For CBT policy to be beneficial to Nigeria, it must;

- Facilitate consensus around specific strategies and objectives for a given destination and define the rules of the game that is, the terms under which CBT operators must function.
- Set out activities and behaviours that are acceptable for visitors and provide a common direction and guidance for all CBT stakeholders within a destination.
- Provide a framework for discussions by the public and private sectors on the role and contributions of the tourism sector to the economy and to society in general and allow CBT to interface more effectively with other sectors of tourism.

- Concentrate on how important and scarce resources could be acquired to meet the needs and provide opportunities for local communities.
- Pay attention to macro-level policies and encourage creativity in order to avoid policies based on stereotyped or outmoded perceptions.
- Permit and facilitate a continuing dynamic social process requiring inputs from multiple sources and associate policies of CBT to those of the total socioeconomic system of destinations or the nation as a whole.
- Have a long-time frame and eliminate the traditional boundaries which render the CBT sector in tourism ineffective.

The cycle is incomplete without implementation. In fact, CBT policy formulation without implementation will be a wasteful venture. The implementation stage consists of strategies which could be for development, promotion and stewardship of the destination, allocation of responsibilities, identification of sources of funding to support tourism programs, specification of timing for implementation of the policy as well as monitoring and evaluation of results (Edgell & Swanson, 2013; Fagbolu & Fallon, 2021; Fallon & Fagbolu, 2021; Moutinho, 2000; Pechlaner & Sauerwein, 2002).

In order to develop a successful CBT policy, the first thing to do is to develop an appropriate strategy for the development, growth, and even diversification of CBT products so as to build capacities at all stages of the CBT intervention and management processes. This chapter proposes eight (8) strategies for CBT destination development. First, matching CBT products with needs. Second, optimally allocating and directing CBT resources. Next, using relevant market intelligence to sense change and to change strategies. Fourth, having greater ability to tailor behaviour and logistics. Fifth, being able to be unique in positioning and differentiating the CBT destination from the competition. Sixth, determining strategies and enlarge the core market. Up next, Organizing frequent traveller programs. Seventh, Organizing frequent traveller programs. Lastly, maintaining consistent CBT product lines and brand integrity.

Planners should endeavour to define the CBT philosophy of the destination and formulate a supportive policy, vision, mission, goals, strategy and tactics (Moutinho, 2000; Tony & Etienne, 2002). These in turn will provide direction and guidance for detailed planning and development initiatives that determine the nature and quality of the experiences the destination is capable of offering. CBT destination management organizations and authorities involved in tourism policy has the obligation to work together towards strategic planning as they control a reasonable portion of event programs, planning, development, and facilities among others.

A workshop could be organized to define strategic interventions for CBT development in order to provide better gains for local communities and to enlighten private and NGO sectors on the roles they can play as collaborators in the tourism industry. The study can be applied in the presentation of evidence-based objectives and interventions to all tourism stakeholders (government: national and local, private sector, international organizations, local communities, etc.) so as to determine their costs, feasibility, impact and return on investment. Government can encourage voluntary involvement of stakeholders in strategic tourism plans in order to ensure participation that is geared towards the achievement of the objectives of tourism projects in the communities.

CONCLUSION

The chapter discusses the current status and challenges of CBT in Nigeria. The chapter also examines the implications of strategic planning for inclusive pro-poor tourism in Uganda for strategic planning for CBT development in the Nigerian context. The CBT strategic planning process is a very critical tool, which will help destination management organizations in Nigeria to know why people choose a destination and to discover tourists' expectations and perceptions. It will assist to ascertain who the real competitors are, the rate of repeat visits, as well as the destination's competitive advantage, target market, the extent of tourism promotion, and levels of tourists' satisfaction (Amalu & Ajake, 2018; Asuk, Ebu & Ifebueme, 2018; Comerio & Strozzi, 2019; Vanhove, 2002).

Strategic planning and a good tourism policy will yield a competitive and sustainable destination (Kozak & Baloglu, 2011). Strategic plans help to guard against failure, which may occur because of poor preparation, hastily formulated goals, lack of sufficient data, and lack of adequate linkage between strategic planning and other control systems. It is critical that public, private and non-governmental organizations in Nigeria rise up to combat the challenges facing CBT by providing funds for strategic planning and development so as to help develop and sustain the sector through research.

Strategic tourism planners and researchers should in turn always endeavour to conduct research so as to formulate policies and programmed that fit their situation. CBT should provide opportunities for the locals. Local people should have autonomy and make decisions that are right for their communities once they are actively and deeply involved in running, leading and sustaining CBT projects in their communities.

REFERENCES

Adetola, B. O. (2015). Ecotourism impacts on support zone communities around Cross River National Park, Nigeria. *Journal of Sustainable Development in Africa, 17*(5), 153–163.

Adiyia, B., & Vanneste, D. (2018). Local tourism value chain linkages as pro-poor tools for regional development in Western Uganda. *Development Southern Africa, 35*(2), 210–224. doi:10.1080/0376835X.2018.1428529

Ahmadu, N. O. (2011). Hospitality and tourism in Nigerian polytechnics and colleges of Technology: Challenges and prospects. *HATMAN Journal of Hospitality and Tourism, 1*(1), 65–70.

Amalu, T. E., & Ajake, A. O. (2018). Developing natural lakes for socio-economic development: The case of Nike Lake, Enugu State, Nigeria. *GeoJournal, 78*(4), 156–172.

Amalu, T. E., Otop, O. O., Duluora, E. I., Omeje, V. U., & Emeana, S. K. (2017). Socio-economic impacts of ecotourism attractions in Enugu State, Nigeria. *Geojournal Spatially Integrated Social Sciences and Humanities*. doi:10.1007/s10708-017-9830-7

Amalu, T. E., Ugwu, O., Idam, N. S., & Oko-Isu, P. E. (2019). Impacts of seasonality and patronage of lakes on the economy of host communities (a case of Nike Lake, Enugu Nigeria). *Journal of Travel. Tourism and Recreation, 1*(1), 7–13.

Amerta, I. M. S. (2017). Community based tourism development. *International Journal of Social Science and Humanity, 1*(3), 97–107.

Aref, F., Gill, S. S., & Aref, F. (2010). Tourism development in local communities: As a community development approach. *The Journal of American Science, 6*(2), 155–161.

Ashley, C., Roe, D., & Goodwin, H. (2001). *Pro-poor tourism strategies: Making tourism work for the poor: A review of experience*. Overseas Development Institute.

Asuk, S. A., Ebu, V. T., & Ifebueme, N. M. (2018). Assessment of community-based ecotourism prospects in southern Nigeria: Case study of Iko Esai community. *International Journal of Agricultural Policy and Research, 6*(7), 111-121. https://doi.org/10.15739/IJAPR.18.013

Bassey, F. O., & Egon, C. I. (2016). Nigerian tourism policy: A critical appraisal and re-appraisal. *Journal of Tourism. Hospitality and Sports., 20*(1), 26–29.

Blackstock, K. (2005). A critical look at community-based tourism. *Community Development Journal: An International Forum*, *40*(1), 39–49.

Briones, Z. B. H., Yusay, R. M. S., & Valdez, S. (2017). Enhancing community-based tourism programs of Gawad Kalinga Enchanted Farm towards sustainable tourism development. *Journal of Economic Development, Management, IT. Finance and Marketing*, *9*(1), 51–60.

Brunet, M., & Aubry, M. (2016). The three dimensions of a governance framework for major public projects. *International Journal of Project Management*, *34*(8), 1596–1607.

Burgos, A., & Mertens, F. (2017). Participatory management of community-based tourism: A network perspective. *Community Development (Columbus, Ohio)*, *48*(4), 546–565.

Burton, S., & Steane, P. (2004). *Surviving your thesis* (1st ed.). New York: Routledge.

Chan, K. L. (2009). The consumption of museum service experiences: Benefits and value of museum experiences. *Journal of Hospitality Marketing & Management*, *18*(2-3), 173–196.

Christie, I. T. (2002). Tourism, growth and poverty: Framework conditions for tourism in developing counties. *Tourism Review*, *57*(1/2), 35–41.

Cole, S. T., & Illum, S. F. (2006). Examining the mediating role of festival visitors' satisfaction in the relationship between service quality and behavioral intentions. *Journal of Vacation Marketing*, *12*(2), 160–173.

Comerio, N., & Strozzi, F. (2019). Tourism and its economic impact: A literature review using bibliometric tools. *Tourism Economics*, *25*(1), 109–131.

David, L. (2011). ICT and digital records management in the Ugandan public service. *Records Management Journal*, *21*(2), 135–144. doi:10.1108/09565691111152062

Di Minin, E., Leader-Williams, N., & Bradshaw, C. J. (2016). Banning trophy hunting will exacerbate biodiversity loss. *Trends in Ecology & Evolution*, *31*(2), 99–102.

Dieke, P. U. C. (2003). Tourism in Africa's economic development: Policy implications. *Management Decision*, *41*(3), 287–295.

Dimoska, T., & Trimcev, B. (2012). Competitiveness strategies for supporting economic development of the touristic destination. *Procedia: Social and Behavioral Sciences*, *44*, 279–288.

Edgell, D. L., & Swanson, J. R. (2013). *Tourism, policy and planning: yesterday, today and tomorrow* (2nd ed.). Routledge.

Emeka, N., Ejim, E. P., & Amaka, O. (2015). Effect of strategy formulation on organizational performance: A study of Innoson manufacturing company Ltd Emene, Enugu. *Word Journal of Management and Behavioral Studies*, *3*(1), 9–20.

Esu, B. B. (2015). Transforming the Nigeria tourism industry through tourism entrepreneurial development. *African Journal of Business Management Review*, *9*(15), 569–580.

Eze-Uzomaka, P. I. (2006). Tourism & community development in issues in tourism planning and development. Afro-Orbis Publishing Co.

Fagbolu, A. O. (2021). Hospitality and tourism management education versus employment in Nigeria. *Global Scientific Journals, 9*(5), 131-139. https://www.globalscientificjournal.com/journal_volume9_issue5_May_2021_edition.ht ml

Fagbolu, A. O., & Fallon, J. (2021). Strategies exploration for academic achievement improvement through organization development intervention (ODI): A study of Kwara state university. *ABAC ODI Journal*, *8*(1), 161–179.

Fallon, J., & Fagbolu, A. O. (2021). Developing possible strategies for academic achievement improvement of hospitality and tourism management students in Nigeria: A study of Kwara state university. *ABAC Journal*, *41*(2), 177–203. http://www.assumptionjournal.au.edu/index.php/abacjournal/issue/current

Fariborz, A., Sarjit, S. G., & Farshid, A. (2010). Tourism development in local communities: As a community development approach. *The Journal of American Science*, *6*(2), 155–161.

Getz, D. (2008). Event tourism: Definition, evolution, and research. *Tourism Management*, *29*(3), 403–428.

Giampiccoli, A., & Mtapuri, O. (2020). Towards a coalescence of the community-based tourism and 'Albergo Difusso' tourism models for Sustainable Local Economic Development. *African Journal of Hospitality, Tourism and Leisure*, *9*(1), 1–20.

Goeldner, C. R., & Ritchie, J. R. B. (2009). *Tourism principles, practices, philosophies* (11th ed.). John Wiley & Sons, Inc.

G¨Ossling, S. (2009). Carbon neutral destinations: A conceptual analysis. *Journal of Sustainable Tourism*, *17*(1), 17–37.

Guo, Y., Jiang, J., & Li, S. (2019). A sustainable tourism policy research review. *Sustainability, 11*(3187), 1–16.

Gwang-Chol, C. (2006). Strategic planning in education: Some concepts and steps. Paris: Division of Educational Policies and Strategies, United Nations Educational, Scientific and Cultural Organization (UNESCO).

Igbojekwe, P. A. (2011). Globalization: Challenges on the development of hospitality and tourism in Nigeria. *HATMAN Journal of Hospitality and Tourism, 1*(1), 42–50.

Ijasan, K. C., & Izobo, M. O. (2012). Assessing community engagement in tourism planning and development in Nigeria: A case study of Arinta Waterfall Tourist Resort, Ipole Iloro Ekiti State. *Transnational Journal of Science and Technology, 2*(4), 11–20.

Islam, M. S. (2015). Study on factors influencing tourism: Way forward for sustainable tourism in Bangladesh. *Journal of Tourism. Hospitality and Sports, 6*(1), 1–13.

Jennings, G. (2010). *Tourism Research.* John Wiley & Sons.

Kozak, M., & Baloglu, S. (2011). *Managing and marketing tourist destinations: Strategies to gain a competitive edge* (1st ed.). Routledge.

Kruczek, Z., Kruczek, M., & Szromek, A. R. (2018). Possibilities of using the tourism area life cycle model to understand and provide sustainable solutions for tourism development in the Antarctic region. *Sustainability, 10*(89), 1–16.

Kwangseh, B. E. (2014). *Community based tourism (CBT) planning –An analysis of opportunities and barriers: A case study of Cameroon* (Doctoral Dissertation). Eastern Mediterranean University, Gazimağusa, Cyprus.

Lansing, P., & De Vries, P. (2007). Sustainable tourism: Ethical alternative or marketing ploy? *Journal of Business Ethics, 72*(1), 77–85.

Lewis, R. C., & Chambers, R. E. (2000). *Marketing leadership in hospitality: Foundations and practices* (3rd ed.). John Wiley & Sons, Inc.

MacKay, K. J., & Campbell, J. M. (2012). Innovative community tourism in Uganda: International tourists, national networks, and local livelihoods. *Innovations in Sustainable Tourism (TEOROS),* 59-64.

Marx, S. (2011). Community-based & pro-poor tourism. *International Trade Forum,* 22.

Mayaka, M., Croy, W. G., & Cox, J. W. (2019). A dimensional approach to community-based tourism: Recognising and differentiating form and context. *Annals of Tourism Research*, *74*(1), 177–190.

Menbere, I. P., & Menbere, T. P. (2017). Opportunities and challenges for community-based ecotourism development: A case study in Dinsho and Goba Woredas, Southeast Ethiopia. *International Journal of Ecology*, *4*(1), 5–16.

Mensah, I. (2017). Benefits and Challenges of Community Based Ecotourism in Park-Fringe Communities: The Case of Mesomagor of Kakum National Park, Ghana. *Tourism Review International*, *21*(1), 81–98.

Mitchell, J., & Ashley, C. (2010). *Tourism and poverty reduction: Pathways to prosperity*. Earthscan.

Moutinho, L. (2000). *Strategic management in tourism*. CABI Publishing.

Mtapuri, O., & Giampiccoli, A. (2016). Towards a comprehensive model of community-based tourism development. *The South African Geographical Journal*, *98*(1), 154–168.

Nickols, F. (2016). *Strategy: Definitions and meanings*. Distance Consulting LLC. Retrieved from http://www.nickols.us/strategy_definitions.pdf

Pechlaner, H., & Sauerwein, E. (2002). Strategy implementation in the Alpine tourism Industry: International perspectives on travel and tourism development. *International Journal of Contemporary Hospitality Management*, *14*(4), 157–168.

Pillay, M., & Rogerson, C. (2013). Agriculture-tourism linkages and pro-poor impacts: The accommodation sector of urban coastal KwaZulu-Natal, South Africa. *Applied Geography (Sevenoaks, England)*, *36*(1), 49–58.

Ritchie, W. B. (2004). Chaos, crises and disaster: A strategic approach to crisis management in the tourism industry. *Tourism Management*, *25*(6), 669–683.

Russell, M., & Scheyvens, R. (2012). Tourism and poverty alleviation in Fiji: Comparing the impacts of small-scale and large-scale tourism enterprises. *Journal of Sustainable Tourism*, *20*(3), 417–436.

Schiffman, L. G., & Kanuk, L. L. (2000). *Consumer behavior* (7th ed.). Prentice Hall/Pearson Education.

Seaton, A. V., & Bennett, M. M. (1996). *Marketing tourism products: Concepts, issues, cases* (1st ed.). International Thompson Business Press.

Sène-Harper, A., & Séye, M. (2019). Community-based tourism around national parks in Senegal: The Implications of colonial legacies in current Management policies. *Tourism Planning & Development*, *16*(2), 217–234.

Snyder, C. R., Harris, C., Anderson, J. R., Holleran, S. A., Irving, L. M., Sigmon, S. T., ... Harney, P. (1991). The will and the ways: Development and validation of an individual-differences measure of hope. *Journal of Personality and Social Psychology*, *60*(4), 570–585.

Stavros, J. M., & Hinrichs, G. (2009). The thin book of SOAR: Building strengths-based strategy. The Thin Book Publishing Co.

Tanjua, A. (2003). Innovative human resource practices and organizational commitment: An empirical investigation. *International Journal of Human Resource Management*, *14*(2), 175–197.

Terencia, N. M. (2018). Community based tourism and development in third world countries: the case of the Bamileke region of Cameroon: World academy of science, engineering and technology. *International Journal of Social and Tourism Sciences*, *12*(1), 26–30.

The Republic of Uganda. (2011). *Office of the Auditor General: Value for money audit report on the management of wildlife conservation by the Uganda Wildlife Authority*. http://www.oag.go.ug/wp-content/uploads/2016/07/uwa-final-report-for-signature.pdf

The Republic of Uganda. (2014). *The Uganda Wildlife Policy: Ministry of Tourism, Wildlife and antiquities*. National Policy on Conservation and Sustainable Development of Wildlife Resources. http://extwprlegs1.fao.org/docs/pdf/uga174672.pdf

The Republic of Uganda. (2019). *The Uganda Wildlife Act*. file:///C:/Users/JM505%20 Computers/Downloads/Uganda%20Wildlife%20Act,%202019.pdf

Theerapappisit, P. (2009). Pro-poor ethnic tourism in the Mekong: A study of three approaches in northern Thailand. *Asia Pacific Journal of Tourism Research*, *14*(2), 20–31.

Thomas, F., Barya, G., & Katongole, C. (2011). *Opportunity study: Inclusive tourism programme: Uganda inclusive tourism*. File:///C:/Users/Jm505%20computers/Downloads/Itc_Opportunity_Study_Uganda.Pdf. Retrieved On 19/08/2021

Tony, B., & Etienne, N. (2002). Tourism as a local development strategy in South Africa. *The Geographical Journal*, *168*(3), 235–247.

Tourism Development International. (2006). *Nigeria tourism development master plan: Institutional capacity strengthening to the tourism sector in Nigeria.* Retrieved from http://www.ibadektyma.com/tourism_masterplan.pdf

Tucker, H. (2011). Critiquing pro-poor tourism – by Scheyvens Regina. *Asia Pacific Viewpoint, 52*(3), 379–380.

Ullah, I., & Kim, D. (2020). A model of collaborative governance for community-based trophy- hunting programs in developing countries. *Perspectives in Ecology and Conservation, 18*(1), 145–160.

United Nations World Tourism Organization (UNWTO). (2002). *Tourism and Poverty Alleviation*. UNWTO.

Vanhove, N. (2002). Tourism policy - between competitiveness and sustainability: The case of Bruges. *Tourism Review, 57*(3), 34–40.

Veal, A. J. (2011). *Research methods for leisure & tourism* (4th ed.). Prentice Hall/ Pearson Education.

World Bank. (2000). *Attacking Poverty: World Development Report 2000/01*. World Bank.

World Tourism Organization (UNWTO). (2016). *Tourism Highlights*. Calle Capitán Haya. https://www.e-unwto.org/doi/book/10.18111/9789284418145

Yang, X., & Hung, K. (2014). Poverty alleviation via tourism cooperatives in china: The story of Yuhu. *International Journal of Contemporary Hospitality Management, 26*(6), 879–906.

Chapter 3
Community-Based Tourism and Local People's Perceptions Towards Conservation:
The Case of Queen Elizabeth Conservation Area, Uganda

Jim Ayorekire
https://orcid.org/0000-0001-6415-3792
Makerere University, Uganda

Francis Mugizi
Makerere University, Uganda

Joseph Obua
Makerere University, Uganda

Grace Ampaire
Makerere University, Uganda

ABSTRACT

Uganda is among the most bio-diverse countries and a competitive wildlife-based tourism destination in the world. Community-based tourism approach has been adopted in the country's conservation areas as a strategy to ensure that local communities benefit and support wildlife conservation. This chapter analyses local communities' perceptions of conservation and the benefits they get from tourism in Queen Elizabeth Conservation Area. The study reveals that local communities were concerned about loss of protected resources and support their conservation irrespective of the benefits they get from tourism in the conservation area. There is need to design conservation programmes that focus on local community-conservation-benefits nexus which take into consideration the perceived conservation values, strategies for benefit sharing and incorporation of indigenous knowledge systems.

DOI: 10.4018/978-1-7998-7335-8.ch003

INTRODUCTION

Community-based tourism (CBT), also known as community-based tourism initiative (CBTI) or rural-based community tourism (RCT) is one of the alternative strategies for empowerment of local communities through enhancement of their capacity to harness opportunities in tourism related enterprises around protected areas (Giampiccoli & Mtapuri, 2012). CBT is intended to promote locally based development processes by building local capacity in creating jobs with positive spin offs for the community including indirect impact by tapping into local supply chains of the informal economy (Giampiccoli & Mtapuri, 2012). It focuses on mechanisms through which communities can collaborate with other actors to meet locally defined community development needs and deliver livelihood and other benefits through tourism enterprises (Mayaka et al., 2019).

For conservation to succeed the local communities need to have positive perceptions towards protected areas and wildlife resources. According to Eaton & Visser (2008), perceptions have predictable influence on behavior. Thus, predicting local community's conservation behavior requires knowledge of their perceptions which can guide conservation area managers and local community leaders on how to mobilize and engage them in community-based tourism programmes. Limited community awareness and negative perception of conservation lead to destructive activities such as poaching and unsustainable harvesting of plants that compromise conservation and jeopardize tourism. Knowledge of local community perceptions is, therefore, crucial in Queen Elizabeth Conservation Area (QECA) which is surrounded by a high population who's positive perceptions of conservation is pivotal in sustainable tourism.

For a long time, community-based tourism has been promoted to meet the social, environmental and economic needs of local communities. According to Weaver (2010), Hall (2005) and WWF (2001), the concept CBT originated in the 1980s, when it was regarded as a development alternative for rural communities. At that time, the World Conservation Strategy linked protected area management with economic activities that would benefit local communities (Mearns, 2003). Community-based tourism was, therefore, considered a viable approach to empower local communities to take control of their land and resources as well as acquire skills to develop themselves (Mearns, 2003). CBT is a development alternative characterized by local community management and conservation of natural resources and economic empowerment (Goodwin & Santilli, 2009; Okazaki, 2008; Manyara & Jones, 2007). The application of community development approach is a resident-responsive strategy for motivating local communities to embrace community-based tourism in QECA and enhance their participation.

Tourism in developing countries provides economic benefits to rural communities. The benefits that are expected to accrue to the local communities in QECA include natural resource conservation through participation which in turn provides economic incentives for improved livelihoods. The benefits of community participation include strengthened capacity in planning and decision-making, ensuring equitable sharing of economic benefits from tourism often retained in the local communities, local control of development and incorporation of resident values (Mensah, 2017).

New paradigms under sustainable development support and encourage self-help, self-reliance and empowerment of communities to engage in different enterprises including community-based tourism. Barriers to community-based tourism include weak policies to support socially inclusive and ecologically sound community-based tourism (Yanes et al., 2019) which the local community leaders and conservation managers in QECA need to bear in mind. When sustainable tourism was introduced as a new global standard, many countries designed policies to reflect the shift. As a result, community participation, empowerment, transparency, fairness, and equity were introduced and incorporated into national regulations and development strategies. However, these concepts have not been fully applied in community-based tourism.

Spenceley & Meyer (2012) defined community-based tourism as collective action by a community to participate in or develop small and medium-scale tourism enterprises while Tolkach & King (2015) and Salazar (2012) refer to community-based tourism as a community-managed enterprise in which local people participate in the planning and development of tourist products to generate income and conserve nature. In addition to the benefits outlined by Mensah (2017), community-based tourism promotes a destination's image and increases tourists' experience through environmental education and interpretation. In the early 1980s, community-based tourism was regarded as the *sine qua non* of alternative tourism (Weaver, 2010) because it provides opportunities for local communities to manage tourism resources and engage in tourism enterprises (Dodds et al., 2016), create jobs and stem rural-urban migration (Anuar et al., 2017). Furthermore, it establishes mutually beneficial relationships between the local community and tourists.

Public interest in community participation has grown in the last two decades with widespread support across the ideological spectrum (Usadolo & Caldwel, 2016). Stukas and Dunlap (2002) postulated theoretical approaches and practices in community involvement which underpins the concept of community-participation. Building on an earlier work by Kraft (1996), Stukas and Dunlap elaborated on the social structure that surrounds community participation and the factors that hinder it. They noted that community participation originated from the social psychology concept of collective involvement that benefits individuals and the community (Bamberger, 1986, Afify, 2011, Omoruyi Osagie, 2016). The aim of community participation is to build social capital networks, provide opportunity for engaging in socio-economic activities and improve local people's livelihoods (Omoruyi Osagie, 2016). It is widely recognized as the vehicle for achieving poverty alleviation, socio-economic transformation, and sustainable development (Pozzoni & Kumar, 2005; The World Bank, 2020).

Social Exchange Theory and Perceived Tourism Benefits

The Social Exchange Theory (SET) has been widely applied in understanding residents' perceptions of tourism development (Byrd et al., 2009; Akarsua, et al. 2020; Gan, 2020; Chang, 2021; Qin et al., 2021). It is one of the approaches that have made considerable theoretical contribution to the study of residents' perceptions of tourism (Fredline & Faulkner, 2000) based on the benefits value exchange (Nunkooa, 2016; Change 2021). Ap (1992), who pioneered application of SET in tourism research (Özel & Kozak, 2017), traced its origin to the general sociological theory concerned with interactive understanding of the exchange of resources between individuals and groups. SET is based on the premise that human behaviour or social interaction is an exchange of activity, tangible and intangible, particularly of rewards and costs (Homans, 1961), that affects patterns of interaction (Molm, 1991). The main hypothesis is that actors will engage in an exchange and interaction if the resulting rewards are of value to them and the perceived costs do not exceed the perceived benefits (Ap, 1992). In tourism, SET suggests that residents perceive the effects of tourism in terms of socio-economic benefits or costs that they expect in return for conserving resources that support tourism (Özel & Kozak, 2017).

Based on SET, this study conceptualizes and analyzes how the exchange and interaction of benefits and costs between the community and QECA have shaped the perceptions of conservation. Tourism's economic benefits to the community are viewed in terms of employment opportunities in the national park and tourist establishments, access to resources such as water and firewood, and availability of market for local produce such as fish sold to the tourist accommodation facilities. The social benefits include construction of schools and increased opportunities for education bursaries. The benefits from tourism from QENP which is part of QECA were analyzed to determine how they influence community perceptions of conservation. The perceptions combined the communities' views and opinions about the importance and value they attached to biodiversity conservation and maintenance of aesthetic landscape values.

CBT Approach, Local Community Perceptions and Benefits

Community-based tourism is linked to the concept of community development that was introduced in the 1960s to support rural development in post-independence of least developed countries, educate the local people and involve them in decision making and development activities (Zdenek & Walsh, 2017). Although there has been a global call for involvement of local communities in tourism to spur rural development and ensure that benefits continuously accrue from natural resources, local communities tend to view the costs as outweighing the benefits (Sindiga, 1995; Higginbottom, 2004). For instance, Mensah (2017) reported a number of disadvantages of CBT including weak connection with the mainstream tourism industry, centralized decision-making processes, lack of skills and knowledge, lack of an appropriate policy framework for the development of community-based initiatives, lack of ownership and low levels of employment, loss of habitats (due to infrastructural and super structural changes), culture, communal land and grazing areas. Community perceptions of tourism are, to a large extent, determined by their level of participation in conservation of resources, tourism activities and benefits they get from it. Given that participation is voluntary, there is need for deliberate action in which individuals establish a common understanding to harness opportunities and participate in community-based tourism (Tosun, 2000). Therefore, community participation should entail control over tourism resources, ownership and generation of benefits (Dodds et al., 2016) in order to establish a strong local community-tourist relationship, conserve environment and culture, inculcate social responsibility and improve livelihoods (Anuar et al, 2017).

The concepts of sustainable tourism and community-based tourism are critical dimensions of tourism that have evolved simultaneously (Dangi & Jamal, 2016) and whose relationship requires deeper understanding. In the last three decades, sustainable tourism debates have focused on the long-term sustainability of the sector and the extent to which community participation influences tourism development and management (Weaver, 2010). CBT approach addresses the dual demands of conservation and community development by linking conservation with local communities' livelihoods through provision of economic alternatives. The success of CBT is grounded on grassroots development approach where community participation and benefit are based on the level of equity and empowerment for decision making and management of community-based tourism enterprises (Saayman & Giampiccoli, 2016). Furthermore, CBT emphasizes capacity building, local community stewardship of natural and cultural resources, and sharing of tourism benefits taking into account the economic disadvantages and social injustice (Dangi & Jamal, 2016). This approach points to the need for community-based tourism to address current and future economic, social, and environmental impacts while meeting visitors' needs.

Tourism Growth, Conservation and Community Based Tourism in Uganda

Before COVID-19 pandemic, tourism was Uganda's highest foreign exchange earner generating USD 1.6 billion, contributing 7.7% of Uganda's GDP and employing 667,600 people (MTWA, 2018). As a result, it was prioritized as a growth sector with a high potential to generate employment (NPA, 2020). Although the pandemic has negatively impacted the tourism sector, it is expected to recover and make significant contribution to the country's economic development. The tourism industry in Uganda is majorly nature-based with over 50% taking place in wildlife protected areas which include 10 national parks, 12 wildlife reserves and 14 wildlife management areas (UWA, 2021). The protected areas share boundaries with local communities who are legally considered as stakeholders and hence eligible to share benefits from conservation.

The history of conservation in Uganda dates back between the late 1880's and 1902, when major ecosystems and wildlife communities were established for conservation and sustainable use (UWA, 2012). Subsequently, the country experienced political instability and insecurity that weakened institutions mandated to conserve natural resources such as national parks and forests that support Uganda's tourism. From 1986 the security improved and conservation institutions such as National Environment Management Authority, Uganda Wildlife Authority, National Forestry Authority, Wetlands Department, and others were established to manage natural resources (Okoth-Ogendo & Tumushabe, 1999). To promote conservation, legal instruments such as National Environment Statute, (1995), Uganda Wildlife Statute (1996), National Forestry and Tree Planting Act (2003), Uganda Wildlife Policy (2014), Uganda Wildlife Act (2019), among others, were enacted.

Several conservation strategies have been implemented in Uganda to reduce local communities' dependence on protected resources and promote sustainable utilization of wildlife. One of the strategies is community-based tourism approach where local communities living adjacent to protected areas are supported to engage in development projects and strengthen partnerships with conservation agencies and local governments. It is clear from tourism and conservation studies that the extent to which local communities living near conservation areas participate in and support conservation depends on their behavioral response to conservation awareness and the benefits thy derive (Vodouhê et al., 2010; Andrade & Rhodes, 2012; Latip et al., 2018). The status of community-based tourism in Uganda is not formally documented despite the existence Uganda Community Tourism Association (UCOTA) that coordinates community-based tourism enterprises in the country. There are scattered reports of isolated initiatives by different organizations involved in promoting community-based tourism. For instance, Uganda Vision Resource Centre is managing the Community Tourism Project in Western Uganda to empower local communities in sustainable development through small-scale tourism enterprises. The project receives few tourists to minimize cultural and environmental impact. The profits are channeled to support community-based projects such as schools, childcare centers, women groups and refugee settlements (Uganda Vision Resource Centre, 2021). Following pro-poor tourism principles of local participation and linking with existing systems, Gorilla Friends Tented Camp is operating in Ruhija village in the vicinity of Bwindi Impenetrable National Park in western Uganda where mountain gorillas have been habituated for gorilla tracking tours. A percentage of profits from Gorilla Friends Tented Camp are returned to support livelihood enhancement initiatives in the village (Mackay & Campbell, 2012).

QECA has over the years adopted a community-based conservation approach and supported community-based initiatives such as Kikorongo Women Community, Agro-tour Walk, Leopard Village, Katwe Tourism Information Center and Nyanz'ibiri Cave Community. However, these efforts have not been fully translated into meaningful benefits for the local communities who continue to bear the cost of living near conservation areas and experience human-wildlife conflicts due to crop raiding, human injuries and sometimes death. This chapter hypothesize that the behavioral response of the community in QECA is determined by their perceptions and benefits they derive from conservation and tourism and the level of mutual interrelationships. In spite of several studies undertaken in QENP on the relationship between conservation and community benefits (Akankwasah et al. 2010, Kabuusu et al. 2018, Driciru et al. 2018), no major study has focused on deepening the understanding of how local communities' perceptions influence their behavior and motivation to participate in conservation and tourism. There is a dearth of knowledge on the local communities' perceptions-conservation nexus that is needed to enhance community participation and increase the benefits of conservation and tourism to the adjacent communities in QECA. This gap in knowledge provides the impetus to examine the afore mentioned nexus. Thus, the objective of the study was to assess ways in which community-based tourism influences local communities' perceptions of conservation and to offer directions for an integrated approach to achieving mutually beneficial interrelationships between conservation and sustainable community-based tourism in QECA. The following questions guided the study: what is the level of local communities' understanding of conservation in QECA? What benefits do they get from tourism? What is the linkage between the benefits from tourism and community perceptions of conservation?

The chapter is organized as follows: an overview of the Social Exchange Theory, perceived tourism benefits, CBT approach, local community perceptions and benefits, tourism growth, conservation and community-based tourism in Uganda is presented in the introduction. The methodology detailing study area, study design and data analysis is presented in the next section. The chapter ends with discussion, conclusions and contribution to knowledge.

METHODOLOGY

Case Study Area: Queen Elizabeth Conservation Area

Queen Elizabeth Conservation Area (QECA) (29° 45'-30° 15' East and 0° 15'-0° 30' South) is located in the Western Rift Valley in southwestern Uganda covering 1,978 square kilometers (Kabuusu et al., 2018) – Figure 1. It comprises Queen Elizabeth National Park (QENP) and the adjoining wildlife reserves, Kyambura (157 km^2) and Kigezi (330 km^2) that harbor 11 fishing enclaves with over 30,000 people rearing livestock that shares water sources and pasture with wildlife (Driciru et al., 2018). QENP (one of the Man and Biosphere Reserves in Africa) was established in 1952 and it is one of the largest in Uganda and borders the Democratic Republic of Congo to the west. The vegetation consists mostly of thickets of small trees, rainforest and of savanna grassland interspersed with acacia trees. The Park is home to over 95 mammal species and over 600 bird species. The park's wildlife includes chimpanzees, leopards, lions, elephants, hippopotamuses, water buffaloes, and several types of antelopes, such as duiker, reedbuck, topi and many species of birds (Uganda Wildlife Authority, 2021).

The study area in Rwenshama region, Bwambara sub-county, Rukungiri District was purposively selected as it is adjacent to QECA area that has experienced increased tourism activities (wildlife tours in adjacent Kigezi Wildlife Reserve and water canoeing on Lake Edward) and increased community collaboration with park management to implement community conservation programmes. The area provided an ideal setting for the study to gain deeper understanding of how the relationship between conservation and benefits from tourism influence community perceptions of conservation.

Study Design and Approach

Guided by an interpretative research design, the study adopted a mixed method approach to investigate the social exchange through which communities interact and benefit from community-based tourism initiatives and how this influences their perceptions of conservation. The mixed methods approach allows concurrent collection, interpretation, and analysis of quantitative and qualitative data (Shorten & Smith, 2017), permits investigation that enriches the results and enables research

Figure 1. Location of Queen Elizabeth conservation area and the study area
Source: Adapted and modified from Oreto, Lemieux & Nobles (2016)

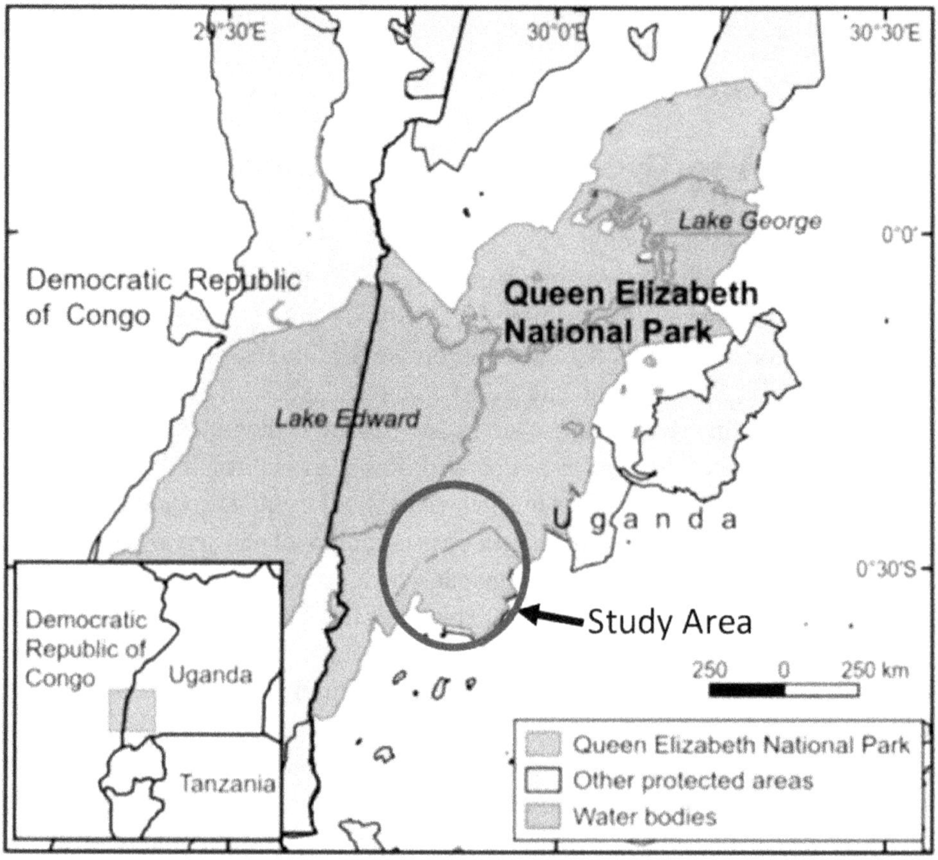

questions to be answered deeply (Tashakkori, 2007) by offsetting the limitations of the different methods when applied individually (Tsegaw & Teressa, 2017).

From a total of 460 households in the study area, a structured questionnaire was administered to 60 randomly selected household heads. Guided by the Social Exchange Theory, information was collected on local communities' perceptions of nature conservation (based on a five-point Likert scale ranging from 'strongly agree' to 'strongly disagree') and the importance that they attached to conservation (based on a five-point Likert scale ranging from 'very important' and to 'very unimportant'). Respondents were also asked to rank the importance of conservation of plants, animals and forests in and around the park on a scale of 1 (highest rank) to 4 (lowest rank). Interviews were held with the manager in charge of community conservation to gain insights into the nature and scope of local community participation in conservation, the benefits obtained from tourism, perceptions of conservation and the role of indigenous knowledge in conservation.

Quantitative data were subjected to chi-square test to determine the relationship between the importance attached to conservation by local communities, the benefits derived from tourism and how these influence their perceptions. Guided by the interpretive approach (Silverman, 1993) the qualitative data gathered from respondents were clustered into themes and sub-themes. Secondary data from district development plan, park records and reports were subjected to content analysis to gain more insights on planning and conservation in QECA and benefits derived conservation and tourism, planning and implementation of conservation and community-based tourism initiatives.

RESULTS

Local Communities' Perceptions of Nature Conservation in Queen Elizabeth Conservation Area

Predetermined statements with corresponding five-point Likert scale were presented to sampled community household heads to measure their perceptions of conservation (Table 1). Results reveal that 35.6% agreed that local community should have moral obligation to conserve biodiversity while the majority 'strongly agreed' (33.39%) and 'agreed' (59.3%) that wildlife should be conserved as tourist attractions. In terms of protecting the park for outdoor recreation, 30% strongly agreed while 53.3% agreed. None of the respondents disagreed with the statement. When asked about the need for conservation of wildlife, 13.3% strongly agreed, 41.7% agreed whereas 11.7% disagreed. Few 'strongly agreed' (3.4%) and majority 'agreed' (84.8%) that it would be unfortunate if plant and animal species became extinct since the plants are critical sources of herbal medicines (73.3% 'agreed') and 43.3% agreed that wild animals should be conserved more than other resources and to support conservation education. About 37% disagreed that wildlife should be conserved even if it does not benefit the community while nearly one third disagreed that wild animals should be conserved more than other resources.

Respondents' ranking of statements indicating levels of conservation in the park based on a five-point Likert scale are presented in Table 2. Results reveal that most community members considered conservation of plants as 'important' (ranked first) and as 'very important' (ranked second). Conservation of animals was revealed as 'important' (ranked first) and as 'neutral' (ranked second), while conservation of forests was considered 'important' (ranked first) and as 'very important (ranked second).

Table 1. Local communities' perception of nature conservation

Statement	SA (%)	A (%)	N (%)	D (%)	SD (%)
Local community should have moral obligation to conserve biodiversity	6.8	35.6	40.7	17.0	0
Forests should be conserved as sources of herbal medicine	20	73.3	1.7	5.0	0
The park should be conserved for outdoor recreation	30	53.3	16.7	0	0
It would be unfortunate if plant or animal species became extinct	3.4	84.8	10.1	1.7	0
Wildlife should be conserved as tourist attractions	33.39	59.3	1.69	5.1	0
It is important to conserve the park's aesthetic values	5	35	58.3	1.7	0
Wildlife should be conserved for posterity and knowledge	3.3	40	55	1.7	0
Wild animals should be conserved more than other resources in the park	10	43.3	11.7	33.3	1.7
It is important to conserve natural resources as part of the environment	8.3	41.7	43.3	6.67	0
Wildlife should be conserved to support conservation education	6.7	43.3	43.3	6.67	0
Irrespective of the benefits, wildlife should be conserved	6.7	33.3	21.7	36.7	1.7
Wildlife should exist and be conserved	13.3	41.7	33.3	11.7	0

Note: SA-Strongly Agree, A-Agree, N-Neutral, D-Disagree, SD-Strongly Disagree

Results of the local community's' concern for loss of biodiversity (plants, animals and natural habitat) in the park based on weighted and ranked responses are presented in Table 3. Results reveal that the local community was 'concerned' about loss of plant species and natural habitats (ranked first) and those that were neutral (ranked second). With regard to loss of animal species, most of the respondents were 'unconcerned' (ranked first) and 'neutral' (ranked second).

Benefits Derived by Local Communities from Tourism

Results on the level and nature of benefits obtained from QECA, adequacy of the benefits and the reasons for not benefiting are presented (Table 4). Sixty two percent of the respondents stated that they benefited from tourism while 38% did not. About 68% revealed that the benefits were adequate while 31.7% indicated that they were inadequate. About 33% indicated that they benefited directly from tourism as they worked as guides and in the lodges while other benefits were schools constructed (16.4%), people employed in the park (10.9%), people can access resources especially water (10.9%), firewood (10.9%), among others. Those who did not benefit attributed it to corruption that scuttled benefits sharing (39.1%), poor benefits sharing mechanism (17.4%), and lack of sensitization of communities on benefit sharing from tourism (13.1%).

Table 2. Importance of conservation of plants, animals and forests

Level of importance for conservation	Likert scale	No.	Weight	Ranking
Importance of Plant conservation				
Very Important	1	18	18	2nd
Important	2	40	80	1st
Neutral	3	1	1	3rd
Unimportant	4	0	0	4th
Very unimportant	5	0	0	4th
Importance of conservation of animals				
Very Important	1	8	8	3rd
Important	2	42	84	1st
Neutral	3	7	21	2nd
Unimportant	4	2	8	3rd
Very unimportant	5	0	0	5th
Importance of conservation of forests				
Very Important	1	20	20	2nd
Important	2	38	76	1st
Neutral	3	1	3	3rd
Unimportant	4	0	0	4th
Very unimportant	5	0	0	4th

Linkage between Benefits from Tourism and Perceptions of Conservation

Chi-square test revealed no significant relationship ($p>0.05$) between the benefits derived from the QECA and the importance local community attach to conservation of animals, plants and forests (Table 5). This implies that local people did not see a link between benefits derived from tourism and conservation of biological resources because they are not adequately sensitized on the contribution of tourism to conservation of wildlife in QECA. Local communities do not seem to appreciate that the financial support they get from QECA comes from tourism hence the inability to see the link.

Results of the chi-square test presented in Tablc 6 also shows no significant relationship ($p>0.05$) between adequacy of benefits got from the park and the importance of conservation of plants and animals. This is possibly because the majority of the respondents did not view conservation in items of environmental services that positively influence their occupation such as farming, fishing, craft

Table 3. Ranking of community concern for loss of biodiversity

Level of concern for biodiversity loss in the park	Likert scale	No.	Weight	Ranking
Concern for plant loss				
Very concerned	1	13	13	3rd
Concerned	2	36	72	1st
Neutral	3	8	24	2nd
Unconcerned	4	3	12	4th
Very Unconcerned	5	0	0	5th
Concern for Animal loss				
Very concerned	1	8	8	4th
Concerned	2	15	30	3rd
Neutral	3	19	57	2nd
Unconcerned	4	18	72	1st
Very Unconcerned	5	0	0	5th
Concern for loss of natural habitat				
Very concerned	1	9	9	4th
Concerned	2	29	58	1st
Neutral	3	12	36	2nd
Unconcerned	4	9	36	2nd
Very Unconcerned	5	0	0	5th

making and petty trade and therefore were not likely to visualize the link between conservation of the park's resources (plants and animals) and the benefits they derive from the park. In addition, the respondents noted that the losses they incur due to crop raids cannot justify the adequacy of the benefits derived from the conservation area.

To gain further insights on the relationship between benefits got from tourism and local people's perceptions tof conservation, the study tested the association between benefits obtained from tourism and the local communities' concerns for loss of the park's biodiversity. Results reveal no significant relationship ($p>0.05$) between local community's concern for loss of biological resources and the benefits got from the park. This implies that local people did not see loss of biological resources from the park affecting the benefits they get from conservation. However, there was a significant relationship between local people's concern for loss of animal species and the benefits they expected from the conservation area ($p<0.05$) (Table 7) because they link tourism to wild animals in the park and loss of the animals would adversely affect tourism that generate the 20% entry fees for revenue sharing.

Table 4. Ways in which local communities benefited from tourism

Benefit parameters	%
Ever benefitted from QECA	
Yes	62
No	38
How Benefitted from QECA	
Bursary for education	1.8
Construction of Schools	16.4
Employment in the Park	10.9
Working as tour guides, employee in lodges etc)	32.7
Accessing water from the park	10.9
Accessing firewood from the park	10.9
Market for fish	1.8
Allowed to fish from the water bodies in the park	9.1
Cutting thatch grass	1.8
Harvesting mat making materials	1.8
Settlement	1.8
Adequate Benefits	
Yes	68.3
No	31.7
If not benefitted, Why?	
Corruption among park and local government staff	39.1
Lack of sensitization	13.1
Limited projects supported	4.4
Poor benefits sharing mechanism	17.4

The above findings on the relationships between benefits obtained from tourism and local communities' concerns over loss of biological resources were corroborated with the views of the conservation manager who noted that local communities and their leaders were engaged in a number of conservation related activities. For instance, local resource user groups were involved in formulation of the general management plan and user agreements to sustainably utilize resources such as papyrus, fish, rattan and others. Through the engagements, it was noted that local communities were concerned about loss of biological resources such as plants and animals from QECA. The level of concern varied with the education background and cultural

Table 5. Relationship between conservation and benefits derived from QECA

	Benefitted from QENP				
Level of importance for conservation	Yes	No	Total		p-value
Importance of plants					
Very Important	17	1	18		
Important	35	5	40		
Neutral	1	0	1	0.770	0.680
Unimportant					
Very unimportant	7	1	8		
Importance of Animals					
Very Important	39	3	42		
Important	5	2	7		
Neutral	2	0	2	3.289	0.349
Unimportant					
Very unimportant					
Importance of Forests					
Very Important	17	3	20		
Important	35	3	38		
Neutral	1	0	1	0.839	0.675
Unimportant					
Very unimportant					

norms. Those who were less educated requested for part of the conservation area to carry out farming while the pastoralists with strong culture and tradition preferred to graze livestock in the conservation area. According to the conservation manager, the local communities that benefited from resources from the park, employment opportunities, and supplying of food items to the hospitality establishments in QECA had positive perceptions of conservation.

DISCUSSION

Results of the study provide critical insights for deeper understanding of opportunities and challenges faced in ensuring sustainable CBT through wildlife resources conservation and utilization. The Social Exchange Theory (SET) provides a general but appropriate sociological framework through which the study provides deeper

Table 6. Relationship between adequacy of benefit got from QECA and importance of conservation

	Are benefits from the park adequate?				
Level of importance for conservation	**Yes**	**No**	**Total**		**p-value**
Importance of plants					
Very Important	7	2	9		
Important	19	10	29		
Neutral	9	3	12		
Unimportant	6	3	9	1.402	0.496
Very unimportant					
Importance of Animals					
Very Important	7	1	8		
Important	29	13	42		
Neutral	4	3	7		
Unimportant	1	1	2	2.089	0.554
Very unimportant					
Importance of Forests					
Very Important	11	2	13		
Important	34	2	36		
Neutral	7	1	8		
Unimportant	2	1	3	3.079	0.380
Very unimportant					

understanding of the interactive exchange of resources between individuals and groups in the study area. Furthermore, the analysis focused on how community perception of conservation in QECA has ve been shaped by the community's social interaction with tangible and intangible tourism benefits and costs.

The long-held view by proponents of inclusive protected area management has been that local communities are most likely to conserve biological resources when they receive direct benefits from them (Stone & Nyaupane, 2016). However, results from this study disagree with this assumption. The local communities in QECA were concerned about conservation of natural resources irrespective of whether they were benefiting or not from tourism. The study raises questions about the effectiveness of community conservation approaches that have traditionally been based on the

Table 7. Relationship between benefits got from QECA and concern for loss of biological resources (plants, animal and natural habitats)

Parameters	Benefitted from QENP				
	Yes	No	Total		p-value
Concern for plant species loss					
Very concerned	11	2	13		
Concerned	34	2	36		
Neutral	7	1	8		
Unconcerned	2	1	2	3.079	0.380
Very Unconcerned					
Concern for animal species loss					
Very concerned	7	1	8		
Concerned	15	0	15		
Neutral	19	0	19		
Unconcerned	13	5	18	10.154	0.017
Very Unconcerned					
Concern for natural habitat loss					
Very concerned	8	1	9		
Concerned	26	3	29		
Neutral	11	1	12		
Unconcerned	8	1	9	0.0627	0.996
Very Unconcerned					

general premise that local communities inherently have negative perceptions of conservation which is untrue (Abukari & Mwalyosi, 2020).

In this study the local communities revealed that, much as they need to benefit from conservation, they had a moral obligation to conserve resources in QECA. This observation suggests the need for protected area managers to design conservation programmes that take into account local communities' conservation values, strategies and indigenous knowledge systems. Indigenous knowledge is a dynamic and important facet of cultural diversity, supports sustainable development (Indigenous Corporate Training, 2020) and informs routine decision-making (UNESCO, 2019). Tourism is one of the socio-economic activities that incorporates use of indigenous knowledge (Timothy, 2017; Nkwanyana, 2018). Through the four 'Hs' (habitat, heritage, history, and handicrafts) that underpin community-based tourism (Smith, 1996; Ringo, 2019), indigenous knowledge and practices can be developed into cultural tourist attractions

(Zeppel, 2006) and generate economic value (Kwanya, 2019, Ayorekire et al., 2019). This implies that community conservation approaches and initiatives can harness the potency of indigenous knowledge to preserve and pass on positive conservation value systems from one generation to another. Cultural festivals, exhibitions, dances, drama and museums are among the most effective ways of transmitting indigenous knowledge and values (Ringo, 2019). Therefore, CBT approach can be effective only when the tourism planners and managers appreciate the role played by social network structures and relationships which affect the tourism development (hang 2021) and conservation of resources on which the tourism industry thrives.

The study is cognizant of the fact that local communities value conservation if it enhances their economic livelihoods (Ayivor et al., 2020), while limited community participation heightens conservation challenges (Benyei et al., 2020; Popp et al., 2019). In Kenya's Amboseli national park, for instance, pastoralists were unhappy with the constraints imposed on their livelihoods by conservation regulations and connived with poachers to kill elephants and lions. When strategies were changed to ensure that local people received tangible benefits from the park, the killings diminished, the pastoralists turned against the poachers and protected wildlife (Western 1994). This reveals that for the local communities to appreciate the benefits of conservation and ecosystem services, they should realize tangible benefits from the conserved biological resources (Austina et al., 2016; Arora, 2018). However, this study reveals that despite having tangible benefits (employment, enhancement of social services, resource access from the park) local community still has limited appreciation of intangible benefits of conservation such as ecosystem services, aesthetic values and others. Thus, there is need for conservation programs to focus on measuring the impacts of the tangible and intangible benefits (Scholtz & Slabbert, 2017) and sensitizing local community about the interconnections between conservation and livelihoods.

Furthermore, attention should focus on mitigating human wildlife conflicts especially from crop raids that affect food security (Akankwasah et al. 2010) and causes negative perceptions of conservation (Tweheyo et al., 2005). The study reveals a worrying situation where local community seems to be less concerned about loss of animals yet on the other hand, they acknowledge that they are key tourism attractions. This perception is most likely because of the long-standing human-wildlife conflict that disrupts livelihoods and amplifies perceived costs of wildlife-based tourism (Gan, 2020).

Thus, there is need to make the local community understand the negative ripple effects of loss of animals on their livelihoods in order to appreciate the mutual relationship between conservation and community-based tourism. However, holistic community development can only be attained when existing community conservation efforts are reinforced by social justice, empowerment, active participation and ownership of tourism processes, as well as redistributive and equitable sharing of benefits (Giampiccoli & Saayman, 2018). Local communities often find themselves in situations where they cannot exercise control over local resources (Lim and Cooper, 2009) leading to unbalanced social exchange, and negative perceptions of tourism and conservation as observed among communities in QECA. Therefore, for more accurate analysis of local communities' perceptions of conservation and tourism development there is need for wider research into the constructs of power and trust which underpin the SET. As noted by Nunkooa (2016), research that integrates the constructs of power and trust provides good theoretical understanding of local communities' perceptions of tourism development.

Furthermore, there is need for policy makers and conservation managers to segregate conservation aspects and analyse local communities' perceptions (Redpath et al., 2013), and affirm that benefits from tourism do not automatically translate into positive perceptions of conservation. Therefore, organisations such as Uganda Wildlife Authority (UWA), need to design conservation strategies to address specific community-based tourism challenges. This implies that the 'one-size fits all' approach to community-based conservation, tourism and benefits to community should be revised to take into consideration the local communities' heterogenous conservation perceptions, socio-cultural values, and indigenous knowledge systems. As reported by the conservation managers, utilization of indigenous knowledge would have meaningful impact if combined with environmental education, in which local people are sensitized about effective conservation practices, equipped with skills to enhance livelihood options and subjected to exposure visits to share best practices in conservation.

CONCLUSIONS AND CONTRIBUTION TO KNOWLEDGE

The chapter advances the argument that if CBT is to motivate protected area conservation, there is need for a policy and strategic focus realignment by conservation managers to enable them design and effectively implement community-based conservation and tourism programmes. Results from QECA reveal that the local communities are concerned about the continued loss of plant species and natural habitats irrespective to whether they are benefiting from tourism or not. This chapter brings to the fore the argument that unless the factors shaping local communities' perceptions of different conservation aspects are comprehensively understood, the CBT approach will encounter numerous challenges. Therefore, there is need to analyse community perceptions of conservation and examine strategies to achieve the dual objective of conservation and sustainable tourism. To this end, UWA should discuss the various aspects of community-based conservation and mechanisms for sharing the benefits from tourism, allow people to ask questions, and recommend conservation strategies from their own perspectives. This will enable the conservation managers to gain deeper insights on the local communities' understanding of conservation and how socio-cultural values and indigenous knowledge systems can be incorporated. However, this should be done cautiously to avoid an over optimistic representation of CBT as a source of livelihood and guard against diminishing the role of tourism in conservation conflict management (Gan, 2020).

The chapter contributes to the existing body of knowledge on conservation-community based tourism nexus by elaborating on local communities' perceptions of conservation hitherto generalized and treated as homogenous (Scholtz & Slabbert, 2017). It challenges the commonly held view by conservation managers that local communities inherently have negative perceptions of conservation when they do not benefit from tourism, which can be turned around by ensuring that they benefit from CBT initiatives. The results from QECA reveal that local communities' perceptions of conservation are not inherently negative as they support conservation even when they do not directly benefit from CBT initiatives. The study reveals that community perceptions are heterogeneous and vary with specific conservation aspects thus guiding conservation managers to design strategies that take into consideration communities interactive exchanges and relationships in and around QECA.

Furthermore, the study deepens understanding of local community concern about loss of protected biological resources, the importance they attach to conservation and how benefits from tourism can influence their perceptions of conservation and community-based tourism. The study also identifies the need for further research to improve understanding of local communities' indigenous knowledge systems and how they can be harnessed to enhance conservation and contribute to sustainable community-based tourism development. Moreover, there is need to apply other theoretical frameworks such as the Social Representation Theory to gain deeper understanding of how varying knowledge systems influence community conservation actions and decisions. Lastly, there is need for further research to examine how local communities' construct the concepts of power and trust, how they shape their interpretation of tourism benefits and costs as well as perceptions of conservation and community-based tourism.

REFERENCES

Abukari, H., & Mwalyosi, R. B. (2020). Local communities' perceptions about the impact of protected areas on livelihoods and community development. *Global Ecology and Conservation*, *22*, 22. doi:10.1016/j.gecco.2020.e00909

Afify, A. M. N. (2011). Challenges and advantages of community participation as an approach for sustainable urban development in Egypt. *Journal of Sustainable Development, 4*(1), 23–36.

Akankwasah, B., Mukadasi, B., Kansiime, F., & Basuta-Isabirye, G. (2010). The Effects of Crop Raiding on Household Food Security in the Albertine Rift: A Case Study of Queen Elizabeth National Park, Western Uganda. *Human Dimensions of Wildlife, 15*(1), 45–54. doi:10.1080/10871200903115070

Akarsua, T. N., Foroudib, P., & Melewarb, T. (2020). What makes Airbnb likeable? Exploring the nexus between service attractiveness, country image, perceived authenticity and experience from a social exchange theory perspective within an emerging economy context. *International Journal of Hospitality Management*, 91.

Andrade, G. S., & Rhodes, J. R. (2012). Protected areas and local communities: An inevitable partnership toward successful conservation strategies? *Ecology and Society, 17*(4), art14. doi:10.5751/ES-05216-170414

Anuar, A. N. A., & Sood, N. A. A. M. (2017). Community Based Tourism: Understanding, Benefits and Challenges. *Journal of Tourism & Hospitality (Los Angeles, Calif.), 6*(01), 1. doi:10.4172/2167-0269.1000263

Ap, J. (1992). Residents' perceptions on tourism impacts. *Annals of Tourism Research, 19*(4), 665–690. doi:10.1016/0160-7383(92)90060-3

Arora, N. K. (2018). Biodiversity conservation for sustainable future. *Environmental Sustainability, 1*(2), 109–111. doi:10.100742398-018-0023-1

Austin, Z., McVittie, A., McCracken, D., Moxey, A., Moran, D., & White, P. C. (2016). The co-benefits of biodiversity conservation programmes on wider ecosystem services. *Ecosystem Services, 20*, 37–43. doi:10.1016/j.ecoser.2016.06.002

Ayivor, J. S., Nyametso, J. K., & Ayivor, S. (2020). Protected Area Governance and Its Influence on Local Perceptions, Attitudes and Collaboration. *Land (Basel), 9*(9), 310. doi:10.3390/land9090310

Ayorekire, J., Obua, J., & Byaruhanga, M. B. (2019). Broadening Uganda's tourism product base through indigenous cultural resource utilization. In M. Mkono (Ed.), *Positive tourism in Africa* (pp. 79–90). Routledge. doi:10.4324/9780429428685-7

Bamberger, M. (1986). *The Role of Community Participation in Development Planning and Project Management*. EDI Policy Seminar Report No. 13. The World Bank.

Benyei, P., Arreola, G., & Reyes-García, V. (2020). Storing and sharing: A review of indigenous and local knowledge conservation initiatives. *Ambio*, *49*(1), 218–230. doi:10.100713280-019-01153-6 PMID:30859400

Byrd, E. T., Bosley, H. E., & Dronberger, M. G. (2009). Comparison of stakeholder perceptions of tourism impacts in rural eastern North Carolina. *Tourism Management*, *30*(5), 693–703. doi:10.1016/j.tourman.2008.10.021

Chang, K. C. (2021). The affecting tourism development attitudes based on the social exchange theory and the social network theory. *Asia Pacific Journal of Tourism Research*, *26*(2), 167–182. doi:10.1080/10941665.2018.1540438

Dangi, T. B., & Jamal, T. (2016). An integrated approach to "sustainable community-based tourism". *Sustainability*, *8*(5), 475. doi:10.3390u8050475

Dodds, R., Alisha, A., & Kelly, G. (2016). Mobilising Knowledge: Determining key elements for success and pitfalls in developing Community Based Tourism. *Current Issues in Tourism*, *21*(13), 1547–1568. doi:10.1080/13683500.2016.1150257

Driciru, M., Rwego, I. B., Asiimwe, B., Travis, D. A., Alvarez, J., VanderWaal, K., & Pelican, K. (2018). Spatio-temporal epidemiology of anthrax in Hippopotamus amphibious in Queen Elizabeth protected area, Uganda. *PLoS One*, *13*(11), 1–21. doi:10.1371/journal.pone.0206922 PMID:30485342

Fredline, E., & Faulkner, B. (2000). Host community reactions: A cluster analysis. *Annals ofTourism Research*, *27*(3), 763–784. doi:10.1016/S0160-7383(99)00103-6

Gan, J. (2020). Uncovering the Environmental and Social Conflicts Behind Residents' Perception of CBT: A Case of Perak, Malaysia. *Tourism Planning & Development*, *17*(6), 674–692. doi:10.1080/21568316.2020.1749121

Giampiccoli, A., & Mtapuri, O. (2012). Community-based tourism: An exploration of the concept (s) from a political perspective. *Tourism Review International*, *16*(1), 29–43. doi:10.3727/154427212X13431568321500

Giampiccoli, A., & Saayman, M. (2018). Community-based tourism development model and community participation. *African Journal of Hospitality, Tourism and Leisure*, *7*(4), 1–27.

Goodwin, H., & Santilli, R. (2009). Community-based tourism: A success. *ICRT Occasional Paper, 11*(1), 37.

Hall, C. M. (2005). *Tourism: Rethinking the Social Science of Obility*. Prentice-Hall.

Higginbottom, K., & Tribe, A. (2004). Contributions of wildlife tourism to conservation. *Wildlife tourism: Impacts, management and planning*, 99-123.

Homans, G. C. (1961). *Social behavior*. Harcourt, Brace, and World.

Kabuusu, R. M., Amuno, J. B., Maseruka, Y., & Macpherson, C. (2018). Incidence and Risk Factors of Wildlife-Associated Human Injuries in Queen Elizabeth National Park, Uganda: A Retrospective Cohort Study. *Journal of Zoonotic Diseases and Public Health*, *2*, 1.

Kraft, R. J. (1996). Service learning: An introduction to its theory, practice, and effects. *Education and Urban Society*, *28*(2), 131–159. doi:10.1177/0013124596028002001

Kwanya, T. (2019). Leveraging tourism in Kenya through indigenous knowledge. Broadening Uganda's tourism product base through indigenous cultural resource utilization. In Positive tourism in Africa (pp. 106-116). Routledge. doi:10.4324/9780429428685-10

Latip, N. A., Rasoolimanesh, S. M., Jaafar, M., Marzuki, A., & Umar, M. U. (2018). Indigenous participation in conservation and tourism development: A case of native people of Sabah, Malaysia. *International Journal of Tourism Research*, *20*(3), 400–409. doi:10.1002/jtr.2191

Lim, C. C., & Cooper, C. (2009). Beyond sustainability: Optimizing island tourism development. *International Journal of Tourism Research*, *11*(1), 89–103. doi:10.1002/jtr.688

Mackay, K. J., & Campbell, J. M. (2012). Innovative community tourism in Uganda. *Téoros - Journal of Tourism Research*, *31*, 3.

Manyara, G., & Jones, E. (2007). Community-based tourism enterprises development in Kenya: An exploration of their potential as avenues of poverty reduction. *Journal of Sustainable Tourism*, *15*(6), 628–644. doi:10.2167/jost723.0

Mayaka, M. W., Croy, G. N., & Cox, J. W. (2019). A dimensional approach to community-based tourism: Recognising and differentiating form and context. *Annals of Tourism Research*, *74*, 177–190. doi:10.1016/j.annals.2018.12.002

Mearns, I. (2003). Community-based Tourism: The key to empowering the Sankuyo community in Botswana. *Africa Insight*, *33*(1), 29–32.

Mensah, I. (2017). Benefits and Challenges of Community-Based Ecotourism In Park-Fringe Communities: The Case of Mesomagor of Kakum National Park, Ghana. *Tourism Review International*, *21*(1), 81–98. doi:10.3727/154427217X14866652018947

Molm, L. D. (1991). Social exchange: Satisfaction in power-dependence relations. *American Sociological Review*, *56*(4), 475–493. doi:10.2307/2096269

MTWA (Ministry of Tourism Wildlife and Antiquities). (2018). *Uganda Tourism Sector Annual Performance Report, Financial Year 2017/2018*. MTWA.

National Environment Statute. (1995). Government of Uganda.

National Forestry and Tree Planting Act. (2003). Government of Uganda.

Nkwanyana, S. (2018). Conserving cultural heritage and the usage of indigenous knowledge: A case study of the Zululand District Municipality, Kwazulu-Natal. *African Journal of Hospitality, Tourism and Leisure*, *7*(2), 1–12.

NPA. (2020). *National Planning Authority – Third National Development Plan (NDPIII) 2020/2021 – 2024/2025*. NPA.

Nunkooa, R. (2016). Toward a More Comprehensive Use of Social Exchange Theory to Study Residents' Attitudes to Tourism. *Procedia Economics and Finance*, *39*, 588–596. doi:10.1016/S2212-5671(16)30303-3

Okazaki, E. (2008). A community-based tourism model: Its conception and use. *Journal of Sustainable Tourism*, *16*(5), 511–529. doi:10.1080/09669580802159594

Okoth-Ogendo, H. W. O., & Tumushabe, G. W. (1999). Governing the environment: political change and natural resources management in Eastern and Southern Africa (No. 10). Acts Press.

Omoruyi Osagie, L. (2016). Significance of community Participation in Rural Development. *International Journal of Agriculture and Earth Science*, *2*, 44–53.

Oreto, W. D. M., Lemieux, A. M., & Nobles, M. R. N. (2016). 'It's in my blood now': The satisfaction of rangers working in Queen Elizabeth National Park, Uganda Article. *Oryx*, 1–9.

Özel, C. H., & Kozak, N. (2017). An exploratory study of resident perceptions toward the tourism industry in Cappadocia: A Social Exchange Theory approach. *Asia Pacific Journal of Tourism Research*, *22*(3), 284–300. doi:10.1080/10941665.2016.1236826

Popp, J. N., Priadka, P., & Kozmik, C. (2019). The rise of moose co-management and integration of Indigenous knowledge. *Human Dimensions of Wildlife*, *24*(2), 159–167. doi:10.1080/10871209.2019.1545953

Pozzoni, B., & Kumar, N. (2005). *A review of the literature on participatory approaches to local development for an evaluation of the effectiveness of World Bank support for community-based and driven development approaches*. Academic Press.

Qin, X., Shen, H., Ye, S., & Lingqiang Zhou, L. (2021). Revisiting residents' support for tourism development: The role of tolerance. *Journal of Hospitality and Tourism Management*, *47*, 114–123. doi:10.1016/j.jhtm.2021.02.010

Redpath, S. M., Young, J., Evely, A., Adams, W. M., Sutherland, W. J., Whitehouse, A., Amar, A., Lambert, R. A., Linnell, J. D. C., Watt, A., & Gutierrez, R. J. (2013). Understanding and managing conservation conflicts. *Trends in Ecology & Evolution*, *28*(2), 100–109. doi:10.1016/j.tree.2012.08.021 PMID:23040462

Ringo, J. E. (2019). The Usage of Indigenous Knowledge in Cultural Tourism Destinations. *Tengeru Community Development Journal*, *6*(1), 86–98.

Saayman, M., & Giampiccoli, A. (2016). Community-based and pro-poor tourism: Initial assessment of their relation to community development. *European Journal of Tourism Research*, *12*, 145–190.

Salazar, N. B. (2012). Community-based cultural tourism: Issues, threats and opportunities. *Journal of Sustainable Tourism*, *20*(1), 9–22. doi:10.1080/09669582.2011.596279

Scholtz, M., & Slabbert, E. (2017). Tourism is not just about the Money: A Comparison of Three South African Communities. *African Journal of Hospitality, Tourism and Leisure*, *6*(3).

Shorten, A., & Smith, J. (2017). Mixed Methods Research: Expanding the Evidence Base. *Evidence-Based Nursing*, *20*(3), 74–75. doi:10.1136/eb-2017-102699 PMID:28615184

Silverman, D. (1993). *Interpreting Qualitative Data: Methods for Analysing Talk, Text and Interaction*. Sage.

Sindiga, I. (1995). Wildlife based tourism in Kenya: Land use conflicts and government compensation policies over protected areas. *Journal of Tourism Studies*, *6*(2), 45.

Smith, V. L. (1996). The four Hs of tribal tourism: Acoma—a pueblo case study. *Progress in Tourism and Hospitality Research*, *2*(3-4), 295–306. doi:10.1002/pth.6070020310

Spenceley, A., & Meyer, D. (2012). Tourism and poverty reduction: Theory and practice in less economically developed countries. *Journal of Sustainable Tourism*, *20*(3), 297–317. doi:10.1080/09669582.2012.668909

Stone, M. T., & Nyaupane, G. P. (2016). Protected areas, tourism and community livelihoods linkages: A comprehensive analysis approach. *Journal of Sustainable Tourism*, *24*(5), 673–693. doi:10.1080/09669582.2015.1072207

Stukas, A. A., & Dunlap, M. R. (2002). Community involvement: Theoretical approaches and educational initiatives. *The Journal of Social Issues*, *58*(3), 411–427. doi:10.1111/1540-4560.00268

Tashakkori, A. (2007). Exploring the Nature of Research Questions in Mixed Research Methods. *Journal of Mixed Research Methods*, *1*, 207–211. doi:10.1177/1558689807302814

The World Bank. (2020). *Community-Driven Development*. The World Bank. https://www.worldbank.org/en/topic/communitydrivendevelopment

The World Bank in Uganda. (2021). *Economic Overview*. https://www.worldbank.org/en/country/uganda/overview

Timothy, D. J. (Ed.). (2017). *Managing heritage and cultural tourism resources: Critical essays, Volume one*. Routledge.

Tolkach, D., & King, B. (2015). Strengthening community-based tourism in a new resource-based island nation: Why and how? *Tourism Management*, *48*, 386–398. doi:10.1016/j.tourman.2014.12.013

Tosun, C. (2000). Limits to community participation in the tourism development process in developing countries. *Tourism Management*, *21*(6), 613–633. doi:10.1016/S0261-5177(00)00009-1

Tsegaw, W. E., & Teressa, D. K. (2017). Tour guiding quality assurance mechanisms and respective tourist satisfaction: Evidence from South Ethiopia. *Journal of Tourism & Hospitality (Los Angeles, Calif.)*, *6*(6), 1–9. doi:10.4172/2167-0269.1000317

Tweheyo, M., Hill, C. M., & Obua, J. (2005). Patterns of crop raiding by primates around the Budongo Forest Reserve, Uganda. *Wildlife Biology*, *11*(3), 237–247. doi:10.2981/0909-6396(2005)11[237:POCRBP]2.0.CO;2

Uganda Vision Resource Centre. (2021). *Community tourism*. Accessed from website https://v2o40rc.wordpress.com/tour-and-volunteer-program/community-tourism/

Uganda Wildlife Act. (2019). Government of Uganda.

Uganda Wildlife Policy. (2014). Government of Uganda.

Uganda Wildlife Statute. (1996). Government of Uganda.

UNESCO. (2019). *Local and Indigenous Knowledge Systems (LINKS).* https://en.unesco.org/links

Usadolo, S. E., & Caldwel, M. (2016). A stakeholder approach to community participation in a rural development project. *SAGE Open*, *6*(1), 2158244016638132. doi:10.1177/2158244016638132

UWA. (2012). *Guidelines for Revenue Sharing between wildlife protected areas and adjacent local governments and communities.* Uganda Wildlife Authority.

UWA. (2021). *Queen Elizabeth National Park at a Glance.* https://www.ugandawildlife.org/explore-our-parks/parks-by-name-a-z/queen-elizabeth-national-park

Vodouhê, F. G., Coulibaly, O., Adégbidi, A., & Sinsin, B. (2010). Community perception of biodiversity conservation within protected areas in Benin. *Forest Policy and Economics*, *12*(7), 505–512. doi:10.1016/j.forpol.2010.06.008

Weaver, D. (2010). Community-based tourism as strategic dead-end. *Tourism Recreation Research*, *35*(2), 206–208. doi:10.1080/02508281.2010.11081635

Western, D. (1994). Ecosystem Conservation and Rural Development: the Case of Amboseli. In Natural Cormections: Perspectives in Community-based Conservation. Academic Press.

WWF (World Wildlife Fund). (2001). *Guidelines for Community-Based Ecotourism Development.* WWF International.

Yanes, A., Zielinski, S., Diaz Cano, M., & Kim, S. I. (2019). Community-based tourism in developing countries: A framework for policy evaluation. *Sustainability*, *11*(9), 2506. doi:10.3390u11092506

Zdenek, R. O., & Walsh, D. (2017). The Background and History of Community Development Organizations. In Navigating Community Development (pp. 21-52). Palgrave Macmillan. doi:10.1057/978-1-137-47701-9_3

Zeppel, H. (2006). *Indigenous ecotourism: Sustainable development and management* (Vol. 3). Cabi. doi:10.1079/9781845931247.0000

Chapter 4
Smiling vs. Frowning Faces:
Community Participation for Sustainable Tourism in Ghanaian Communities

Ama Amponsaa Birch Freeman
Takoradi Technical University, Ghana

Ishmael Mensah
https://orcid.org/0000-0002-8068-5446
University of Cape Coast, Ghana

Kwabena Barima Antwi
University of Cape Coast, Ghana

ABSTRACT

Participatory development has become an integral strategy for achieving sustainable tourism in view of the varying impacts brought by mass tourism. Community participation ensures local control of the pace of tourism development and the creation of a more personalised tourism product. However, some studies have stated tension between governments and locals, which has hindered tourism development. This study, thus, sought to explore the extent to which communities are involved in decision-making and how useful the outcome has been. To achieve the study objectives, two tourism sites in Ghana implemented based on participatory principles, namely Kakum National Park and Afadjato Mountain, were compared. Results of the study suggest there is a fair degree of community involvement at the Gbledi Traditional Area (Afadjato) leading to satisfaction (smiling-faces), rather than Abrafo-Odumase (Kakum), which exhibited dissatisfaction (frowning-faces). The policy and research implications relate mainly to the challenge of achieving sustainable tourism development at the community level.

DOI: 10.4018/978-1-7998-7335-8.ch004

INTRODUCTION

The varying impacts brought about by mass tourism has made the issue of sustainable tourism more popular. Community participation and sustainability has been a concern to major international bodies such as the World Tourism Organization (WTO), United Nations (UN) and other international bodies. This has also been the main focus of Ghana's 15-year National Tourism Development Plans i.e., 1996-2010 and 2013-2027. According to Ghana's 1996-2010 tourism national policy framework, tourism was to be developed on a sustainable basis and the socio-economic benefits well distributed in the country and society, giving emphasis to community-based tourism especially in rural areas. Subsequently, one of the main objectives of the Ghana National Tourism Development Plan (2013-2027) is to 'promote sustainable and responsible tourism in such a way as to preserve the historical, cultural and natural heritage' of the country (p.5). These objectives are in line with the UNWTO sustainable development goals 1, 12 and 15, which focus on developing tools that alleviate poverty, create jobs, promote local culture and products as well as conserving and preserving biodiversity in relation to tourism. Additionally, the national tourism plan emphasizes community involvement in the tourism sector in that, communities should be represented in decision-making processes that impact their lives.

According to Mugizi, Ayorekire and Obua (2017), tourism has many social, economic and environmental benefits however, in order to realize the full benefits of tourism it is important to involve the local people and incorporate their expectations into tourism planning and development. Community-based tourism is an essential approach to ensuring sustainable tourism development as exemplified by the Banteay Chhmar community in Cambodia where there was active participation due to free flow of information and collaboration (Ellis & Sheridan, 2015). Mensah (2017) argues that misunderstandings and conflicts among host communities, tourists and park authorities are reduced when there is community participation. Additionally he argues that through community-based tourism a more sustainable way of managing natural resources is achieved by involving local communities to ensure conservation, which in turn provides economic incentives to the communities. Chang et al. (2018) mention that although tourism development can bring about positive impacts, it can also lead to the adulteration of local culture, damage to the natural environment and conflicts with tourists. Yang, Ryan & Zhang (2013) assert that tourism development basically kindles social conflict as it brings together various stakeholders leading to a change in culture and social structure in communities. They state that conflicts occur between the indigenous community and external entrepreneurs over economic benefits derived from tourism. This has hindered tourism development in certain communities due to the negative impacts experienced in these developing countries.

Tourism development in rural areas has often been aimed at alleviating poverty and enhancing the well-being of local residents. Ibanescu et al. (2018) argue that tourism activities create jobs, provide rural female employment, generate an increase in revenues, modernise facilities, and improve standards of living for residents. They demonstrate that rural tourism is perceived as a sustainable 'cure-all' solution for various problems of rural economies. This explains why community-based tourism initiatives aim at increasing local people's involvement in tourism as a useful component of pro-poor tourism (PPT). Lin, Chen & Filieri (2017) also agree that tourism contributes to the overall growth in the quality of life of residents and supports rural sustainable development. According to Wen, Cai & Li (2021) tourism has long been recognized as a major contributor to economic development and can contribute significantly to poverty reduction in developing countries, and this has been acknowledged by major organizations in the world. Bagus et al. (2019, p.71) state that 'Community-based tourism development is expected to be a model of sustainable tourism development in accordance with government policies in the field of tourism'. However, if tourism development only considers the welfare of the government but overlooks the socio-economic interests of the community, CBT will not run as planned (Nurhidayati, 2015 cited in Bagus, et al., 2019).

Tourism is the fourth highest foreign exchange earner for the country after Cocoa, Gold and Oil & Gas. According to the World Travel & Tourism Council (2021) Ghana's total contribution to tourism employment in 2019 was 742,500 jobs representing 6% of total employment but this reduced to 514,800 jobs in 2020, representing 2% of total employment. The total domestic visitor spending in 2019 was USD 1,733.8 million representing 56%, with its equivalent international visitor spending at USD 1,378.6 million which represents 44% of total visitor spending. Domestic visitor spending however dropped in 2020 to USD 1,063.4 million representing 74% with international visitor spending dropping further to USD 372.6 million, representing 26% of total visitor spending for the year 2020. Prior to the COVID-19 pandemic, there was an increase in total number of jobs (direct and indirect jobs) created by the Ghana tourism sector from 550,000 in 2017 to 602,425 in 2018; out of which direct jobs provided by the tourism sector rose from 135,000 in 2017 to 158,231 in 2018 (Ministry of Tourism, Arts and Culture, 2019).

The government of Ghana has through the Wildlife Division of the Forestry Commission and some non-governmental agencies endorsed some community-based tourism projects using various interventions (Mensah & Adofo, 2013. The Wildlife Division oversees all protected reserves in Ghana including the Kakum National Park and the Afadjato in the Agumatsa range. The Ghana Heritage Conservation Trust (GHCT), managers of the Kakum National Park in an interview with the Ghana News Agency (GNA), disclosed that the facility has within the space of four months lost in excess of GH¢400,000.00 in revenue due to the COVID-19 outbreak which culminated in its closure (Abdullah, 2020). The park attracts an estimated 100,000 visitors on an annual basis and revenue generation will certainly be affected. The Ghana Tourism Authority has already indicated that the Ghanaian tourism industry is expected to lose a revenue of about GH¢ 400,000 due to the novel COVID-19 outbreak which has seen countries shutting their borders (Antwi, 2020).

Statistics released by the Ghana Heritage Conservation Trust indicates that patronage of the site has significantly reduced over the past years. In 2014, a total of 439,323 tourists visited the Park but the figure reduced significantly to 270,779 in 2015. The document said 12,209 Ghanaian adults visited the Park as compared to 12,601 in the previous year, with 9,919 non-Ghanaian adults visiting last year as against 10,582 in 2016. Patronage saw a further drastic reduction in 2017 as a total of 113,830 tourists visited the park as against the 141,053 recorded in the previous year. This was attributed to the deteriorating nature of the 30 kilometre road from Cape Coast leading to the site could extend the driving time to over an hour instead of about 30 minutes before one could access the facility (Ghana News Agency, January, 2018). The road has since been repaired but the impact of COVID-19 has kept tourist arrivals on a downward spiral. In the case of Afadjato, 10,578 domestic and international tourists visited, spending an equivalent of GH¢9,300.00 in 2010. These visitor numbers increased to 12,456 with and an equivalent spending of GH¢18,122.5 in 2011 (Ackuayi et al., 2014).

Mount Afadjato and the Kakum National Park (KNP) have attracted a number of studies. The studies on Afadjato have been conducted by Owusu, Ekpe & Asamoah (2005), Owusu (2008), Ofori, Owusu & Attuquayefio (2014), Wiafe & Arku (2014), and Siakwah (2018). These studies on Afadjato have mainly been on forest conservation (Owusu, 2008), diversity of species (Wiafe & Agyei, 2013) and perception of the community towards conservation (Owusu, 2001; Owusu, 2008). In the same vein, there have been various studies on KNP (Abane et al., 1999; Akyeampong, 2011; Amoah & Wiafe, 2012; Appiah-Opoku, 2011; Cobbinah, Black & Thwaites 2015; Dei, 1996; Eshun, 2014; Fiagbomeh & Bürger-Arndt, 2015; Mensah, 2016; Nanang & Owusu, 2010; Twerefou & Ababio, 2012). These include studies on small mammal diversion and population sizes of elephants (Eggert, Eggert & Woodruff, 2003 and Yeboah, 1998) and tourist motivation and satisfaction (Agyeman, Aboagye & Ashie; 2019; Amuquandoh, 2017).

However, most of the studies have been on community livelihoods (Abane et al., 1999; Akyeampong, 2011; Appiah-Opoku, 2011; Cobbinah, Black & Thwaites 2015 and Mensah, 2017). Generally, these studies on livelihoods have shown that the majority of community members' livelihoods have not significantly improved as a result of tourism. However, the satisfaction of community members with regard to their involvement in decision-making through community participation have not been examined in a comparative study. Additionally, majority of these studies have used the quantitative method thereby discounting the intricate details of the personal experiences of community members. It is against this background that this paper seeks to explore the extent to which two rural communities were involved in ecotourism development as well as the socio-economic benefits they derived from it. To achieve this, the level of participation in an already established attraction (Kakum National Park) and an emerging attraction (Afadjato Mountain) will be analysed. These two projects are considered because whilst the Afadjato Conservation Project was initiated by the chiefs and people of the Gbledi Traditional Area, Kakum was undertaken by the government with funding from NGOs. Community-based tourism in both communities were developed based on the principles of sustainable tourism development.

CONCEPTUAL ISSUES

Sustainable Tourism Development

The Concept of sustainable development was first introduced globally by the World Conservation Strategy, which was issued by the International Union for the Conservation of Nature and Natural Resources (ICUN) in 1980; which produced a document linking conservation of nature for human development (Schwass, 2011). The definition of sustainable development as a concept is still unclear and is also unclear what constitutes a sustainable period (Onchwati, Somerville & Brockway, 2010). However, the most widely quoted definition is the one by the World Commission on Environment and Development (WCED) in the Brundtland Report of 1987, as the 'development that meets the needs of the present without compromising the ability of future generations to meet their own needs' (United Nations General Assembly, 1987, p. 43). This definition has however been widely criticized as basically illogical, as it allows for a wide range of interests to support it and is generally impossible to fully implement due to its lack of practicality (Butler, 2018). He further explains that the definition is ambiguous and open to varying interpretations due to its contradictory views on goals, failure to measure outcomes and ensure suitable controls on development.

Lu & Nepal (2009) suggest that Sustainable tourism (ST) was initially regarded as a resolve for negative tourism impacts, with the long-term sustainability and welfare of destinations focusing on minimizing negative impacts. Butler (2018) points out that there is an unfortunate connection with the assumption that mass tourism is automatically unsustainable and, for that reason, has nothing to do with sustainable development. Lu and Nepal (2009, p.9) on the other hand suggest that sustainable tourism research "has shifted its focus from niche tourism to mass tourism however, the most recent position is one of convergence which suggests that ST is a goal which is applicable to all forms of tourism regardless of scale". Thus the initial trend was a paradigm shift in the industry from mass tourism and other unsustainable tourism practices to sustainable tourism development. The concept of sustainable development has since evolved and is defined by the United Nations World Tourism Organisation as "Tourism that takes full account of its current and future economic, social and environmental impacts, addressing the needs of visitors, the industry, environment and host communities" (UNWTO, 2013, p.17).

From the updated definition of sustainability, the underpinning factors are that sustainability requires appropriate consideration of the long-term economic, environmental, socio-cultural and political well-being of all stakeholders, which can be attained by engaging all stakeholders involved in the whole process (Fletcher et al., 2018). Sustainable tourism is often linked to the conservation of ecosystems, generational equity, the promotion of human welfare and community participation in decision-making (Bramwell, 2015). According to Bramwell, Higham, Lane & Miller (2017), sustainable tourism goals are generally now seen as adaptable depending on the context and changing circumstances and have over time been seen as relevant for all forms of tourism, irrespective of its scale. Some tourism scholars have argued that with regards to developing countries a "sustainable livelihood approach" is more relevant and appropriate than a Sustainable Tourism approach (Tao & Wall, 2009). In effect, the industry must ameliorate the impacts of tourism on the environment, and reduce the consumption of natural resources for Sustainable Tourism Development to be achieved. Kiper (2013) maintains that sustainable tourism embroils strong political leadership and informed participation of all related stakeholders to bring about consensus building and extensive participation. This requires the involvement of the community in decision-making, implementation, monitoring impacts and exacting corrective actions at different levels.

Community Participation

Pacione (2009, p. 356) defines a community as "a group of people who share a geographic area and are bound together by common culture, values, race, or social class". These descriptions imply a group of people living in the same area with similar characteristics and interests which could be economic, social, cultural, environmental and political. There is however no one agreement on the definition of community, and in recent times a community cannot be described by merely geographic area, commonalities and physical interaction as diverse people could live in the same community or virtually. Geol (2014) points out that technology has made it possible to cross physical boundaries of place by connecting people through new emerging communities such as Facebook and other social media. For the purpose of this paper and from and African perspective, communities may be described 'as clans or tribes with traditional heads such as the *Ewe*, *Ga* and *Akan* tribes of Ghana. The Abrafo-Odumase community is made up of *Akans* whilst the people of Gbledi are *Ewes*. The emerging argument has been that community participation in tourism will help create entrepreneurial opportunities for the local people and engage them in developmental decision-making as well as benefit host communities and thus leading to community policies (Dei, 2000). Community-based ecotourism ensures the community has total control over tourism development in terms of its scale and activities (Amoako-Atta et al. 2020). Wang et al (2021) support this point that the community is entitled to partake in planning activities that impact their everyday lives. Yanes et al. (2019) assert that, participation revolves between two extremes: people either have the power to influence decisions or are just observers of the process.

Community participation is a term that describes various forms of participation at different levels and under specific situations; and is likely to be embraced by the community if they have been invited to participate in the decision-making process (Mak et al., 2017). Furthermore, the community participation concept is portrayed as a means to achieving sustainable tourism as a driving force for economic and socio-cultural development (Tosun, 2006). In an earlier writing, Tosun (2000) argued that the participatory approach would facilitate the implementation of sustainable tourism development principles, thereby creating better opportunities for the local community and thus gaining more benefits from tourism development in their area. In effect, locals will develop more positive attitudes to tourism development, by conserving their resources (Inskeep, 1994 quoted in Tosun, 2006). However, despite all the arguments for the participatory approach, there are limits to community involvement in tourism development at the structural, operational and cultural levels (Tosun, 1998). The argument is that ''the extent of local entrepreneurial involvement is usually very limited, owing to the fact that the local indigenous groups are rarely adequately pre-adapted to the business culture in tourism'' (Din, 1988, p.563). To buttress this, Akama (2011) and Lee and Jan (2019) postulate that local communities should organise themselves to plan and manage tourism development in their areas by taking control over decision-making and determining how their resources will be used and managed..

Tosun (1999) proposed the Typologies of Community Participation which establishes that different stakeholders expect different types of community participation to achieve their own aims which may conflict with each other. Tosun's model is categorised into three typologies - spontaneous, induced and coercive participation. Spontaneous participation is a type of community participation in tourism that gives authority and total managerial responsibility to the host community.

Induced community participation is a type of community participation in tourism in which the local community has a voice in tourism. Their views are taken and they are told what has been planned but they do not have the power to enforce their interests. Other powerful stakeholders such as multinational companies, government, etc. take final decisions and only seek the approval of the local community (Tosun, 1999). It is thus a top-down approach where the host community does not partake in decision-making but shares in the benefits and implementation of tourism, which is common in developing countries (Tosun, 2006).

Coercive participation in Tosun's model is also a top-down approach which is rather manipulative. With this strategy, when taking certain decisions, the local leaders may be consulted in order to reduce the socio-political threats to tourism development. Thus some basic needs of the local community may be taken into consideration although the real aim is to forge for the requests and needs of policymakers, tourists and tourism operators (Tosun, 2006). Tosun's typology compares to Arnstein's (1971) and Pretty's (1995) typologies of community participation. Arnstein's refers to citizen participation as the redistribution of power which allows the have-nots to be consciously included in tourism development in future as well as share in the benefits of the rich in society (Tosun, 2006). Pretty (1995) on the other hand gives seven levels of community participation. Pretty's model has manipulative participation at the bottom and self-mobilization at the top thus showing the power relationship between citizens and authority. Tosun's typology however relates directly to the tourism industry while Arnstein and Pretty's typologies are not skewed to any economic sector but are for general discourse in development studies. Generally, typologies of participation in the literature suggest that when diverse social actors engage in community projects, the outcome can range from coercion to co-learning (Arnstein, 1969; Pretty, 1995; Keen and Mahanty (2005) with co-learning seen as a more sustainable form of participation (figure 1).

METHODOLOGY

Profile of Study Areas

Kakum National Park

The Kakum Conservation Area is of high national importance in addition to its normal role of conserving biodiversity, tropical rain forest ecosystem and being the catchment area of the water supply of Cape Coast Municipality. The Kakum Conservation area consists of the Kakum National Park and the Assin Attandanso Resource Reserve and was formally gazetted as a Wildlife Protected Area in 1992 by Legislative Instrument 1525. This was the official transfer of two areas which had been established as a forest reserve between 1925-26 and 1935–36 respectively

Figure 1. Typologies of community participation
Source: *Adapted from Arnstein (1969); Pretty (1995) Tosun (1999); and Keen, and Mahanty (2005)*

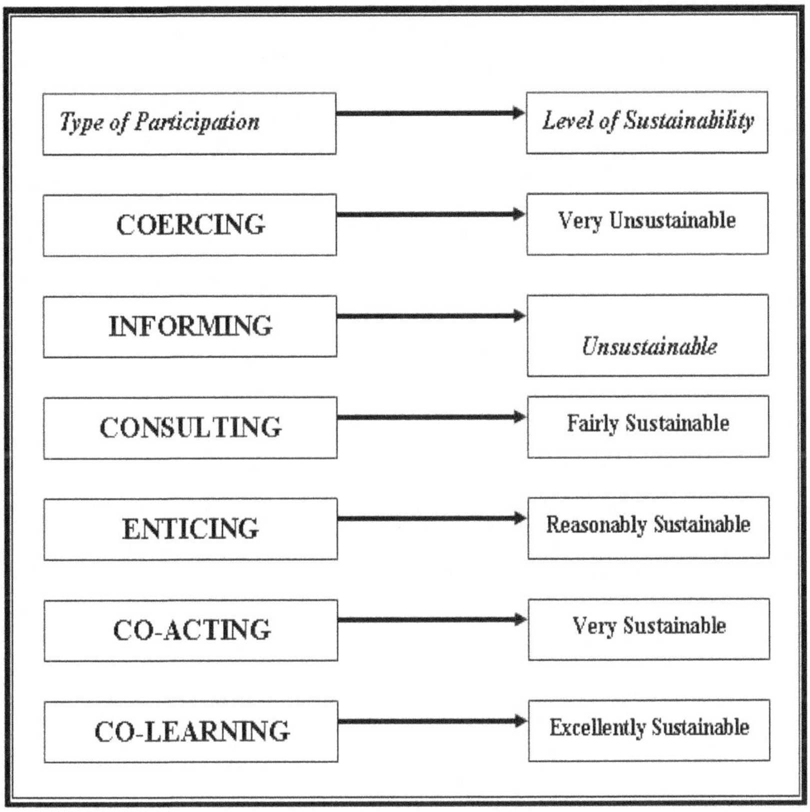

to the Wildlife Division of the Forestry Commission, formerly Wildlife Department on behalf of the Assin Attandanso, Denkyira, Twifu Heman, Denkyira and Abura traditional areas (International Union for Conservation of Nature and Natural Resources, 2010). These are represented by 52 communities in the immediate environs of the conservation area. The reserve served as a source of commercial logging until 1989, when it came under the management of the Wildlife Division and the Central Regional administration, under the auspices of its tourism development scheme. The canopy walkway and visitor centre was built under the direction of Conservation International (CI) and with funding support from USAID and managed by the Ghana Heritage Conservation Trust (GHTC), an NGO. The Kakum forest reserve covers an area of 350 sq. km and stretches from Abrafo-Odumase (the research area) to Twifo Praso in the Twifo-Heman-Lower-Denkyira District (Figure 2).

The forest is a source of food and livelihood for the people who live around and within the area. They derive fuelwood, herbal medicine, alcoholic beverages, snails,

Figure 2. Map of Kakum conservation area

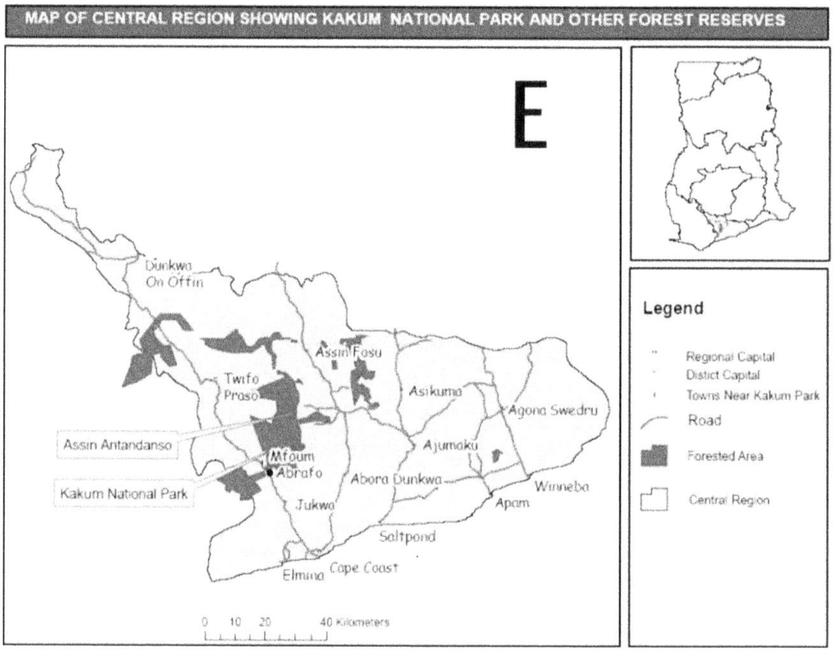

mushroom, game, chewing sticks and a lot more from the forest (Abane et al., 1999). The park was officially opened in 1994 to tourists but a visitor centre and a canopy walkway were opened in 1997.

Afadjato Community Forest Conservation Project

The project is located within the Gbledi Gborgame Traditional Area in the Hohoe District of the Volta Region. The Project area is 300 kms from the nation's capital, Accra and 24 km southeast of Hohoe, the District capital (see Figure 3). Mount Afadjato is the highest peak in Ghana at 885 metres above sea level. The project was initiated by the chiefs and people of the Gbledi Traditional Area and Fodome Ahor and implemented through a partnership with the Ghana Wildlife Society (an NGO). This collaboration is in line with the Ministry of Lands and Natural Resources (2012) Ghana Forest and Wildlife Policy, which promotes community participation in natural resource management. The project is funded by the Netherlands Government through the Netherlands Embassy in Accra, with an approved budget of about 3.7 million Dutch Guilders over a five-year period (1998-2003).

Figure 3. Map of Afadjato conservation area

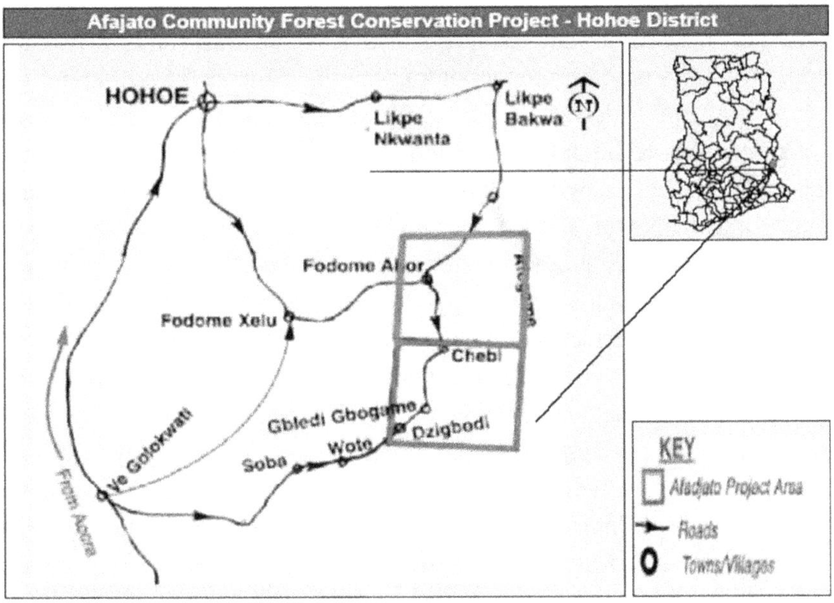

The project was initiated in response to concerns over the degradation of the environment through farming on the slopes, tree felling and bush burning. These activities posed a serious threat to the ecological functions and socio-cultural values of the forest. The objective of the project is for the local community to conserve the forest and its biodiversity as well as manage it sustainably as an alternative source of income for improving local livelihoods. To achieve this the project seeks to use the natural attractions of the habitat and the Agumatsa ranges to promote community-based eco-tourism as an alternative to farming and tree felling in the forest. The tourism enterprise is expected to provide employment opportunities for the people and adequate revenue for community development.

Data Collection and Analysis

The paper compared two different cases from two communities in Ghana, Kakum in the Central region and Afadjato in the Volta region. The Central Region is not only of historical importance but is referred to as the 'tourism heartbeat' of Ghana. The Kakum National Park which is about 30kms north of Cape Coast, the regional capital is the most visited attraction in the whole country. The Volta Region is a mountainous area with beautiful vegetation, wildlife and waterfalls, and the choice of Mount Afadjato at Gbledi Gborgame is significant because Afadjato is the highest peak of the Country.

Case study research design 'scientifically investigates into a real-life phenomenon in-depth within its environmental context, and is a means of advancing theories by comparing similarities and differences among cases' (Ridder, 2017, p.2). This type of research is commonly associated with a location, such as a community or organisation as is the case with this study. The case study was considered appropriate since the topic deals with social issues and thus, the qualitative method was found to be suitable for bringing out the information relevant for answering the research questions. Besides, it allows researchers to employ multiple sources of evidence; direct observation, interviews as well as secondary data in the form of documents. The target population consisted of tourism professionals, government officials and local people. Data was gathered from both secondary and primary sources. The former included published and unpublished data and reports of government institutions. The latter were data solicited from residents using interview schedules, in-depth interviews, telephone interviews, e-mails and focus group discussions (FGDs).

As a triangulatory study, a sample of forty community members were targeted. This was made up of twenty respondents from each study area. In practice, however, 30 (15 for each community) quality interview results could be used for the analysis. The purposive sampling technique was used to target government officials, officials of NGOs, tour operators, tourism developers and traditional heads, whilst the convenient sampling technique was used to select the community members for the study. The instruments used included an observation checklist, and a semi-structured interview guide. The semi-structured questions were mainly open-ended questions with a few closed-ended ones. Apart from individual bio-data, the themes explored included issues on the level of involvement in decision-making on tourism projects, level of satisfaction with participation, benefits and problems associated with community-based tourism and future steps to sustain tourism in the study area.

Data was collected at both Kakum and Afadjato with the help of Field Assistants who were conversant with the study area in each case. The interviews were kept within a maximum of 45 minutes each, recorded and transcribed. Thematic content analysis was employed to analyze the data, hence the results could be said to be fairly reliable in enhancing understanding of the issues investigated. However, readers have to tread cautiously in the interpretation of the findings by placing it in the individual case study context, which implies the need for analytical rather than statistical generalisation.

RESULTS AND DISCUSSION

Level of Community Participation: Kakum vs. Afadjato

Ecotourism development in a rural setting usually has implications for landuse. In the case of Abrafo-Odumase and Gbledi-Gborgame, the communities have lost their sources of livelihood. This is because the residents of these areas are predominantly farmers and hunters, therefore land given up for ecotourism could have been used for farming, hunting and served as a source of fuelwood, herbal medicines and other natural resources. With the designation of these places as protected areas and tourist attractions, alternative income-generating projects as shown in Table 1 were undertaken to discourage poaching and encroachment.

Table 1. Income generating activities of community participation projects

Abrafo-Odumase	Gbledi Gborgame
	Bee Keeping
Grasscutter rearing	Crop farming
Snail rearing	Cassava processing
Mushroom Cultivation	Animal Husbandry
Beekeeping	Wood carving
Traditional soap making	Palm oil processing
Handicraft making	Traditional soap making
	Rice Farming
	Vegetables

In the case of Abrafo-Odumase, the project has been partially successful or beneficial due to the token involvement of the locals in decision-making, preference for bushmeat and a lack of demand for locally produced handicrafts. With Gbledi Gborgame, the project failed initially when it was undertaken by groups, but when individual projects were later encouraged and local concerns factored into the decision-making process, it proved to be fairly successful. There was apathy towards the initial project when the people realised that the benefits were to accrue to the entire community, but when the projects were individualised, there was recognition that individuals stood to benefit or fail to benefit and hence, management promoted projects undertaken by individuals.

Results analysed from various sources suggest that the people of Abrafo-Odumase have not been adequately involved in decision making in tourism development as ten (67%) of the fifteen respondents interviewed at Abrafo-Odumase confirmed this. This could result in people planning ways to sabotage the project and they become indifferent about participation in such projects and programmes (Afenyo & Amuquandoh, 2014; Lee & Jan, 2019), As expressed by one trader at Abrafo '*our mouths hurt when we talk about Kakum Park because what we say falls on deaf ears, and we're tired of talking to researchers because we don't get results'*.

Somehow, the people of Kakum felt worse-off because they do not have access to the forest which was their source of livelihood. According to the Chief of Abrafo-Odumase, '*before we could get food from the forest, now we've lost all that. Our women could sell snails from the forest and our men could hunt...without money you cannot do anything'*. This finding contravenes the 'pro-poor tourism concept which according to Lin, Chen & Filieri, (2017) contributes to the overall growth in the quality of life of residents and supports rural sustainable development. This finding suggests the local residents would naturally carry 'frowned-faces', instead. In other words, tourism project intervention that would have brought 'smiles' to the faces of the poor in the Abrafo-Odumase somehow worsened their plight.

Comparing the two case scenarios, it emerged that in the case of Gledi Gborgame; almost all respondents confirmed their involvement in decision-making because the conservation project was their own initiative and not the government's. The level of community members involvement notwithstanding, the people were not entirely satisfied in terms of tourism meeting all their felt needs, but at least they could put up 'smiling-faces' even if not to the full.

Another problem is the ratio of local people gainfully employed in the tourism industry. For instance, the people of Abrafo claim that only a few of them are employed in the tourism industry as compared to those from the regional capital, Cape Coast. This sentiment was shared by a trader at the Abrafo market. According to her sometimes they (officials from the Kakum National Park) *'by-pass us and go for people in Cape Coast (the central regional capital)'*. An interview with some of the employees at the park revealed that quite a number were not from the community and that the majority who were from the community did the menial jobs. Unfortunately, the chief felt helpless as he felt lack of education accounts for the exclusion of his subjects from tourism employment. He stated that *'I don't really blame the government for putting my people in low jobs...I can only plead with my subjects to complete their education.* In effect, better training and skills could help communities to take charge of their own tourism ventures, thus empowering them to achieve sustainable tourism. However, the director of GHTC indicated that 19 tour guides had been trained from the surrounding communities but only work on a part-time basis when there was a shortage of Wildlife Division staff.

With the Afadjato Conservation Area, the majority of the staff were from the locality (See Table 2). As a result of this, poaching was less pronounced. Despite the introduction of grasscutter rearing at Kakum, the communities still prefer the authentic *bush meat* from the forest and this was confirmed by most respondents at Abrafo- Odumase. However, in Afadjato, poaching has reduced as natives were allowed to hunt in areas outside the conservation zone. People who choose to hunt in the conservation area are punished to serve as a deterrent to others. As confirmed by the director of GHTC '*we need to use the traditional system...not the European structure which has not worked and will never work here: therefore we must find a localized system'*. This implies that adopting western environmentalism may not yield good results and to an extent affirms Yang, Ryan & Zhang's (2013) assertion that tourism development brings together various stakeholders, leading to a change in the culture and social structure of communities which could kindle social conflict. Thirteen percent who responded that they did not believe community participation would lead to sustainable tourism development were all women. Perhaps this is because few women participate in decision-making. However of the 30 respondents only seven were women making it difficult to conclude that this answer represents the view of women.

The extent to which revenue accruing from the attractions is shared between the government and the local community also influences the degree of community participation or support for the management of natural resources (Archabald and Naughton-Treves, 2001). From the inception of the park the government was taking about 50% of proceeds from the Kakum National Park although USAID sponsored the project whereas in the case of the Afadjato conservation Area, only five percent of proceeds is paid to the government although the Netherlands embassy sponsored the project. More money is therefore retained in the Gbledi Gborgame community to embark on development projects and alternative income generating activities to improve the livelihoods of the community and to reinforce conservation practices. This is evident in the visitor spending which has had a positive impact; as the funds generated, was used to acquired gravels and sand to level the untarred road leading to the attraction to make it easily accessible, reconstruct a track to the tourist site and to refurnish the only community guest house. Additionally, with the help of an NGO, Stepping Stones for Africa they have promoted a Green and Clean Campaign to improve sanitation and create jobs (GraphicOnline, November, 2019).

This has prompted the GHCT to appeal to the government to reduce its share of proceeds from the Kakum National Park to 10%. In 2009 the then Minster of tourism announced that per the formula drawn in the year 2006, proceeds from the park were distributed as follows; central government of Ghana (10%), district assemblies (20%), traditional authorities and communities (20%), with the remaining 50% going into a Maintenance Fund towards the management of the Kakum park. However, according to the chief of Jukwa, since the inception of the Kakum National Park in 1995 royalties had not been paid to the land owners from the revenue accrued (Modern Ghana, July 2009).

Table 2. Native and non-native staff employed at Afadjato conservation area.

Year	Local	Non-Locals	Total
1998	6	3	9
1999	6	2	8
2000	6	6	12
2001	10	7	17
2002	10	8	18
2003	10	10	20
2004	9	4	13
2005	9	4	13
2006	6	1	7
2007	6	1	7
2008	6	1	7
2009	6	1	7
2010	6	2	8
2011	7	2	9
2012	7	1	8
2013	4	1	5
2014	4	1	5
Total	**118**	**55**	**173**
Percentage	**68.2**	**31.8**	**100**

Source: Ghana Wildlife Society, Gbeldi Gborgame (2014)

CONCLUSION, POLICY AND RESEARCH IMPLICATIONS

From the findings, it can be deduced that on one hand, community participation in Afadjato relates to Tosun's spontaneous participation typology due to the direct participation of communities in decision-making processes. On the other hand, community participation in Kakum could be linked to Tosun's induced participation typology. In the case of Afadjato, the local residents are actors in the project (Figure 1) and this is very sustainable however, in the case of Kakum, some opinion leaders were only informed at the inception of the project. They were not given the chance to make inputs and this is unsustainable. This observation is based on the fact that although some local views were listened to, other powerful stakeholders such as the government took the final decisions. It can be concluded that community participation could lead to sustainable tourism development granted that the needs and expectations of the local community are met.

However, sustaining the interest of the communities is equally important as what sustains the participation, is what will keep tourism development going. There is also the need to ensure that the community is well represented and not just a faction taking decisions on the community's behalf. This echoes Simpson's (2001, p.4) assertion that 'theoretical best practices may not necessarily be reflected in the nature of tourism planning in practical situations'. Participation is not an easy task as some people tend to believe, given that all stakeholders try to satisfy their individual self-interests, a characteristic of every 'homo-economicus' (economic-man). However, with collaboration and consensus-building community participation could be achieved and conflicts minimised to the barest minimum. Therefore, communities should be supported and encouraged to embrace the concept of pro-poor sustainable tourism.

Researchers may wish to carry out a similar research by employing quantitative methods to cover a wider range of respondents. There is also the need to research into land ownership and its effect on community-based tourism. This is because locals have insisted they have a right to make their own decisions simply because the land on which tourism is developed belongs to them.

REFERENCES

Abane, A. M., Awusabo-Asare, K., & Kissi, A. K. (1999). In who's Interest? Individual and Societal needs in the Creation of Forest Reserves: The Case of Kakum in Ghana. *Bulletin of the Ghana Geographical Association*, *2*, 21–30.

Abdullah, A. (2020, August 14). Covid-19 Pierces "Heartbeat" of Ghana's Tourism Hard. *Ghana News Agency*. https://www.gna.org.gh/1.18646544

Ackuayi, E. D., Kotoku, G., Dzeto, G. K., & Bonsu-Owu, A. (2014). Contribution of the Hospitality and Tourism Sector to the Development of Local Communities in the Hohoe Municipality, Ghana. *Journal of Economics and Sustainable Development*, *5*(23), 209–219.

Afenyo, E. A., & Amuquandoh, F. E. (2014). Who Benefits from Community-Based Ecotourism Development? Insights from Tafe Atome, Ghana. *Tourism Planning & Development*, *11*(2), 129–190. doi:10.1080/21568316.2013.864994

Agyeman, Y. B., Aboagye, O. K., & Ashie, E. (2019). Visitor Satisfaction at Kakum National Park in Ghana. *Tourism Recreation Research*, *44*(2), 178–189. doi:10.1080/02508281.2019.1566048

Akama, J. (2011). Efficacy of Tourism as a Tool for Local Community Development. A case Study of Mombasa Kenya. *Journal of Social Sciences*, *1*(1), 1–16.

Akyeampong, O. A. (2011). Pro-poor Tourism: Residents' Expectations, Experiences and Perceptions in the Kakum National Park Area of Ghana. *Journal of Sustainable Tourism*, *19*(2), 197–213. doi:10.1080/09669582.2010.509508

Amoah, M., & Wiafe, E. D. (2012). Livelihoods of Fringe Communities and the Impacts on the Management of Conservation Area: The Case of Kakum National Park in Ghana. *International Forestry Review*, *14*(2), 131–144. doi:10.1505/146554812800923381

Amoako-Atta, E., Dayour, F., & Bonye, S. Z. (2020). Community Participation in the Management of Wechiau Community Hippo Sanctuary, Ghana. *Ghana Journal of Development Studies*, *17*(1), 1–24. doi:10.4314/gjds.v17i1.1

Amuquandoh, F. E. (2017). Tourists' Motivation for Visiting Kakum National Park, Ghana. *Ghana Journal of Geography*, *9*(1), 152–168.

Antwi, L. S. (2020, March 21). Covid-19: Kakum National Park Closes Down. *Graphic Showbiz*. https://www.graphic.com.gh/entertainment/covid-19-kakum-national-park-closes-down.html

Appiah-Opoku, S. (2011). Using Protected Areas as a Tool for Biodiversity Conservation and Ecotourism: A Case Study of Kakum National Park in Ghana. *Society & Natural Resources*, *24*(5), 500–510. doi:10.1080/08941920.2010.495108

Archabald, K., & Naughton-Treves, L. (2001). Tourism Revenue-sharing around National Parks in Western Uganda; Early Efforts to Identify and Reward Local Communities. *Environmental Conservation*, *28*(2), 135–149. doi:10.1017/S0376892901000145

Arnstein, S. R. (1969). A Ladder of Citizen Participation. *Journal of the American Institute of Planners*, *35*(4), 216–224. doi:10.1080/01944366908977225

Arnstein, S. R. (1971). A Ladder of Citizen Participation in the USA. London: Royal Town Planning Institute.

Bagus, S. I., Imade, S. U., Nyoman, S. I. A., & Putu, W. S. N. (2019). Community Based Tourism as Sustainable Tourism Support. *RJOAS*, *10*(94), 70–78.

Bramwall, B. (2015). Theoretical Activity in Sustainable Research. *Annals of Tourism Research*, *54*, 204–218. doi:10.1016/j.annals.2015.07.005

Bramwall, B., Higham, J., Lane, B., & Miller, G. (2017). Twenty-Five Years of Sustainable Tourism and the Journal of Sustainable Tourism. *Journal of Sustainable Tourism*, *25*(1), 1–9. doi:10.1080/09669582.2017.1251689

Butler, R. (2018). Sustainable Tourism in Sensitive Environments: A Wolf in Sheeps's Clothing? *Sustainability*, *10*(1789), 1-11.

Chang, K. G., Chien, H., Cheng, H., & Chen, H. (2018). The Impacts of Tourism Development in Rural Indigenous Destinations: An Investigation of the Local Residents' Perception Using Choice Modeling. *Sustainability*, *10*(4766), 1–15. doi:10.3390u10124766

Cobinnah, P. B., Black, R., & Thwaites, R. (2015). Ecotourism Implementation in the Kakum Conservation Area, Ghana: Administrative Framework and Local Community Experiences. *Journal of Ecotourism*, *14*(2-3), 223–242. doi:10.1080/14724049.2015.1051536

Dei, L. A. (2000). Community Participation in Tourism in Africa. In P. U. C. Dieke (Ed.), *Political Economy of Tourism in Africa* (pp. 285–298). Cognizant Communication Corporation.

Din, K. H. (1988). Reports: Social Cultural Impacts of Tourism. *Annals of Tourism Research*, *15*(4), 563–566. doi:10.1016/0160-7383(88)90054-0

Eggert, L. S., Eggert, J. A., & Woodruff, D. S. (2003). Estimating population sizes for elusive animals: The forest elephants of Kakum National Park, Ghana. *Molecular Ecology*, *12*(6), 1389–1402. doi:10.1046/j.1365-294X.2003.01822.x PMID:12755869

Ellis, S., & Sheridan, L. (2015). The Role of Resident Perceptions in Achieving Effective Community Based Tourism For Least Developed Countries. *International Journal of Tourism and Hospitality Research*, *26*(2), 244–257.

Fiagbomeh, F. F., & Burger-Arndt, R. (2015). Prioritisation of Strategies for protected area management with local people using the hybrid SWOT-AHP analyses: The case of Kakum Conservation Area, Ghana. *Management Science Letters*, *5*(5), 457–470. doi:10.5267/j.msl.2015.3.008

Fletcher, J., Fyall, A., Gilbert, D., & Wanhill, S. (2018). *Tourism Principles and Practice* (6th ed.). Pearson Education Ltd.

Geol, K. (2014). Understanding Community and Community Development: Defining the Concept of Community. *ResearchGate*. file:C:/Users/Ama/Downloads/1. UnderstandingCommunityandCommunityDevelopment.pdf

Ghana News Agency. (2009, July 7). *Chief Appeals for Royalties from Kakum Park*. Modern Ghana. https://www.modernghana.com/news/225929/chief-appeals-for-royalties-from-kakum-park.html

Ghana News Agency. (2018, January 30). Tourism at Kakum National Park Drops. *Modern Ghana*. https://www.modernghana.com/news/831799/tourism-at-kakum-national-park-drops.html

GraphicOnline. (2019, November 7). *Poor Roads Hamper Ecotourism Promotion*. Daily Graphic. https://www.graphic.com.gh/news/general-news/ghana-news-poor-roads-hamper-ecotourism-promotion.html

Ibanescu, B., Stoleriu, O. M., Munteanu, A., & Oatu, C. (2018). The Impact of Tourism on Sustainable Development of Rural Areas: Evidence from Romania. *Sustainability*, *10*(3529), 1–19. doi:10.3390u10103529

International Union for Conservation of Nature and Natural Resources. (2010). *Parks and Reserves of Ghana: Management Effectiveness Assessment of Protected Areas*. https://portals.iucn.org/library/sites/library/files/documents/2010-073.pdf

Keen, M., & Mahanty, S. (2005). Collaborative Learning: Bridging Scales and Interests. In M. Keen, V. A. Brown, & R. Dyball (Eds.), *Social Learning in Environmental Management: Towards a Sustainable Future* (pp. 1–21). EARTSCAN.

Kiper, T. (2013). Role of Ecotourism in Sustainable Tourism Development. *IntechOpen*, *31*, 773–802.

Lee, J. H., & Jan, F. H. (2019). Can Community-Based Tourism Contribute to Sustainable Development? Evidence from Residents' Perceptions of Sustainability. *Tourism Management*, *70*, 368–380. doi:10.1016/j.tourman.2018.09.003

Lin, Z., Chen, Y., & Filieri, R. (2017). The Role of Resident's Perceived Tourism Impacts and Life Satisfaction. *Tourism Management*, *61*, 436–442. doi:10.1016/j.tourman.2017.02.013

Lu, J., & Nepal, S. (2009). Sustainable Tourism Research: An analysis of Papers Published in the Journal of Tourism. *Journal of Tourism*, *17*(1), 5–16.

Mak, K. L., Cheung, L. T. O., & Hui, D. L. H. (2017). Community Participation in the decision making process for sustainable tourism development in rural areas of Hong Kong, China. *Sustainability*, *9*(6975), 1–13. doi:10.3390u9101695

Mensah, I. (2016). Effects of socio-demographic characteristics and perceived benefits of tourism on community participation in tourism in the Mesomagor Area of the Kakum National Park, Ghana. *Athens Journal of Tourism*, *3*(3), 211–230. doi:10.30958/ajt.3-3-3

Mensah, I. (2017). Benefits and Challenges of Community-Based Ecotourism in Park-Fringe Communities: The Case of Mesomagor of Kakum National Park, Ghana. *Tourism Review International*, *21*(1), 81–98. doi:10.3727/154427217X14866652018947

Mensah, I., & Adofo, E. (2013). Community Participation in Ecotourism: The Case of Bobiri Forest Reserve and Butterfly Sanctuary in Ashanti Region of Ghana. *American Journal of Tourism Management*, *2*(1), 34–42.

Ministry of Lands and Natural Resources. (2012). *Ghana Forest and Wildlife Policy*. https://www.documents.clientearth.org/wp-content/uploads/library/2012-01-01-ghana-forest-and-wildlife-policy-2012-ext-en.pdf

Ministry of Tourism. (2012). *Arts and Culture*. Ghana National Tourism Development Plan.

Ministry of Tourism, Arts and Culture. (2019). *Medium Term Expenditure Framework for 2019-2022*. Accra: Republic of Ghana.

Modern Ghana. (2009, July 9). *Formula for Sharing Proceeds from Kakum to be revisited*. Ghana News Agency. https://www.modernghana.com/news/ 226414/formula-for-sharing-proceeds-from-kakum-park-to-be-revisited.html

Mugizi, F., Ayorekire, J., & Obua, J. (2017). Factors that Influence Local Community Participation in Tourism in Murchison Falls Conservation Area. *Journal of Environmental Science & Engineering*, *A6*, 209–223.

Nanang, D., & Owusu, E. H. (2010). Estimating the Economic Value of recreation at the Kakum National Park, Ghana. *Nova Science Publishers*, *12*, 247-267. https://www.researchgate.net/publication/284031218_Estimating_The_Economic_Value_of_Recreation_at_the_Kakum_National_Park_Ghana

Nurhidayati, S. E. (2015). The Evaluation Study of Implementation as Community-Based Tourism (CBT) in Supporting Sustainable Agritourism. *Society. Cultural Politics*, *28*(1), 1–10.

Ofori, B. Y., Owusu, E., & Attuquayefio, D. K. (2014). Ecological Status of the Mount Afadjato-Agumatsa Range in Ghana After a Decade of Local Community Management. *African Journal of Ecology*, *53*(1), 116–120. doi:10.1111/aje.12165

Onchwati, J., Sommerville, H., & Brockway, N. (2010). Sustainable Tourism Development in the Masai Mara National Reserve, Kenya, East Africa. *WIT Transaction on Ecology and the Environment*, *139*, 319–330. doi:10.2495/ST100281

Owusu, E. A. (2008). Ecotourism as a Conservation Tool – A Case of Afadjato-Agumatsa Convertion Area, Ghana. *Journal of Science and Technology (Ghana)*, *28*(3), 166–177.

Owusu, E. H. (2001). *Community Based Conservation in Ghana: The Potential of the Afadjato and Agumatsa Range for Ecotourism* [Unpublished PhD Thesis]. University of Kent at Canterbury.

Owusu, E. H., Ekpe, E. K., & Asamoah, A. A. (2005). Large Mammal Fauna of Afadjato and Agumatsa range in Ghana: An Important Bird Area. *West African Journal of Applied Ecology*, *8*(1), 18–28.

Pacione, M. (2009). *Urban Georgraphy: A Global Perspective* (3rd ed.). Routledge. doi:10.4324/9780203881927

Pretty, J. (1995). Participatory Learning for Sustainable Agriculture. *World Development*, *23*(8), 1247–1263. doi:10.1016/0305-750X(95)00046-F

Ridder, H. (2017). The Theory Contribution of Case Study Research. *Business Research*, *10*, 281-305. https://link.springer.com/article/10.1007/s40685-017-0045-z

Schwass, R. D. (2011). World Conservation Strategy of International Union for the Conservation of Nature and Natural Resources (IUCN). In D. V. J. Bell & Y. A. Cheung (Eds.), *Introduction to Sustainable Development* (pp. 147–171). Eolss Publishers.

Siakwah, P. (2018). Tourism Geography and Spatial Distribution of Tourism Sites in Ghana. *African Journal of Hospitality, Tourism and Leisure*, *7*(1), 1–19.

Simpson, K. (2001). Strategic Planning and Community Involvement in Contributors to Sustainable Tourism Development. *Current Issues in Tourism*, *4*(1), 3–41. doi:10.1080/13683500108667880

Tao, T. C. H., & Wall, G. (2009). Tourism as a Sustainable Livelihood Strategy. *Tourism Management*, *30*(2), 90–98. doi:10.1016/j.tourman.2008.03.009

Tosun, C. (1998). Roots of Unsustainable Tourism Development at the Local Level: The Case of Ürgüp in Turkey. *Tourism Management*, *19*(6), 595–610. doi:10.1016/S0261-5177(98)00068-5

Tosun, C. (1999). Towards a Typology of Community Participation in the Tourism Development Process. *Anatolia, 10*(2), 113–134. doi:10.1080/13032917.1999.9686975

Tosun, C. (2000). Limits to Community Participation in Tourism Development Process in Developing Countries. *Tourism Management, 21*(6), 613–633. doi:10.1016/S0261-5177(00)00009-1

Tosun, C. (2006). Expected Nature of Community Participation in Tourism Development. *Tourism Management, 27*(3), 493–504. doi:10.1016/j.tourman.2004.12.004

Twerefou, D. K., & Ababio, D. K. A. (2012). An Economic Valuation of Kakum National Park: An Individual Travel Cost Approach. *African Journal of Environmental Science and Technology, 6*(4), 199–207.

United Nations General Assembly. (1987). *Report of the world commission on environment and development: Our common future.* Oslo, Norway: United Nations General Assembly, Development and International Co-operation *Environment.*

United Nations World Tourism Organisation. (2013). *Sustainable Tourism for Development Guidebook: Enhancing Capabilities for Sustainable Tourism for Developing Countries.* https://www.e-unwto.org/doi/pdf/10.18111/9789284415496

Wang, M., Jiang, J., Xu, S., & Guo, Y. (2021). Community Participation and Residents' Support for Tourism Development in Ancient Villages: The Mediating Role of Perceptions of Conflicts in Tourism Community. *Sustainability, 13*(2455), 1–15.

Wen, S., Cai, X., & Li, J. (2021). Pro-poor Tourism and Local Practices: An Empirical Study of an Enormous County in China. *SAGE Open*, 1–11. https://journals.sagepub.com/doi/pdf/10.1177/21582440211022740

Wiafe, E. D., & Agyei, D. (2013). Spices Richness, Diversity and Distribution of Amphibians Along Elevation Gradient on Mountain Afadjato Ghana. *Eurasian Journal of Forest Science, 1*(2), 68–76.

Wiafe, E. D., & Arku, F. S. (2014). The Role of Women in Community-Based Resource Management in the Afadjato Mountain Ecosystem, Ghana. *Applied Research Journal, 1*(1), 1–11.

World Travel & Tourism Council (2021). *Ghana 2021 Annual Research Key Highlights, Economic Impact Report.* Wttc.org.

Yanes, A., Zidinski, S., Cano, M. D., & Kim, S. (2019). Community Based Tourism in Developing Countries: A Framework for Policy Evaluation. *Sustainability*, *11*(2506), 2–23. doi:10.3390u11092506

Yang, J., Ryan, C., & Zhang, L. (2013). Social Conflict in Communities Impacted by Tourism. *Tourism Management*, *35*, 82–93. doi:10.1016/j.tourman.2012.06.002

Yeboah, S. (1998). Small Mammal Diversity in the Kakum National Park in Ghana. *Journal of Science*, *38*, 25–32.

Chapter 5
Distributive Justice Issues in Community-Based Tourism

Victor Mawutor Agbo
University of Waterloo, Canada

ABSTRACT

Conventional tourism and its attendant challenges for nature and communities have necessitated the need for tourism to be more sustainable, equitable, and responsible, hence the introduction of community-based tourism (CBT). CBT was developed as a model and a viable instrument for poverty reduction, offering opportunities for conservation and rural economic development. However, despite the potential of CBT to generate welfare for communities, many initiatives have failed to deliver on their promises. Since tourism and justice issues cannot be separated from each other, concerns over the discussion of justice related issues have emerged as a community concept which requires critical investigation. This chapter presents a theoretical exploration of how justice is conceptualized, with an emphasis on distributive justice in the context of CBT, and how it is shaping the production of CBT. It also explores some critical approaches to tourism studies and practice and how justice for local communities is conceptualized in CBT contexts.

INTRODUCTION

Justice as a concept cannot be ignored in any discussion on tourism because many justice issues are embedded in tourism (Fennel, 2006). For this reason, concerns over justice related issues, in the context of Community-based tourism (CBT) have recently become of critical importance to practitioners, policymakers, tourism businesses, and academics. When tourism is produced and marketed from local communities

DOI: 10.4018/978-1-7998-7335-8.ch005

many critical issues become note-worthy. It is important to be aware of the impacts that potentially result from the relationship that tourism development has with local community well-being. Many tourism resources of the world including endangered species exist in natural habitats and vulnerable communities, and the ever-growing desire of many travellers to experience and explore the world puts most of these resources at risk. Conventional tourism, and its attendant challenges for nature and communities have raised many alarms over the world. For that matter, many scholars (e.g., Campbell, Gray, & Meletis, 2008; Harris, 2009; Ross & Wall, 1999), have suggested that there is an urgent need for a re-think about tourism, one that ensures the long term economic, social and environmental sustainability, especially in the Global South. Out of this desire came, the concept of community-based tourism (CBT), with its beginnings dating back to the 1980s. CBT was developed as a model and a viable instrument for poverty reduction, conservation and rural economic development. However, despite the potential of CBT to generate benefits for communities, many initiatives have failed to deliver on their promises. According to the World Commission on Environment and Development [WCED] (1987), justice and other similar concepts, such as equity and fairness, were the core goals of many pioneers of sustainability; however, in developing new management approaches and principles, many seem to have lost focus on justice-related fundamentals in sustainable tourism. Jamal and Camargo (2014) also lamented that for many local communities who share the bulk of tourism's negative impacts, justice and equity discussions have been slow to address these issues.

This chapter presents a theoretical exploration of how justice is conceptualised, with an emphasis on distributive justice, in the context of CBT, and how it is shaping the planning and development of CBT at the destination level. It also identifies the principles that need to influence policy and guideline formulation, tackle critical issues such as benefit-sharing, and aid in the evaluation of programmes and institutional structures that facilitate sustainable CBT development, management and marketing for destinations. The last part of the chapter explores some critical approaches to tourism studies and practice, and how justice for local communities fit into the production and management of CBT.

COMMUNITY-BASED TOURISM (CBT)

Community-based tourism is a concept that is difficult to define. This is mainly because of many stakeholders involved in it have made attempts to define it in ways that reflect their respective interests. No matter how it is defined, there are a number of common characteristics in all the definitions. To understand this concept, there is the need to examine what 'community-based' is. Van Bibber (1997) defined 'community-based' as a philosophical approach that allows communities to have an active role and participate in highlighting and addressing the issues that matter to them. From this understanding, CBT can be described as a type of tourism that gives the residents of a community substantial control over, and active involvement in the production and management of tourism. Additionally, this type of tourism ensures that a major part of the benefits that accrue from tourism remains within the community. Mensah (2017) posited that CBT projects, especially in developing countries, have been used to achieve two things: improved livelihoods for residents of host communities and natural resource conservation. He referred to this as a two-edged sword. One thing is clear about CBT projects: they enable community members to be active stakeholders in the process of decision-making. This positions them for just, equitable, and fair distribution of power (Dei, 2000; Tosun, 2000). Li (2006) believes that in order for communities to partake in the benefits of tourism, for the distribution of the benefits to be fair and just and to ensure the sustainability of tourism resources, there is the need for active local participation in decision-making.

Hall (1996) pointed out that under CBT, the host community's involvement is at the core of planning and maintaining tourism development. This is a necessary prerequisite for creating a more sustainable tourism industry. In another vein, Pearce (1992) noted that CBT makes it possible for local communities to have control over development, to be represented in decision making, and to have equitable access to benefits that accrue from the tourism industry. Despite these promising potentials of CBT, many scholars have also highlighted many ways in which CBT has failed to be just, fair, and equitable to local communities. Blackstock (2005), in her fieldwork in Port Douglas, a North Queensland tourism destination, found that CBT examples do not carry the transformative and empowering promises of development for communities. She explained that this was because for almost all the time, CBT is projected as a catalyst for the survival of the supply side of tourism, specifically tourism business, instead of focusing on just, equitable, and fair treatment of local communities. Blackstock (2005) continued to note that local communities, are more often than not, represented as homogeneous units, not having any internal power struggles, or many different and competing interests and values. This is what Mayo, as quoted in Craig (2003:3) called "an example of community development 'imposter' driven by economic imperatives and neo-liberal agenda, rather than values of empowerment and social justice".

CRITICAL APPROACHES TO CBT

There have been many changes in epistemological and methodological thought in the social sciences. Aitchison (1999) observed some of these notable changes as 'new cultural and sociological studies' (e.g., Atkinson, 1990; Du Gay, 1997; Long, 1997), 'new human and cultural geographies (e.g., Jackson, 1993; Mansvelt, 2005; Massey, 2005) and 'new studies in leisure'. The word 'new' here does not suggest new sub-disciplines. It also does not imply a total rejection of pioneer 'traditional thinkers'; instead, its use is a general reference to the diverse work that goes beyond disciplinary boundaries to knowledge construction.

Tourism studies have been influenced by many disciplines; however, only recently have critical concerns been introduced in the methodological and philosophical engagements (e.g., Franklin & Crang, 2001; Hollinshead, 1999; Phillimore & Goodson, 2004; Tribe, 1997). Having examined the available literature on the positive and negative impacts of tourism, Britton (1982) noted that many scholars engage in debates on tourism without necessarily paying attention to theories of political economy concerned with the ever-growing inequalities that exist between and among many nations. As a way to critically assess the impacts of tourism, he stressed the importance of asking why tourism frequently deepens economic problems, social tensions, and regional and class inequalities even though it has also contributed to positive outcomes in many developing countries.

Tribe (2007) noted that ideas such as utopia, hope, and the 'good life for tourism' are important concepts which are critical tourism scholars as well. He also explained that key concepts such as justice, power, and equity are key interest areas for critical tourism researchers. In the CBT context, scholars would seek to reveal how justice is served to local communities, and which of their interests and values are honoured. This becomes important in explaining how justice is conceptualized in CBT contexts. Jamal (2019) pointed out the importance of discussing the extent to which discourses of sustainability address justice-related concerns for lower-income, marginalized groups, and the disadvantaged, who are mostly left out of decision-making about their own identities, resources, and culture.

Ateljevic, Morgan and Pritchard (2013:1) also emphasized that critical approaches to tourism move the study of tourism beyond the production, governance, and marketing to consider 'reclaiming the world for the whole of humanity'. This includes the understanding of concepts like justice, equity, power, and sustainability in the study of tourism and how they, in one way or the other affect community wellbeing. What this approach to tourism does is that it puts distributive justice into perspective, hammering on actions that carry many great prospects for tourism in communities.

Another critical approach to tourism was adopted by Grimwood, Mair, Caton and Muldoon (2018) to critically assess tourism and wellness. In their book entitled, *Tourism and Wellness: Travel for the Good for All?* they advanced tourism knowledge by situating the tourism and wellness nexus within a sophisticated interdisciplinary terrain, underscored by a broad social justice mandate. The authors pulled together cutting-edge, critical tourism analyses from scholars in the field to disrupt and reconfigure how wellness is understood in tourism, and by whom and for whom. They established that if wellness, as an end, is about optimizing the various dimensions of individuals' well-being, then tourism plays a major role and carries a great promise. To further emphasize the need for their book, they made a critical point: "that wellness and tourism are not just about individuals. Both are also about collectives and contexts, that is, social, cultural, ecological, and epistemic" (Grimwood et al., 2018: xvi). They however observed that on one hand, the utility of wellness as a metaphor attending to the consequential dynamics of tourism at scales above and beyond the individual (i.e., the tourist), has received little consideration or critique. On the other hand, the relationships between wellness and tourism have been insufficiently situated in time, space, or the moral conceptions and enactments that shape the making of our worlds (Haraway, 2016), hence the urgent need to cast the nexus of tourism and wellness in such light (Grimwood et al., 2018).

Just like any other critical tourism agenda, the core mandate of their book shares some common grounds with bell hooks' 2003 book *Teaching Community: A Pedagogy of Hope*. Pritchard et al. (2011) reviewed the book and found that it provides deep insights into ways to model critical education arenas in order to dismantle any form of ethnic, gender, racial, national lines, and class oppression and strive for communities that are socially just. Situating broad questions of justice, well-being, health, and equity within the scaffolds of critical tourism studies allows for the enhancement and extension of the analysis of tourism as a social force (Higgins-Desbiolles, 2006; Hollinshead, 2007). To make tourism studies and practice more critical, many approaches continue to emerge to cater to the changing dynamics underlying tourism development in communities. To ensure that CBT is carried out in a manner that is just, equitable, and sustainable to tourism communities, and all other stakeholders, there is the urgent need to rethink the development and management of CBT, having in mind those who are affected the most by tourism development and are at the bottom of the global capitalist hierarchy – in most cases, local communities.

IDEAS OF JUSTICE

Justice is a fundamental concept that has been examined by many scholars in leisure sciences. Early pioneers of sustainability had justice, equity, and fairness as a major part of their campaigns, but many of them have now lost sight of issues of justice in sustainable tourism (Jamal & Camargo, 2014). Higgins-Desbiolles (2018) explained that justice issues come from the realm of philosophy. She noted that this is the domain where for a very long time, ideas of what creates a just community and a good life have been contemplated. She argued that from a contemporary position, modern day tourism is seen as an individualistic, consumerist, commercial, and hedonistic pursuit which should not be considered a conducive terrain for what makes a just society and a good life (Higgins-Desbiolles, 2018). There were, however, scholars such as Alain de Botton, who shared different thoughts. In his book '*The Art of Travel*' (2003), he projected tourism and travel as phenomena that allow people to live a meaningful life, for that matter, worthy of philosophical contemplation. The question, however, is 'a meaningful life for whom?"

According to Sandel (1998), justice is served when people get what they deserve. Among the main philosophical traditions commonly raised in the discussion of justice, grounded in comprehensive moral doctrine are utilitarianism, Kantian deontology, and Aristotle's virtue ethics (Jamal & Camargo, 2014). Different theories of utilitarianism address different types of utility, which include consequences, or happiness, for the majority. For instance, government policy generally draws upon utilitarian perspectives to frame policy and action based on maximizing utility. Jamal (2019) explained that policymakers may favour this approach in seeking to maximize benefits to the majority of their constituents. However, the question of what about those who fall into the minority and who do not benefit from the policies passed still stands.

It is for this critique that, Kant's (1998) influential deontological theory, stands in contrast, to focus on duty or obligation. Based on the idea that every moral action is guided by a categorical imperative, Jamal and Menzel (2009) consider this as highly formal and abstract. The humanity formulation of the categorical imperative is: "Act in such a way that you treat humanity, whether in your own person or in the person of another, always at the same time as an end and never simply as a means" (Kant, 1998: 432). Kant's approach has been influential in informing human rights discourses and development views that emphasize human dignity (e.g. Nussbaum, 2011).

Aristotle's virtue ethics is the third perspective that tourism studies draw upon significantly. It is community-based, with a central idea that humans belong to a political community, and through participation in the community, the citizen develops the virtues and character conducive to a good life. His ideology was that all good things such as health, friends, community, that one chooses to pursue are pursued for the sake of the general, highest good of living well, also seen as the good life (Jamal & Camargo, 2014).

Justice can also be considered as *retributive*, one which is oriented toward blame and punishment in the criminal justice system. It can also be seen as *restorative*, commonly defined as a justice approach that addresses crime, and the harm it causes while making the offender take responsibility for their actions. By extension, it also focuses on providing an opportunity for the communities, offenders and victims to identify and address their needs in the aftermath of a crime. Another type of justice is *procedural*. This form of justice examines the extent to which the processes used by people in authority to arrive at specific outcomes or decisions demonstrate fairness. Research suggests that when citizens make overall judgments about the legitimacy of those in positions of authority, they are more concerned about procedural fairness, that is, how they are treated, than they are about the outcome of the encounter (Wright & Boese, 2015). Jamal (2019) also pointed out that conflicts over the distribution of goods and the claims made to them may be resolved by calling on justice to determine what people are properly entitled to. Also, one can think about justice for groups with respect to the distribution of state resources among various groups of residents. This is called *distributive* justice. "Justice here requires that the resources available to the distributor be shared according to some relevant criterion, such as equality, desert, or need" (Miller, 2017:252). Even though there is more to justice, this chapter only discusses distributive justice as conceptualized in CBT.

DISTRIBUTIVE JUSTICE IN CBT

Distributive justice points to people's assessment of fairness regarding how the benefits of tourism are allocated to people. Miller (2017) explained that at the core of distributive justice, how individuals are treated and how resources are distributed are crucial. Deutsch (1975), while summarizing distributive justice, identified three fundamental principles that people consider as fair depending on the particular social goals that are relevant to a particular relationship or social context. These principles are equality, need, and equity (Wright & Boese, 2015). *Equality* represents the distribution of resources in equal proportions. Under the principle of *need*, there is an unequal distribution of resources in such a way that larger proportions are given to people with the most need. In communities where responsibility and social welfare are the main goals, the principle of need is seen as the most acceptable. The principle of *equity* also calls for an equal distribution of resources and that the ratio of any individual's inputs to his/her outcomes should be equivalent to the same ratio of the person with whom the individual is being compared. In this case, every person should receive resources commensurate to his/her contributions.

A critical take on the concept of justice in discussions around CBT development is crucial as it tackles the various ways in which tourism development has been critiqued. It also allows for the development of critical, and evaluative directions in order to assess whether or not the approaches used by policymakers, governments, marketers, and tourism organisations to help with community wellbeing and the fair treatment of the natural and cultural goods that are shared by residents with tourists are morally justifiable (Jamal, 2019). In the context of CBT, justice is conceptualised as a societal concern. Jamal (2019) refers to this type of justice as social justice.

She further explained that at the core of distributive justice is the discussion of what kinds of social goods there are, and how they are allocated. In practice, justice has shaped many aspects of CBT, focusing on the fair distribution of power and the benefits that come out of tourism in communities where it is practised. This all-encompassing nature of justice as fairness, does not only focus on the fair and equitable distribution of power, opportunities, and honour to host communities, but also to tourists who travel to these communities. In other words, this is distributive justice to both hosts and guests.

Higgins-Desbiolles (2018) noted that Hultsman was one of the earliest scholars in the academy to examine justice in the context of tourism. He developed the idea of "just tourism", which he described as representing "fair" or "proper" tourism (Hultsman, 1995: 560). Hultsman's (1995) idea of just tourism is the suggestion that an ethical framework for tourism must be developed to guide every form of CBT. For him, this framework needs to focus not only on the tourists' experience of tourism but must also provide guidelines for developing a 'principled' practice and 'ethicality' in tourism to guide the activities of all tourism stakeholders in the conduct of their activities and its impacts on local communities. Lee and Jamal (2008) referred how the incorporation of a comprehensive equity and justice framework into planning and development of tourism has started to emerge. They argued that Hultsman's approach to the study of justice in tourism "provides important direction and guidance for addressing injustices related to human - environmental relationships, particularly with respect to disadvantaged, low-income, and minority communities" (Lee & Jamal 2008: 44).

A key factor in ensuring a successful CBT is allowing for local communities' ability to self-determine what happens in their own communities, and the exercise of their human rights to be at the center of every tourism development. This should include, but is not limited to the right to meaningful participation and consultation including free, prior, and informed consent on whether tourism should take place, to what extent it should be developed, and in what form it takes place (Jamal, 2019). In some CBT projects, the application of justice frameworks has helped in the widespread and fair distribution of socio-economic benefits. This includes improving community wellbeing, local prosperity, and social equity. For instance, Afenyo and Amuquandoh (2014), in their quest to explore the benefits that have accrued from an ecotourism project, and to examine the distribution process in Tafi Atome, a rural community in Ghana, observed that there was enough evidence of collective benefits for the community. Their study established that infrastructural development, the provision of alternative employment opportunities, and increased environmental consciousness among community members were some outcomes of the project. To be clear, this is what Jamal (2019) calls distributive justice in tourism contexts.

Jamal (2019) again pointed out that justice in CBT contexts also means accountability to local communities. Her idea is that since governments and industry operators utilize land and cultural resources belonging to local communities to the benefit of businesses and tourists, they must be accountable to the local community. Kiss (2004) warns against the potential of tourism developers ignoring dynamics that exist within a local community when engaging in tourism development. To this end, Afenyo and Amuquandoh (2014) stressed the need for tourism actors to strive to balance the needs of the different power groupings that exist within local communities in order to achieve the objectives of CBT projects. They concluded that this move helps to ensure that all members of local communities, especially "powerless" groups benefit from the projects' outcomes.

Overall, the discussion about justice in CBT helps to clearly determine the rights and responsibilities of communities, governments, tourism businesses, and tourists. Also, justice and equity discussions allow for the prioritization of marginalised and vulnerable groups including minorities, children, women, indigenous people, and all manner of people who are in one way or the other affected by tourism, or work in the tourism industry (Jamal, 2019).

IMPORTANCE OF DISTRIBUTIVE JUSTICE TO CBT

To understand the ways in which justice is favourably shaping CBT in many parts of the world, the chapter revisits Hultsman's (1995) idea of just tourism, which is the ethical framework for tourism planning and development. His view is that the concept of justice in CBT studies allows for the development of a 'principled' practice and 'ethicality' in tourism to guide the activities of all tourism stakeholders. This does justice to the many challenges of CBT by clearly laying out what needs to be done, who needs to do what, and who benefits from what. Local community members, who are mostly ignored, find a place to share their ideas and benefit from tourism. Lee and Jamal (2008) stressed the importance of this ethical framework. They argued that it makes room for tackling injustices that happen as a result of human - environmental relationships, especially those that are meted out to minorities, low-income, and disadvantaged communities (Lee & Jamal, 2008).

Rami Kassis, the Director of the NGO, the Alternative Tourism Group of Palestine outlined his conceptualization of justice through tourism from his experience of using tourism to effect change in Palestine (Higgins-Desbiolles, 2018). He observed that globalization of tourism has been responsible for the norm of cultural domination, especially to the detriment of local communities, however, prioritizing justice in tourism development is a social and cultural response to this cultural domination (Kassis, n.d.). Also, Scheyvens (2002) discussed how tourism can be used as a force for community empowerment. She explained that justice and equity are shaping CBT in many destinations. She described just tourism as "both ethical and equitable" and claimed it has the following attributes: ability to build solidarity between hosts and guests and the power to promote symbiotic relationships. She described these relationships as respectful, equitable, carrying the potential to maximize local economic, cultural, and social benefits, and supporting self-sufficiency and self-determination of local communities (p. 104).

Additionally, justice has been seen as a driver for building solidarity and synergetic relationships between host communities and guests which is based on a caring attitude and understanding for one another, mutual respect marked by equality of relations (Dangi, 2018). Other scholars (e.g. Fennell, 2006; Lovelock & Lovelock, 2013) have attempted to define tourist-resident interactions by the application of the lens of ethics of care. They noted that justice has usefully shaped CBT development and management through the fostering of emotional solidarity, described by Dangi (2018) as a feeling of closeness or bonding that individuals experience with one another.

Again, the concept of justice has helped in the promotion of respect and preservation of the environment. Defining justice for stakeholder groups and integrating them in the production of CBT has enabled residents to revitalize activities that are productive to them, engage in reclaiming historical practices, highlight their cultural resources, and highlight the unique aspects of their locality, such as architecture, cuisine, climate, handicrafts, and topography. For example, Taylor (2009) pointed to the reduction in resource degradation in Zimbabwe through the Communal Areas Programme for Indigenous Resources (CAMPFIRE) project. This project assigned the *de facto* rights to occupiers of titled land as custodians of wildlife, fish, and plants by the legally mandated authority responsible for wildlife management in the country. This gave the local community members complete control over the project, resulting in mutual respect for each group and success in the CAMPFIRE project. This is seen as one of the most successful CBT tourism projects in Zimbabwe. Dangi (2018) explained that tourism communities who feel that their needs are catered for, and their voices represented have participated in sustainable collaborative efforts and respect for all stakeholders.

In some CBT contexts, justice concerns have helped to fairly distribute economic and social benefits to local communities, including improving community health and well-being, local prosperity, and social equity. Jamal (2019) noted that justice also means that governments and operators in the tourism industry must provide timely and fair accountability to local communities because it is their land and cultures that are being utilized for the benefit of tourism businesses and tourists. This is shown in many areas where CBT is practised. For instance, Mensah (2017), in exploring the benefits and challenges of community-based ecotourism in Ghana's park-fringe communities found that community participation, which is a product of justice, offered economic gains in the development of infrastructure such as roads, classrooms and health facilities for the Mesomagor community. Overall, the integration of justice in CBT projects have helped to clearly determine the rights and responsibilities of communities, governments, tourism businesses, and tourists. As the discussion of marginalised and vulnerable groups is mostly limited, justice and equity discussions allow for the prioritization of local community concerns (Jamal, 2019).

Lastly, justice and equity have shaped policies and guidelines, raised appropriate practical questions such as benefit-sharing, and helped evaluate institutional structures and programs set up to facilitate sustainable CBT development and marketing for destinations. For example, Harris (2009), in assessing CBT in Bario, Malaysia observed that there were some conditions that help facilitate CBT as an effective tool for community development. He indicated that policies and guidelines were intentionally designed to enable communities to derive the best from tourism. The key guiding principle that was behind CBT in Bario, according to him, has consistently been how CBT could be made useful to community members. He noted that the approach was evident in the various forms of entrepreneurial activities that tourism utilized. Some of the activities included homestays for the occupants of the longhouse with a lot of space, and the engagement of farmers as tour guides when they were not busy in their fields (Harris, 2009).

The prioritization of justice and equity at the core of CBT development, provides the potential for a *Just Destination*, which is, a destination where tourism planning, policy, and practices enable the fair treatment of both tangible and intangible social, cultural, and environmental resources and facilitate the well-being of host communities.

CONTEMPORARY ISSUES IN RELATION TO DISTRIBUTIVE JUSTICE IN CBT

Even though the integration of justice in the development and management of CBT has many potentials, many tourism destinations are still confronted with many challenges. In an age of broader contemporary issues such as environmental degradation and community health and well-being, there is the need for innovation in the planning and management of CBT that clearly reflects how justice is served. This section provides an overview of some contemporary issues and the innovations needed to ensure successful CBT.

Environmental degradation continues to be a major concern for many countries. The United Nations International Strategy for Disaster Reduction (2009) defines environmental degradation as the reduction of the capacity of the environment to meet social and ecological objectives and needs. It is important to note that no matter how tourism is planned, it still has the potential of putting pressure on the environment. Increased demand for travel resulting in pressure on fossil fuels, forests, wetlands, and wildlife have become major concerns for humanity. This alarming extent of environmental degradation caused by many anthropogenic factors including tourism, requires innovative approaches to CBT development.

Community health and well-being which is broadly defined as "a function of the actual conditions of life, and what a person or community makes of those conditions" (Michalos, 2008, pp 349 – 350) is another issue that confronts humanity. Studies have shown that a community's health can be influenced by tourism either directly or indirectly. Even though community health and wellbeing have been discussed by many tourism scholars, many criticisms of tourism production and development point to the fact that the existing literature has mainly prioritized the wellbeing of tourists to the neglect of host communities' wellbeing (Genc, 2012). Unfortunately, only recently have studies started to examine the links between tourism and local communities' health and wellbeing. Aside from the increased competition for resources such as drinking water, existing health care services and other common resources, tourism can also displace local people and destroy traditional livelihoods (Neef, 2019). Realising that these are becoming prevalent, the question is: what innovations are needed in the production of CBT given these contemporary issues?

IMPROVING DISTRIBUTIVE JUSTICE IN CBT PROJECTS

Integration of Sustainable Development Goals (SDGs) for Improving Distributive Justice in CBT

There is the need to integrate the SDGs in the development and management of CBT. The SDGs represent 17 global goals that were designed to be a blueprint for achieving a better and more sustainable future for all (UNDESA, 2017). These goals have been seen as an urgent call for action by both developed and developing countries in a global partnership. Tourism only appears specifically in the targets of Goals 8, 12 and 14, however, it still connects directly or indirectly with all 17 of these goals with its outreach and impact (UNWTO-UNDP, 2017).

The fact that these goals cover a wide range of issues including justice, equity, power, and sustainability means that developing and managing CBT in line with them provides the incentive for tackling broader contemporary issues. It is important to note, however that, the goals in themselves are not enough for successful CBT development and management. Boluk et al. (2019) emphasized that the prospects carried by these goals and what the UN claims they could achieve need to be complemented with analysis from a diversity of perspectives and approaches, coupled with critical thinking. Therefore, given the right conditions, and appropriate partnerships, CBT projects stand the chance of being successful even in the face of contemporary issues. Even though only three of the goals speak directly to tourism, prioritizing all 17 goals carries a promise for just, equitable, and sustainable CBT. Boluk et al. (2019) pointed out that tourism ranges from the local to the global, from the individual to the societal, and from the immediate impacts of activities to the longer-term impact of environmental change. This implies that using the SDGs as a framework for developing CBT requires a critical consideration of every concern within its scope. This is why Hall (2008) recommended systems thinking as an approach to building understanding and enabling better CBT management. Figure 1 shows SDGs 8, 12 and 14 as they relate to tourism.

Table 1. Showing SDGs 8, 12 and 14 and their direct connection with tourism development

8 DECENT WORK AND ECONOMIC GROWTH	**SDG 8** – Promote sustained, inclusive, and sustainable economic growth, employment, and decent work for all. Tourism, as services trade, is one of the top four export earners globally, currently providing one in ten jobs worldwide. Decent work opportunities in tourism, particularly for youth and women, and policies that favour better diversification through tourism value chains can enhance tourism positive socio-economic impacts.
12 RESPONSIBLE CONSUMPTION AND PRODUCTION	**SDG 12** – Ensure sustainable consumption and production patterns. The tourism sector's plans to adopt sustainable consumption and production modes, accelerating the shift towards sustainability. Tools to monitor sustainable development impacts for tourism including for energy, water, waste, biodiversity, and job creation will result in enhanced economic, social, and environmental outcomes.
14 LIFE BELOW WATER	**SDG 14** – Conserve and sustainably use the oceans, seas, and marine resources for sustainable development. Coastal and maritime tourism rely on healthy marine ecosystems. Tourism development as part of Integrated Coastal Zone Management in order to help conserve and preserve fragile marine ecosystems and serve as a vehicle to promote a blue economy, contributing to the sustainable use of marine resources.

Source: UNWTO – UNDP (2017)

Over the years, more responsible tourists have emerged. These are tourists that demand services and products which support the environmental protection and the welfare of local communities. Boluk et al. (2019) referred to this as 'ethical consumption'. In their assessment of the SDGs, they found that these ideas of ethical production and consumption mean that tourism is no longer just a business and development sector but gradually also becoming a moral sphere. They, therefore, proposed that all tourism stakeholders should have ethical responsibilities that need to be evaluated. The integration of the SDGs into the planning, development, and management of CBT serves as the starting point for this campaign. This understanding makes it possible to maximise the potentials of CBT, while minimizing the challenges that are associated with it. It also emphasizes the importance of developing and strengthening existing partnerships between local stakeholders and tourism authorities, government and the private sector, as well as partnerships among the global tourism community.

Interactive Governance Approach for Enhancing Distributive Justice in CBT Projects

Another approach to CBT in the wake of contemporary issues is an all-inclusive governance approach to the planning, development, and management of CBT. Kooiman and Bavinck (2005: 17) refer to this approach to managing community-based projects as an 'interactive governance approach'. Even though this approach has its roots in the discipline of fisheries, its applicability in CBT contexts is highly possible. It is a model that can help curb degradation (Agbo, 2019; Calderon & Nawir, 2006; Pandit et al., 2009), improve community justice, equity, power, and participation at all levels, and ensure sustainability of CBT projects (Agbo, 2019). Just like any community-based project, CBT has many actors and stakeholders at various levels and scales. These actors and stakeholders have different interests, and this is revealed in the different values they attach to resources and projects (Agbo, 2019). In fact, no matter the number of interests, what ultimately matters in terms of action in most community-based projects is the value attached to the projects by these stakeholders and actors, and that is what determines to what extent projects are supported (Gregersen & Contreras, 1992). This approach to governance allows for all these values to be clearly understood and managed.

Justice should be at the core of every governance or management approach. That is to say, that creation of societal opportunities must include, but is not limited to an all-inclusive approach to decision making. This gesture makes all stakeholders feel valued by one another, hence the possibility of co-operation. According to Kooiman and Bavinck (2005), the most important feature of governance is interaction. Crucial in the interactive governance model is attention to three main elements - *structure, actors, and interaction*. According to them, actors operate within frameworks and guidelines. These frameworks and guidelines are referred to as structures. Structures include the law, technical and material possibilities, culture, and agreements. *Actors*, in their model refer to all social units including individuals, households, and all other stakeholders that possess power of action. *Interaction* is a specific form of action, undertaken by actors to remove obstacles and tread new pathways (Kooiman & Bavinck 2005: 17).

Three components can be identified in the interactive governance model; the governing system, the system-to-be-governed and governing interactions, which is the mediating component between the two. These altogether ensure the governability of the system. The governing system is a 'social and man-made system consisting of institutions and steering instruments and mechanisms' (Jentoft, 2007: 360). In the CBT context, this includes the institutions mandated to develop and implement policies for tourism at the community level, Destination Management Organisations (DMOs), and local and central governments. Chuenpagdee and Jentoft (2009) identified two sub-systems within the system-to-be-governed, namely the socioeconomic and natural systems. The natural system represents an ecosystem and the resources it contains, while the resource users and stakeholders represent the socioeconomic system. Both the governing system and the system-to-be-governed are characterized by diversity, dynamics, and complexity due to the linkages and interdependencies among its components (Kooiman et al., 2008). Meeting the needs of and understanding the interests of local community members goes a long way in improving community health and wellbeing. The third component, the governance interactions, represents the day-to-day relationships between the governing system and the system-to-be-governed. The outcomes of these interactions determine the degree of governability of the system, that is to say if governance outcomes are effective or ineffective. In the context of CBT, the outcomes of the day-to-day interactions between hosts and tourists, and local communities and governing bodies determine the success or failure of CBT projects.

The interactive governance model is designed to utilize community leaders, maximize local economic benefits, work with existing community initiatives, and employ residents so that income generated from tourism stays in the community. This, in effect takes care of justice concerns, which in many instances, are neglected.

FUTURE RESEARCH DIRECTIONS

There are many concerns over the potential of CBT as a tool for reducing the challenges facing local communities due to the uncertainties over its applicability. For this reason, and in order to guarantee justice and equity, there is the need for a re-thinking and a re-assessment of CBT's dominant ideas. The point is that tourism, in any form is predominantly colonizing in nature. Jamaica Kincaid, in her book entitled *'A Small Place'* depicts the idea that tourism makes the tourist create separation by "othering" the locale and the individuals that inhabit it (Kincaid, 1988). A critical look at this, and its relevance to tourism studies, presents the argument that despite the many advances in tourism practice, research and scholarship, knowledge about tourism is still predominantly 'colonial' (Chambers & Buzinde, 2015). This means that there is still a privileging of Western epistemologies and western tourists' behaviour at destinations, which continuously marginalize the locale and the individuals that are at the receiving point of tourism. Unfortunately, the tourism industry is part of a global economic system that ultimately does not always translate into benefits for the local communities who enable it. The result therefore is that destinations and their locals are recolonized by the industry and are put at the subservient and disenfranchised position in the global economic system which mostly fails to serve them.

This chapter places emphasis on providing justice for all stakeholders, especially local communities, a core component of any CBT effort. It also highlights the many potentials this approach to the production of CBT carries. It establishes that CBT is a unique lens into understanding what it means for local communities to be put at the heart of every development effort. Justice also explains practically what it means to engage in good actions as a tourist or host, and to fairly treat all stakeholders involved in CBT. It is imperative to understand that since tourism has colonial entanglements and exhibits these legacies in many forms, there is the need for future research to address these issues. Using the postcolonial and decolonial lenses, future research should revisit, remember, and interrogate the colonial past and its aftermath in today's CBT context. It should critically analyze the experiences of colonialism and their current manifestations in how CBT is organised and delivered in communities. Finally, it is important that future research deliberately puts the perspectives of marginalised groups at the forefront, while decentring dominant cultures. This helps in new knowledge construction and allows for local communities to be active participators in CBT. Gibson (2019) reviewed progress in tourism studies in three broad essays (Gibson, 2008, 2009, 2010) and found that among the pressing concerns in an emergent critical thread of tourism studies and practice in the 1990s and 2000s were the rise of ethical tourism, (post)colonial power relations, and uneven development and local livelihoods. Against this backdrop, tourism scholars' engagements with critical theory in understanding the challenges of CBT should be broadened and deepened.

CONCLUSION

The global pandemic has slowed down tourism all over the world. However, considering how governments, destinations, businesses and tourism communities are slowly recovering from the effects of the pandemic, there is enough evidence that tourism will get back on track. Some major concerns, post-pandemic, will be how to secure the social and ecological integrities of destinations in a world that is increasingly interdependent and stressed by growing consumerism. This chapter has assessed the importance of placing justice issues at the core of every CBT effort. It has emphasized the role justice plays in minimizing the challenges of CBT in many communities. This notwithstanding, the tendencies of tourism, as a capitalist endeavour to usurp what constitutes justice for tourism communities suggests that intentionality is required in the design and production of CBT to ensure success. Even though there are uncertainties over CBT carrying the transformative and empowering promises of development for communities, incorporating justice and the SDGs in the planning, development, governance, and management of CBT could help to maximise the potentials of CBT, while minimizing the challenges that are associated with it. It is also important to note that the success or failure of CBT projects is not only dependent on justice. Actually, several other equally critical issues such as equity, power, and sustainability are necessary considerations in any CBT project. However, one thing is clear, and that is there are choices to be made by scholars, practitioners, students, industry players, private and public institutions, in order to transform societies through a more just, equitable, and sustainable CBT.

REFERENCES

Afenyo, E. A., & Amuquandoh, F. E. (2014). Who benefits from community-based ecotourism development? Insights from Tafi Atome, Ghana. *Tourism Planning & Development*, *11*(2), 179–190. doi:10.1080/21568316.2013.864994

Agbo, V. M. (2019). *From conflict to collaboration: Atewa Forest governance* (Master's thesis). University of Waterloo.

Aitchison, C. (1999). *New cultural geographies: The spatiality of leisure, and evolution*. Academic Press.

Ateljevic, I., Morgan, N., & Pritchard, A. (Eds.). (2013). *The critical turn in tourism studies: Creating an academy of hope* (Vol. 22). Routledge. doi:10.4324/9780203806586

Atkinson, J. M. (1990). Ethnomethodological approaches to socio-legal studies. *Ethnomethodological Sociology*, 450-472.

Blackstock, K. (2005). A critical look at community-based tourism. *Community Development Journal: An International Forum*, *40*(1), 39–49. doi:10.1093/cdj/bsi005

Boluk, K. A., Cavaliere, C. T., & Higgins-Desbiolles, F. (2019). A critical framework for interrogating the United Nations Sustainable Development Goals 2030 Agenda in tourism. *Journal of Sustainable Tourism, 27*(7), 847–864. doi:10.1080/09669582.2019.1619748

Britton, S. G. (1982). The political economy of tourism in the Third World. *Annals of Tourism Research, 9*(3), 331–358. doi:10.1016/0160-7383(82)90018-4

Calderon, M. M., & Nawir, A. A. (2006). *An evaluation of the feasibility and benefits of forest partnerships to develop tree plantations: case studies in the Philippines. Center for International Forestry Research.* CIFOR.

Campbell, L. M., Gray, N. J., & Meletis, Z. A. (2008). Political ecology perspectives on ecotourism to parks and protected areas. In K. S. Hanna, D. A. Clark, & D. S. Slocombe (Eds.), Transforming parks and protected areas: Policy and Governance in a Changing World (pp. 200–221). Academic Press.

Chambers, D., & Buzinde, C. (2015). Tourism and decolonisation: Locating research and self. *Annals of Tourism Research, 51*, 1–16. doi:10.1016/j.annals.2014.12.002

Chuenpagdee, R., & Jentoft, S. (2009). Governability assessment for fisheries and coastal systems: A reality check. *Human Ecology, 37*(1), 109–120. doi:10.100710745-008-9212-3

Craig, G. (2003). Editorial introduction. *Community Development Journal: An International Forum, 38*(1), 1–5. doi:10.1093/cdj/38.1.1

Dangi, T. B. (2018). Exploring the intersections of emotional solidarity and ethic of care: An analysis of their synergistic contributions to sustainable community tourism development. *Sustainability, 10*(8), 2713. doi:10.3390u10082713

Dei, G. J. S. (2000). Rethinking the role of indigenous knowledges in the academy. *International Journal of Inclusive Education, 4*(2), 111–132. doi:10.1080/136031100284849

Deutsch, M. (1975). Equity, equality, and need: What determines which value will be used as the basis of distributive justice? *The Journal of Social Issues, 31*(3), 137–149. doi:10.1111/j.1540-4560.1975.tb01000.x

Du Gay, P. (1997). Organizing identity: making up people at work. *Production of Culture/Cultures of Production*, 285-344.

Fennell, D. A. (2006). *Tourism Ethics* (Vol. 30). Channel View Publications. doi:10.21832/9781845410360

Franklin, A., & Crang, M. (2001). The trouble with tourism and travel theory? *Tourist Studies*, *1*(1), 5–22. doi:10.1177/146879760100100101

Genç, R. (2012). Subjective aspects of tourists' quality-of-life (QOL). In *Handbook of Tourism and Quality-of-life Research* (pp. 149–167). Springer. doi:10.1007/978-94-007-2288-0_9

Gibson, C. (2008). Locating geographies of tourism. *Progress in Human Geography*, *32*(3), 407–422. doi:10.1177/0309132507086877

Gibson, C. (2009). Geographies of tourism: Critical research on capitalism and local livelihoods. *Progress in Human Geography*, *33*(4), 527–534. doi:10.1177/0309132508099797

Gibson, C. (2010). Geographies of tourism:(un) ethical encounters. *Progress in Human Geography*, *34*(4), 521–527. doi:10.1177/0309132509348688

Gibson, C. (2019). Critical tourism studies: New directions for volatile times. *Tourism Geographies*, 1–19. doi:10.1080/14616688.2019.1647453

Gregersen, H. M., & Contreras, A. (1992). *Economic Assessment of Forestry Project Impacts* (Vol. 106). Food and Agriculture Organization.

Grimwood, Caton, & Muldoon. (2018). Tourism and Wellness: Travel for the Good of All? Lexington Books.

Hall, C. M. (2008). *Tourism Planning: Policies, Processes and Relationships* (2nd ed.). Pearson Prentice Hall.

Hall, S. (1996). Who needs identity? *Questions of Cultural Identity*, *16*(2), 1–17.

Haraway, D. J. (2016). *Staying with the Trouble. Making Kin in the Chthulucene*. Duke University Press. doi:10.1215/9780822373780

Harris, R. W. (2009). Tourism in Bario, Sarawak, Malaysia: A case study of pro-poor community-based tourism integrated into community development. *Asia Pacific Journal of Tourism Research*, *14*(2), 125–135. doi:10.1080/10941660902847179

Higgins-Desbiolles, F. (2006). More than an "industry": The forgotten power of tourism as a social force. *Tourism Management*, *27*(6), 1192–1208. doi:10.1016/j.tourman.2005.05.020

Higgins-Desbiolles, F. (2008). Justice tourism and alternative globalisation. *Journal of Sustainable Tourism*, *16*(3), 345–364. doi:10.1080/09669580802154132

Higgins-Desbiolles, F. (2018). Sustainable tourism: Sustaining tourism or something more? *Tourism Management Perspectives*, *25*, 157–160. doi:10.1016/j.tmp.2017.11.017

Hollinshead, K. (1999). Surveillance of the worlds of tourism: Foucault and the eye-of-power. *Tourism Management*, *20*(1), 7–23. doi:10.1016/S0261-5177(98)00090-9

Hollinshead, K. (2007). 'Worldmaking' and the transformation of place and culture: The enlargement of Meethan's analysis of tourism and global change. In I. Ateljevic, N. Morgan, & A. Pritchard (Eds.), *The Critical Turn in Tourism Studies: Innovative Research Methodologies* (pp. 165–193). doi:10.1016/B978-0-08-045098-8.50016-7

Hooks, B. (2003). *Teaching Community: A Pedagogy of Hope* (Vol. 36). Psychology Press.

Hultsman, J. (1995). Just tourism: An ethical framework. *Annals of Tourism Research*, *22*(3), 553–567. doi:10.1016/0160-7383(95)00011-T

Jackson, P. (1993). *Towards a Cultural Politics of Consumption*. NP.

Jamal, T. (2019). *Justice and Ethics in Tourism*. Routledge. doi:10.4324/9781315162942

Jamal, T., & Camargo, B. A. (2014). Sustainable tourism, justice and an ethic of care: Toward the Just Destination. *Journal of Sustainable Tourism*, *22*(1), 11–30. doi:10.1080/09669582.2013.786084

Jamal, T. & Menzel, C. (2009). Good actions in tourism. *Philosophical Issues in Tourism*, 227-243.

Jentoft, S. (2007). Limits of governability: Institutional implications for fisheries and coastal governance. *Marine Policy*, *31*(4), 360–370. doi:10.1016/j.marpol.2006.11.003

Kant, I. (1998). *Kant: Religion within the boundaries of mere reason: And other writings*. Cambridge University Press. doi:10.1017/CBO9780511809637

Kincaid, J. (1988). *A small place*. Farrar Straus Giroux.

Kiss, A. (2004). Is community-based ecotourism a good use of biodiversity conservation funds? *Trends in Ecology & Evolution*, *19*(5), 232–237. doi:10.1016/j.tree.2004.03.010 PMID:16701261

Kooiman, J., & Bavinck, M. (2005). The governance perspective. *Fish for Life: Interactive Governance for Fisheries*, *3*, 11. doi:10.1017/9789048505326.002

Kooiman, J., Bavinck, M., Chuenpagdee, R., Mahon, R., & Pullin, R. (2008). Interactive governance and governability: An introduction. *The Journal of Transdisciplinary Environmental Studies*, *7*(1), 1–11.

Lee, S., & Jamal, T. (2008). Environmental justice and environmental equity in tourism: Missing links to sustainability. *Journal of Ecotourism*, *7*(1), 44–67. doi:10.2167/joe191.0

Li, W. (2006). Community decision-making participation in development. *Annals of Tourism Research*, *33*(1), 132–143. doi:10.1016/j.annals.2005.07.003

Long, E. (Ed.). (1997). *From sociology to cultural studies: New perspectives*. Wiley-Blackwell.

Lovelock, B., & Lovelock, K. (2013). *The ethics of tourism: Critical and applied perspectives*. Routledge. doi:10.4324/9780203854532

Mansvelt, J. (2005). Geographies of consumption. *Sage (Atlanta, Ga.)*.

Massey, D. (2005). Backfire at the Border: Why Enforcement without Legalization. *Trade Policy Analysis, 29*.

Mensah, I. (2017). Benefits and challenges of community-based ecotourism in park-fringe communities: The case of Mesomagor of Kakum National Park, Ghana. *Tourism Review International*, *21*(1), 81–98. doi:10.3727/154427217X14866652018947

Michalos, A. C. (2008). Education, happiness, and wellbeing. *Social Indicators Research*, *87*(3), 347–366. doi:10.100711205-007-9144-0

Miller, D. (2017). Fair trade: What does it mean and why does it matter? *Journal of Moral Philosophy*, *14*(3), 249–269. doi:10.1163/17455243-46810053

Neef, A. (2019). *Tourism, Land Grabs and Displacement: A Study with Particular Emphasis on the Global South*. Academic Press.

Nussbaum, M. C. (2017). Capabilities, entitlements, rights: Supplementation and critique. In *Justice and the Capabilities Approach* (pp. 173–187). Routledge. doi:10.4324/9781315251240-4

Pandit, B. H., Albano, A., & Kumar, C. (2009). Community-based forest enterprises in Nepal: An analysis of their role in increasing income benefits to the poor. *Small-scale Forestry*, *8*(4), 447–462. doi:10.100711842-009-9094-2

Pearce, D. G. (1992). *Tourist organizations*. Longman Scientific & Technical.

Pritchard, A., Morgan, N., & Ateljevic, I. (2011). Hopeful tourism: A new transformative perspective. *Annals of Tourism Research*, *38*(3), 941–963. doi:10.1016/j.annals.2011.01.004

(2004). Progress in qualitative research in tourism: Epistemology, ontology, and methodology. InPhillimore, J., & Goodson, L. (Eds.), *Qualitative Research in Tourism* (1st ed., pp. 21–23). Routledge.

Ross, S., & Wall, G. (1999). Ecotourism: Towards congruence between theory and practice. *Tourism Management*, *20*(1), 123–132. doi:10.1016/S0261-5177(98)00098-3

Sandel, M. J. (1998). Religious liberty: Freedom of choice or freedom of conscience. *Secularism and Its Critics*, 73-93.

Scheyvens, R. (2002). *Tourism for development: Empowering communities*. Pearson Education.

Taylor, R. (2009). Community based natural resource management in Zimbabwe: The experience of CAMPFIRE. *Biodiversity and Conservation*, *18*(10), 2563–2583. doi:10.100710531-009-9612-8

Tosun, C. (2000). Limits to community participation in the tourism development process in developing countries. *Tourism Management*, *21*(6), 613–633. doi:10.1016/S0261-5177(00)00009-1

Tribe, J. (1997). *Corporate strategy for tourism*. International Thomson Business Press.

Tribe, J. (2007). Critical tourism: Rules and resistance. In I. Ateljevic, N. Morgan, & A. Pritchard (Eds.), *The Critical Turn in Tourism Studies: Innovative Research Methodologies* (pp. 51–62). Routledge. doi:10.1016/B978-0-08-045098-8.50007-6

United Nations Department of Economic and Social Affairs (UNDESA). (2017). *The 17 Goals: History*. Retrieved from https://sdgs.un.org/goals

United Nations – UNISDR, UNDP, & ISDR. (2009). *Making Disaster Risk Reduction Gender Sensitive: Policy and Practical Guidelines*. United Nations. Retrieved from https://www.unisdr.org/files/9922_MakingDisasterRiskReductionGenderSe.pdf

Van Bibber, M. (1997). *It takes a community: A resource manual for community-based prevention of Fetal Alcohol Syndrome and Fetal Alcohol Effects*. Aboriginal Nurses Association of Canada.

World Commission on Environment and Development. Special Working Session. (1987). World commission on environment and development. *Our Common Future, 17*(1), 1-91.

World Tourism Organization and United Nations Development Programme. (2017). *Tourism and the Sustainable Development Goals – Journey to 2030*. Retrieved from https://www.e-unwto.org/doi/epdf/10.18111/9789284419401

Wright, S. C., & Boese, G. D. (2015). Meritocracy and tokenism. In International Encyclopedia of the Social and Behavioural Sciences. doi:10.1016/B978-0-08-097086-8.24074-9

Chapter 6
Cultural Tourism and the Wellbeing of Local Citizens:
Examining the Mediating Role of Cultural Conservation in Northern Pakistan

Saranjam Baig
https://orcid.org/0000-0003-2407-3776
Sultan Qaboos University, Oman

Arifa Shabbnum
Colorado State University, USA

Ahmad Arslan
University of Oulu, Finland

ABSTRACT

Cultural tourism is an increasingly visible trend in the tourism industry. The chapter is one of the first academic studies to specifically analyze the possible impacts of cultural tourism on the local culture itself and the wellbeing of the host community while highlighting cultural conservation as a mediator. The study utilizes primary data collected from rural areas in the Himalayan Gilgit Baltistan region in Northern Pakistan. This region has experienced a significant rise in cultural tourism due to China-Pakistan Economic Corridor (CPEC). As a result, there has been a significant and visible development of touristic facilities in the region. Hence, there is a visible shift from earlier tourism, which was mostly linked to mountaineering adventure, to more relatively mass cultural tourism. The results suggest that cultural conservation serves as a partial mediator and that cultural tourism tends to positively and significantly influence the host community's wellbeing, and cultural conservation partially mediates this relationship.

DOI: 10.4018/978-1-7998-7335-8.ch006

INTRODUCTION

Tourism is frequently viewed as a threat to the preservation of local cultures and cultural sites. Tourism is, in reality, a platform and a vehicle for exposing heritage to the masses, protecting it, and ensuring its economic and social sustainability. As a result, tourism is frequently a balancing mechanism that preserves and protects the culture of a society.

Cultural tourism has clear economic advantages. It is of crucial importance for many mountain communities in developing countries, where there is a lack of economic opportunities due to harsh geographic and climatic conditions (Wester et al., 2019). Tourism has emerged as one of the key sources of economic activities for these mountain communities, as building large industrial infrastructure is not viable in these regions in most cases (Nepal, 2020). Cultural tourism is quickly expanding in the Himalayan Gilgit Baltistan region in Northern Pakistan, and this has been strongly linked to culture conservation. On the one hand, cultural heritage is a fundamental resource for attracting visitors in cultural tourism. Cultural tourism, on the other hand, can provide the financial base for the protection of culture.

Cultural tourism has been referred to as one of the largest and fastest-growing industries in the world tourism market (UNWTO, 2005). Some scholars have found that cultural tourism has both positive and negative socio-cultural, economic and environmental impacts on residents of host communities (Fang, 2020; Richards, 2007; Richards 2018). The commodification of cultural resources for attracting tourists and consumption by these tourists has burgeoned in the recent years. However, not all communities are able to translate the benefits of tourist arrivals into development; though, tourism being an economic activity can be an effective strategy for regional economic development (Fang, 2020; Yang & Wall, 2009). In their study, Perdue et. al. (2010) stressed that the usefulness of cultural tourism is that it expedites and sustains wellbeing such as reducing poverty in host communities, the renaissance of heritage and community culture and the preservation and protection of both cultural and natural resources. In this context, sustainable cultural tourism has been stressed as being useful in avoiding immoderate abuse of resources and encouraging preservation of culture and heritage for the next generation (e.g., Boniface, 2013; Asmelash and Kumar, 2019).

Against this backdrop, the current chapter focuses on a mountainous region in the north of Pakistan that borders China's West (Gilgit Baltistan). The region is rich in cultural heritage in the form of arts and crafts, archaeological heritage, as well as the world-famous rock carvings and inscriptions discovered in the area, which date back to prehistoric times.

Currently, the Gilgit-Baltistan region is experiencing an exponential increase in tourist arrivals (and especially cultural tourists) due to the improvements in infrastructure under the multi-billion dollar China Pakistan Economic Corridor [CPEC] (Baig & Zehra, 2020; Baig et al., 2020). Tourism plays an important role in the local economy as the majority of the population is solely dependent on tourism for their livelihood. Due to poor infrastructure and security conditions, this industry has never flourished in the past as a viable commercial venture. Previously, most of the tourism in this region was based on relatively few mountaineering and nature adventurers. Along with famous mountains like K-2, Nangaparbat, Gasherbrum and Rakaposhi, the region is also home to National parks like Khunjerab National Park, Deosai National Park and the Central Karakorum National Park, which have attracted such adventure tourists from both within Pakistan and abroad. However, in the recent years, cultural tourism has been on the rise, which makes the analysis of its influences on local communities interesting from both academic and practical perspectives. Cultural tourists are attracted to monuments like Baltit Fort in Karimabad, the Altit Fort in Altit, Old Ganish settlements in Ganish or the Nagar Fort and an old mosque in the Nagar district, Shiger Fort in Shiger Baltistan, and the Khaplu Fort in Baltistan. Apart from these cultural sites and monuments, cultural tourists are also attracted to the old silk route settlements along the modern Karakorum Highway. Despite this significant importance and increased visibility of cultural tourism in the Gilgit-Baltistan region, the literature review reveals that specific influences of this rise in cultural tourism on local communities in this specific context are significantly underexplored academically. Due to this visible contextual gap in the extant literature, as well as potential practical implications, this chapter aims to: i). analyze the effect of cultural tourism on the conservation of the culture (language, lifestyle, food, values, social practices, and values) of the host region; ii) specifically examine the role of cultural conservation on the wellbeing of the residents of this region; iii) analyze the mediating role of cultural conservation in the impact of CPEC influenced cultural tourism on the wellbeing of the residents in Gilgit-Baltistan.

This study is one of the first attempts to examine the direct and indirect impact of CPEC influenced cultural tourism on the wellbeing of the residents in selected districts of Hindu Kush Himalayan Gilgit-Baltistan region of Northern Pakistan. We contribute to the extant literature by establishing a direct link between cultural-tourism and wellbeing of the local residents in the Pakistani context. Secondly, we introduce cultural conservation as an important mediating variable that influences the relationship between cultural tourism and the local residents' wellbeing.

The rest of the chapter is organized as follows. The next section presents a literature review followed by a presentation of conceptual framework and methodological details. After that, results are presented. The paper concludes with a presentation of implications, limitations, and future research directions.

LITERATURE REVIEW

Concepts of Cultural Tourism and Cultural Conservation

Many existing studies have tried to define the concept of cultural tourism. For instance, according to Liu and Shu (2020) cultural tourism is distinct from recreational tourism in that it aims to obtain a better knowledge or appreciation of the nature of the destination. Likewise, Mousavi et al. (year, p…) argue that the "cultural tourism involves the consumption of a wide range of cultural manifestations like heritage, art, folklore etc. by tourists. Cultural tourism can be generally seen as a sociocultural relationship between people which is promoted, moderated and mediated by a range of various actors including, planners, politicians, researchers, marketing professionals, travel agencies and so on". Cultural tourism, according to this definition, encompasses not just "heritage tourism," which is the consumption of historical cultural items, but also modern culture, or the "way of life" of the people or place (Liu and Shu, 2020).

In the line of agreement with Liu and Shu (2020), Robinson and Picard (2006), suggest that tourism brings individuals and human communities together and has an important role in facilitating dialogues between cultures, which in turn enables the preservation of human heritage, improves living conditions, and reduces poverty. The linkage between cultural tourism and cultural conversation merits attention because historic monuments and places are essential resources for attracting visitors; and therefore, cultural conservation and tourism are inextricably linked.

Effects of Cultural Tourism on Wellbeing of Host Communities

However, tourism can have a positive or negative long lasting effect on natural heritage and culture, as well as on the environment, cultural diversity and creativity, and the balance of societies. Zadel and Bogdan (2013) added that it is an arduous task, to capture the economic impacts of cultural tourism because most countries do not have statistical systems to analyze and monitor the individual cultural tourism factors. In cultural tourism, cultural resources are transformed into cultural-tourism products. Pandey et al. (1995) attempted to show the effect of tourism on Nepal's environment, economy and culture. Since tourism is an important sector in Nepal's economy; it serves as a source of foreign exchange, generates employment and contributes to GDP. The tourism sector can play a major role in eradicating poverty by generating employment in developing countries (Fang, 2020).

It is an established fact in the tourism literature that tourism directly affects the local economy in multiple ways (Cheer & Lew, 2017). According to many scholars, this direct effect is linked closely to the wellbeing of local populations (e.g. Kim,

2002; Wood, 2017). It has been argued that a key element of CPEC is to improve the economic condition and wellbeing of people in Pakistan, and specifically Gilgit-Baltistan, which lies at an important intersection of this infrastructure development plan (Malik, 2018). Scholars have stressed the factors pertaining to quality of life and material wellbeing like income levels and stability, educational opportunities in the community, living conditions of the residents, the quality of the natural and social environment, and safety. In particular context of tourism, Kim (2002) specified five attributes of wellbeing as material wellbeing, community wellbeing, health wellbeing, safety wellbeing, and emotional wellbeing. These aspects have been linked to people's overall quality of life in touristic regions (e.g. Aref, 2011; Mathew & Sreejesh, 2017). This paper incorporates all these elements of wellbeing in the analysis of cultural tourism in the Gilgit-Baltistan region

Brohman (1996) indicated that the developing world's tourism industry has been growing fast, but it has also been encountering many issues that are shared with other outward-oriented development approaches, which includes the formation of separate areas, unnecessary foreign dependency, and also the fortification of dimensional inequalities and the social and economic factors, increasing cultural isolation and environmental devastation. To escape such issues, institutional structures need to be built to boost vigorous state and community participation in tourism strategies. Similarly, Soutsas et al. (2006) argued that tourism activities and leisure that are perceived to be a part of some specific area's developmental procedures can add to their economic growth at varying degrees of intensity so that they can be achievable and desirable.

Dwyer, Forsyth, and Dwyer (2010) argue that there are direct, indirect, and induced economic impacts of cultural tourism. Money spent by tourists that constitute direct effects include expenditure on food, transportation, and communication services. Likewise, Saarinen and Manwa (2008) suggest dividing the impact created when tourists encounter social and cultural experiences. The impact of tourism on culture can be termed as changes in material culture, for example, crafts, and changes in non-material culture that is tradition, customs, and the acculturation processes. On the other hand, the acculturation process can be termed as the indirect long-term change caused by tourist action in the host culture. In contrast, social impacts can be divided into different categories like individual and social relations, religion, language, moral and ethical values, and health.

Cultural Tourism and CPEC

In 2015, Pakistan and China signed a multi-billion-dollar economic agreement under the aegis of the China-Pakistan Economic Corridor (CPEC). Analysts and researchers consider this mega project as a lifeline between the two all-weather strategic partners (Baig & Zehra, 2020). As the megaproject includes investments on infrastructure and energy, it has been argued that the investments under CPEC will transform Pakistan's economy and help overcome its power shortage. In the years after the initiation of CPEC projects especially the expansion of Karakoram Highway (KKH) and construction of seven tunnels along the KKH has resulted in unprecedented inflow of tourists to the northern region of Pakistan. It has been argued that the CPEC has created an environment that fosters tourism in the study region that has a 450km alignment along the CPEC route (Khan, 2019).

Analyzing cultural tourism in the specific context of the CPEC initiative, Khan et al. (2018) identified the potential development of cultural Tourism in Pakistan and, more specifically, in Peshawar. They recommended that it should be used as a tool to improve local people's socio-economic conditions through job creation and business opportunities. CPEC has already been termed as an economic and geopolitical game-changer for the whole region (Malik, 2018). It has already been mentioned that the Gilgit-Baltistan region's location along the main corridor of the CPEC and its proximity to China has increased its importance, resulting in significant development and plans for further development of infrastructure (Nigar, 2018). These developments have made this region accessible to many cultural tourists for whom arriving in this mountainous region in the past was difficult due to connectivity issues and lack of touristic facilities. Therefore, the change from limited mountaineering and adventure tourism to cultural tourism is significantly visible now in the region.

Studies that were undertaken in different parts of the world and contexts, have found that cultural tourism elements include beliefs and faith of local communities and associated touristic attractions (Olsen, 2011). Some other scholars have referred to the elements of lifestyle (Boniface, 2013; Richards, 2007), local cuisine and food (Fuste-Forne, 2019), and architectural sites (Boniface, 2013), as being depictions of the cultural attractions of a destination. Finally, the works of local artisans including handicrafts (Ratten et al., 2019), music and art (Boniface, 2013) and local ethical traditions (Smith, 2009) have been stressed by some scholars as being important in relation to cultural tourism. Due to the importance of all these elements in the context of cultural tourism, this chapter aims to analyze them in the specific context of CPEC influenced cultural tourism in the GB region of Pakistan, as further depicted in the conceptual model in the next section.

Cultural Tourism and Cultural Conversation

It has further been found in some studies that cultural tourism specifically results in the improvement of cultural elements in local societies, as preserving them becomes an economic need (Boniface, 2013). Even though some researchers have been arguing that mass cultural tourism can negatively influence the culture of a destination, UNWTO (2005) suggests that this can be avoided by properly planning and managing the whole cultural tourism experience. The extent to which the host community's local culture is incorporated into the experience of the tourist has been debated significantly by scholars from a range of perspectives. It has further been highlighted that some indigenous communities may not wish to share their culture with the tourists and isolate themselves from the tourists (Ruhanen & Witford, 2019). On the other hand, it is also possible that many of such communities may not even know the economic worth of sharing their culture with tourists. Tourists are more interested in knowing the culture of host communities. It seems impossible for communities to isolate themselves from others and avoid other cultural invasions in this highly globalized world. When such indigenous communities allow tourists to visit them and share their unique culture, they can earn their living by commodifying their cultural products and attracting culturally curious tourists. This can result in significant economic and wellbeing outcomes for those communities (Boniface, 2013; Richards, 2007). UNWTO (2005) highlighted the importance of cultural conservation to the promotion of cultural tourists by stating that "culture needs to be rooted in the community in order to be authentic". In this context, cultural conservation has been highlighted as an important factor related to cultural tourism. Some scholars argue that an increase in cultural tourism can potentially affect the conservation of cultural assets in the specific tourist destination including both tangible and intangible aspects (Richards, 2007). These cultural products act as a pull factor in attracting culturally curious tourists who are interested in experiencing and learning different cultures. Therefore, it can be argued that it is important to incorporate cultural conservation in any analysis of cultural tourism, as it is strongly linked with the long-term viability of such cultural tourism in an area. This paper also incorporates cultural conservation while analyzing CPEC influenced cultural tourism in Pakistan's Gilgit-Baltistan region, as depicted in the conceptual framework in the next section.

METHODOLOGY

The Region

Gilgit-Baltistan is a quasi-province in the north of Pakistan. It is endowed with natural beauty, highest peaks, glaciers, fresh water, lakes, and valleys. The region is also home with to several cultures and ethnicities. It has been a center of attention for tourists from around the globe. Moreover, Gilgit-Baltistan is located at a junction of neighboring countries including Central Asian countries (e.g Tajikistan) and China. Besides its natural attractions, historically, the region has been strategically important, being the gateway to Central Asia. Today, as the gateway to CPEC, the region is witnessing improved road infrastructure and better communication systems and this has led to an exponential increase in tourist arrivals as argued by Baig et al. (2021).

Conceptual Framework

Based on literature review presented in the preceding section, a conceptual framework for the study is presented in figure 1.

Figure 1. Conceptual framework

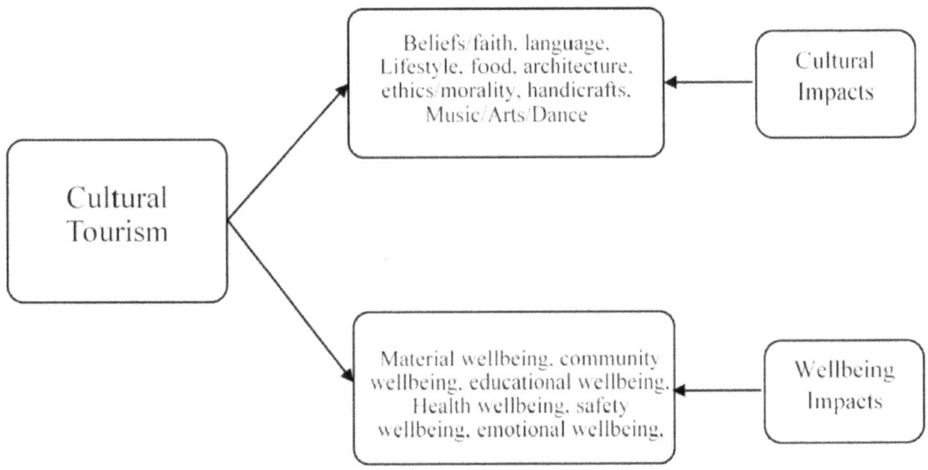

Many studies including Borowiecki and Castiglione (2014), Cellini and Cuccia (2013), and Zieba (2016) have found the existence of a strong relationship between tourism flows and cultural participation in Italy and Austria. In the same line of argument, this research presents the interaction of culture, tourism, and well-being as presented in figure 1. During the interaction of these three components: cultural tourism, cultural impacts, and wellbeing of residents, a by-product is produced due to each component's interaction. The by-product produced due to the interaction of cultural impacts and cultural tourism can result in either cultural conservation or cultural degradation. According to the available literature these cultural changes include changes in the lifestyle, beliefs/faith, language, food, architecture, ethics/morality, handicrafts, and arts/music/dance. It means that when cultural tourists visit a tourist destination, they can influence the host community's culture either positively or negatively. These cultural tourists either influence the residents to conserve their culture in order to attract more culturally curious visitors or adopt the culture of tourists. Furthermore, in the next stage culture tourism influences the wellbeing of the residents of the host community resulting in either wellbeing or ill-being of the residents as a result of the interactions cultural tourists have with the people of that region (Ahebwa et. al., 2016; Mathew & Sreejesh, 2017). When cultural tourists arrive in a region, they need services and facilities such as transportation, food, accommodation, and handicrafts. The host community's engagement by providing tourists with these services, in turn, provides employment opportunities and potentially increases their wellbeing by increasing their income. At the same time, cultural tourists can also cause harm to the local residents' wellbeing by invading their personal or community life as well as degrading the environment, which we term as ill-being.

Although Kim (2002) suggests five indicators of wellbeing fostered by tourism as discussed in the preceding sections, this research considers only economic measures of well-being which include indicators on material wellbeing and educational wellbeing following the tradition of Alatartseva and Barysheva (2015) and Mathew and Sreejesh (2017). Since the scope of this study is limited to economic wellbeing, we examine the influence of culture tourism on employment and income as well as on education. Unlike Kim (2002), limitation of resources and time compelled the authors to consider only two measures of wellbeing: material wellbeing, and education wellbeing. These two indicators were chosen according to their visible influence on the host communities.

Model Construction

Following the conceptual framework proposed in figure 2, this research uses mediation analysis explained by the following two models in figure 3. Model A in this path analysis shows the perceived total effect of CPEC-Culture-Tourism on the wellbeing of the residents while Model B shows both the perceived direct and indirect effect of CPEC-Culture-Tourism on Wellbeing of Residents using Cultural Conservation as a mediator.

This model has three variables; an independent variable (IV) which we denote by CT, a dependent variable (DV) which is denoted by WB and a mediating variable (MV) denoted by CC. CC plays a mediating role between cultural tourism (CT) and wellbeing (WB). Figure (3) further explains the paths leading to the dependent variable: Model A depicts the total impact of the independent variable represented by the coefficient βc Further in Model B, the direct effect of independent on dependent variable represented by the regression coefficient βd, there is another path showing the relationship between independent variable and mediator represented by the regression coefficient βa a d the path showing the relationship between mediator and the dependent variable represented by the regression coefficient βb.

Figure 2. Conceptual framework of cultural tourism, culture conservation and wellbeing along with their components from which they were derived.

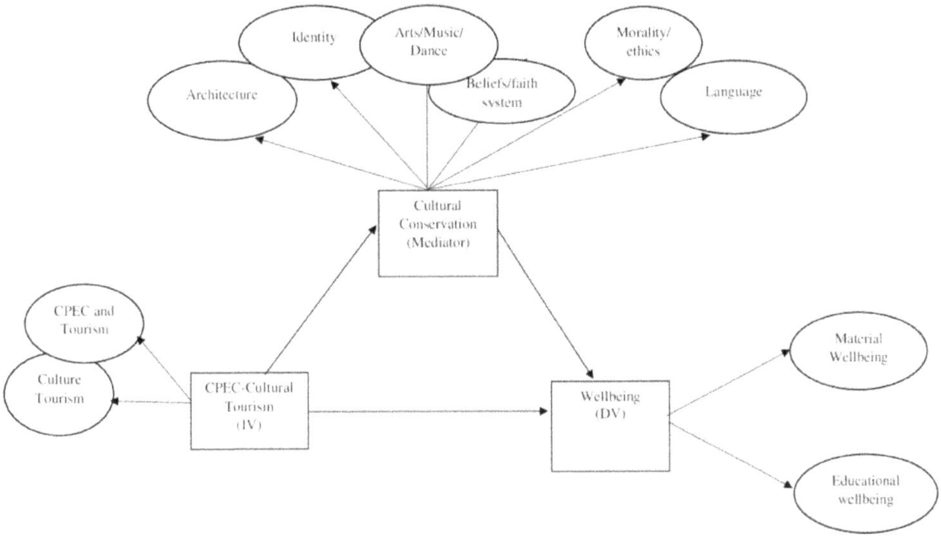

According to Baron and Kenny (1986) the following are the criteria for a variable to function as a mediator: (i) variations in the mediating variable are accounted for significantly by the variations in the independent variables (ii) variations in dependent variable are elucidated by the variations in independent variable (iii) variation in the mediator significantly accounts for variations in the dependent variable when controlling for independent variable, and the impact of the independent variable on the dependent variable decreases greatly once the mediator is engaged with the independent variable as a predictor of the dependent variable. Further, Kenny et.al. (1998) explained that the criterion (iii) will be satisfied only if the first two are satisfied. Hence referring to Model A and model B these prescribed criteria require to be significant whereas should be less than . Nonetheless, some authors have rejected the idea of having a significant total effect of an independent variable on the dependent variable to be a necessary condition for the mediation to occur (Shrout & Bolger, 2002). Henceforth, the regression equations following these three criteria are listed below:

$$WB = \alpha_i + \beta_c CT$$

$$CC = \alpha_{ii} + \beta_a CT$$

Figure 3. Mediation analysis for cultural tourism, cultural conservation and wellbeing.

Model A

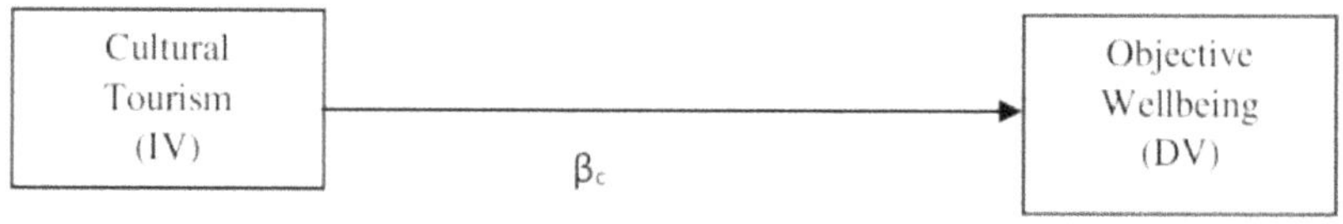

Model B

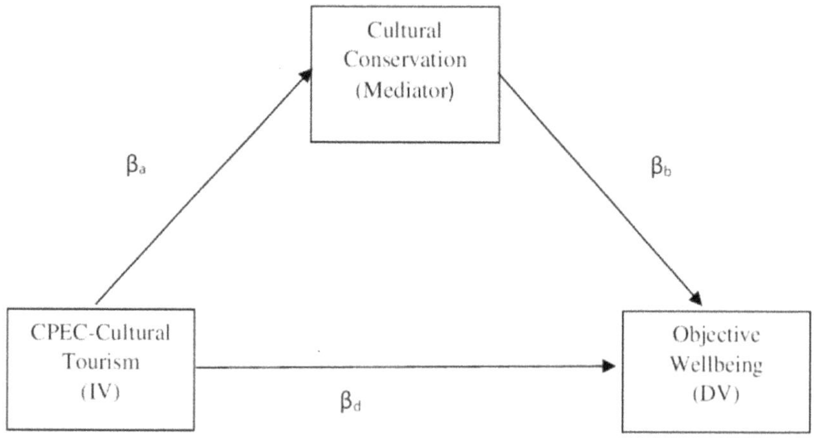

$$WB = \alpha_{iii} + \beta_d CT + \beta_b CC$$

Where, αi, αii and αiii are the intercept terms and β's is the regression coefficient from the respective regression equation. The succeeding section shall describe the descriptive statistics of the socio-economic and demographic indicators of respondents as well as the detailed results and discussion of the relationship among CPEC-cultural-tourism and objective wellbeing of residents of these respective regions i.e. mediation analysis. Also, it describes the role of Culture Conservation as a mediator in effecting the relationship between the dependent and independent variables.

Data Collection and Sampling

This research uses primary data collected in the form of a questionnaire from local residents, where residents have been classified into different categories. The questionnaire consists of questions about socioeconomic and demographic indicators and questions regarding the variables of culture, wellbeing, culture of youth (aged 15-25) and CPEC fostered tourism with reference to previous studies by Gursoy and Rutherford (2004), Ratz and Michalko (2011), Aref (2011), Allis (2016), Zadel and Bogdan (2013). Following Soontayatron, (2013), the stratified random sampling technique was used. The population was divided into different subgroups (strata) and then respondents were randomly selected from each group. The selected strata for this research are:

Strata 1: local people having direct and regular contact with tourists, who depend on tourism for their incomes (e.g. tour guides, tour operators, hotel management).

Strata 2: local people having regular contact with tourists arriving at their localities but their income is not dependent on tourism activities (Management of fort, guide etc.). as well as those who had an indirect or no continual contact with tourists and their income is partially dependent on tourism. (Shop keepers, restaurant).

Strata 3: local people having no contact with tourists (teachers, workers, etc.).

The sample was collected from the areas including Gilgit city, Danyore, Sultanabad, Hoper, Karimabad, Aliabad, Altit, Ganish, Gulmit, Passu, Chillas city, Raikot, Jaghlot, Guro. A questionnaire survey was undertaken to evaluate the impacts of CPEC influenced cultural tourism on the well-being of the residents and the role of Cultural Conservation in mediating this relationship. A sample size of 385 was determined with a 95% confidence interval and 5 percent margin of error. The target population was approximately 50,000 individuals. The following formula was used to determine the sample size:

$$n = N^*X / (X + N - 1),$$

where,

$$X = Z_{\alpha/2}{}^2 * p^*(1-p) / MOE^2,$$

Where $Z_{\alpha/2}$ is the critical value of the Normal distribution at $\alpha/2$ (e.g. for a confidence level of 95%, α is 0.05 and the critical value is 1.96), MOE is the margin of error, p is the sample proportion, and N is the population size. After taking out the incompletely filled questionnaires 336 useful responses remained. The field survey was completed in three months from December 2018 to February 2019. The questionnaire was structured using a 5-point Likert scale with 1 representing "strongly disagree" and 5 "strongly agree". Before conducting the actual survey, a pilot survey was performed in November 2017. Data was collected from Gilgit city, involving a sample size of 40. This data was used to check the reliability of the questions in the questionnaire using Cronbach's Reliability test. After checking for the reliability, the questionnaires were filled by the residents of all four districts. All segments of the questionnaire had a Cronbach alpha value greater than 0.55 recommended by Yurdugul (2008). According to Yurdugul (2008), for sample sizes between 30 and 50 the alpha value should be greater than 0.55. The reliability test was run again on the data collected for the analysis and the results are shown in table 1.

Table 1. Results for Chronbach's alpha for the reliability of data

Sr. No.	Instruments	Cronbach's alpha	No. of items
1	Culture Tourism	.55	3
2	Culture	.720	31
3	Wellbeing	.799	26
4	Culture of Youth	.670	9
5	CPEC	.771	21

Source: Primary data

RESULTS AND DISCUSSION

Variables Construction

The reliability analysis on the data from the 336 questionnaires comprised of the first part of the index construction. Principal Component Analysis is used to reduce the number of indicators to a single index. The first index that is CPEC-Cultural Tourism is obtained by using PCA, extracted from two principal components. These Principal components were derived from the rescaled rotated component matrix loadings multiplied by the responses of their respective indicators. Then these principal components were multiplied by the total variance explained by the extracted sum of square loadings to provide the final index value as suggested by Suhr (2005) and Abeyasekera (2005). Also, the Kaiser-Meyer-Olkin (KMO) value of 0.671 shows that the sample for the indicators of cultural tourism is 67.1% consistent and adequate. Table 2 shows the indicators, their mean and standard deviation for the respective components. Whereas, the variable Culture Tourism is perception about the trends in culture tourism to their region and also the perception regarding the impact of CPEC on the cultural tourism. Residents were asked questions about these respective issues to help construct the index which has been termed as CPEC influenced cultural tourism.

Table 2. Descriptive statistics of indicators of-cultural tourism

Indicators of Cultural Tourism	Mean	Std. Deviation
Tourists appreciate the culture (lifestyle, food, music, dance, language etc.) more than any other object.	3.76	1.23
Tourist arrivals to this region has increased only because of cultural attractions.	2.89	1.30
The cultural attractions of this region are unique to tourists than any other region in Pakistan.	4.03	1.14
CPEC helps promote the culture of GB through tourism and advertisements	3.74	1.12
CPEC helps promote tourism in GB.	3.85	1.11
CPEC helps facilitate tourism through tourism information desks	3.74	1.18
CPEC helps make this place an easy and safe to go tourist destination.	3.58	1.27

Similarly, table 3 shows the list of indicators, their mean and standard deviation used to obtain cultural conservation index. Cultural conservation index was derived from 5 components derived through a PCA using fifteen indicators. Similarly, the Kaiser-Meyer-Olkin's (KMO) of 0.63 shows that the dataset is 63% consistent and adequate which is fairly good according to the previous literature. Further, these impacts of tourism on culture were derived from previous researchers including Brunt and Courtney (1999) and Besculides et. al. (2002).

In the same manner, the index for the wellbeing of residents was derived by using PCA and following the procedure as described above. Estimating the index from 9 principal respective components explaining the maximum variation in the data. Table 4 shows the mean and standard deviations of the indicators which derive the principal components of the wellbeing index. Similarly, the KMO value of 0.72 shows that the data is 72% sufficient and consistent for this study which is higher than the benchmark value of 0.60.

Finally in deriving a framework for this study based on the theoretical framework in the previous section and hence describing the relationship between CPEC-cultural-tourism, cultural conservation and objective wellbeing using mediation analysis. Following the framework explained in the previous section, the path diagram (figure 2) shows the details of the components of the indexes from which they were derived, hence providing a complete picture of the relationships among the variables.

Table 3. Descriptive statistics of indicators of cultural conservation

Indicators of Cultural Conservation	Mean	Std. Deviation
Tourism helps us strengthen our belief/faith	2.78	1.391
It helps us understand belief/faith of others too	3.21	1.380
Tourism is promoting the native/local language of the region	3.35	1.380
Tourism is promoting the arts and crafts of the region	4.04	.939
It helps handicraft evolve over the time according to the need of tourists.	3.77	1.031
It is bringing positive moral values and ethics to the region.	3.03	1.301
Tourism is helping in the promotion of traditional music of the region.	3.95	.991
It improves and encourages residents to show off their traditional dance.	3.92	.997
Tourists are more attracted to objects like arts/music/dance.	3.96	.982
Tourism helps improve the architectural structure.	3.77	1.059
It is encouraging local residents to adopt modern architecture and the traditional or cultural architecture will soon be extinct.	3.63	1.122
It gives us a sense of pride regarding our culture	3.69	1.087
It helps improve the image of our community and country	3.84	.987
It gives us a stronger sense of ethnic identity	3.74	1.016
It helps eliminate stereotypes/ bad image of other cultures and countries and ours too.	3.64	1.165

Table 4. Descriptive statistics of indicators of wellbeing

Indicators of Wellbeing	*Mean*	*Std. Deviation*
Tourism has increased your income at your current job	4.12	1.051
It has increased employment opportunities	4.41	.817
It gives economic security of your job	4.14	1.112
Your family income has improved	4.13	1.078
You get more pay and fringe benefits	3.91	1.115
It improves the educational standard of the region	3.80	1.169
It brings awareness and competition which makes the people of this region more diverted toward education	3.79	1.091
It helps bring innovative ideas specifically the entrepreneurial education, and gives a platform to practically experiment them	3.80	1.102

Results

Baron and Kenny's (1986) approach was adapted to test for mediation of cultural conservation to impact the relationship between CPEC-Cultural-tourism and wellbeing of residents. Several regression analysis and bootstrapping were used to estimate the result for the relationship and the mediation effect. Hitherto, the most common technique used is these strategies was described by Baron and Kenny (1986). The results of the mediation analysis through bootstrapping (by using process macro in SPSS introduced by a psychologist Andrew F. Hayes) is discussed in the table 5. It reveals both standardized and unstandardized regression coefficients, standard error, t-value and the BootCI (the boot generated upper and lower confidence intervals).

Table 5. The regression results for mediation analysis

	Regression Coefficient β				
Regression	**Unstandardized regression coefficient β**	**Standardized regression coefficient β**	**SE**	**t-value**	**BootLLCI/ BOOTULCI**
β_c (OWB-CT)	0.30***	0.22***	0.05	4.03	0.19/0.43
β_a(CC-CT)	0.17***	0.32***	0.03	5.45	0.11/0.23
β_b(OWB-CC)	0.62***	0.33***	0.10	6.25	0.43/0.82
β_d(OWB-CT. CC)	0.20***	0.18***	0.06	3.37	0.08/0.32

Note: OWB, objective wellbeing; CT, CPEC-Culture Tourism; CC, Culture conservation. The sign * shows the level of significance (** shows significance at p<0.05 & *** shows significance at p<0.01)

The results from mediation analysis is explained in both table 5 above as well as in the figure (4) below.

These results based on a 5000 bootstrapped sample (as suggested by Preacher & Hayes, 2008) show that the total effect of Culture Tourism on objective wellbeing was positive and significant showed in step 1 of table 5 (total effect=0.17 se=0.03, p<0.01). Step 2 and 3 of the regression analysis in table 5 show a positive and significant relationship of independent variable with the mediator (IV-> M=0.32, se= 0.03, p<0.01) consequently, showing a positive and significant relationship of mediator to the dependent variable (M->DV=0.62, se=0.10, p<0.01). Meanwhile, the direct effect (indicated by step 4 in table 5) was not significant (direct effect =0.20, se=0.06, p<0.01). Hence giving an insight that cultural conservation partially mediated the relationship between culture tourism and objective wellbeing of residents (step 4 in table 5). In addition, with reference to Baron and Kenny (1986), complete

Figure 4. Results of the mediation analysis.

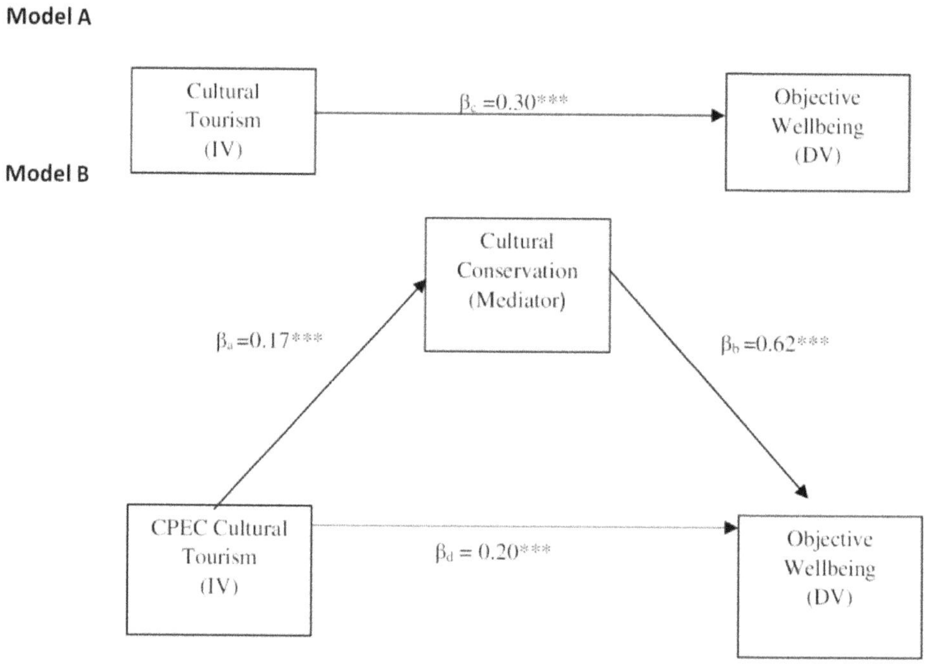

mediation happens when the path from the independent variable to dependent variable is reduced in absolute size and is insignificant when the mediating variable is introduced otherwise it partially mediates the effect.

Discussion

This study shows that increased cultural tourism fostered by CPEC has a significant and positive impact on the wellbeing of the residents in the Gilgit-Baltistan region of Pakistan. Hence, our findings support that of previous studies that linked cultural tourism with residents' wellbeing (Aref, 2011; Kim, 2002; Mathew and Sreejech, 2017). These results further support the argument that increased cultural tourism is linked with the demand for tangible and intangible depiction of the local culture (Richards, 2007; UNWTO, 2005). The study further shows that the market for local arts and crafts, entertainment/music/dance, cultural food, architecture, and museums is linked with increased cultural tourism.

The results further show that there is a positive and significant relationship between cultural tourism and cultural conservation. Therefore, our study supports the notion of positivity in this relationship (Boniface, 2013; Richards, 2017), rather than concerns about potential negative impacts as feared by some scholars (Besculides et al., 2002).

These results suggest that cultural tourism encourages the cultural conservation as well as interest in preserving that specific culture (Bec et al., 2019; Richards, 2017). It is further noted that as culture tourism increases, the residents of the tourist destination try to grab these tourism opportunities by promoting their indigenous culture, hence commodifying both tangible and intangible objects associated with their culture (Robinson and Picard, 2006; Young and Markham, 2020).

The study also reveals that a positive and significant indirect impact of CPEC influenced cultural tourism on residents' wellbeing is mediated by culture conservation. This finding is in line with some prior studies analyzing similar relationships in different contexts (eAllis, 2016; Ahebwa et. al., 2016 UNWTO, 2005). Culture conservation playing the role of mediator states that cultural tourism plays a positive role in increasing the host community's objective wellbeing in the presence of cultural conservation. Since cultural assets, including tangible and intangible objects as well as expression of a society's values, play an important role in attracting tourists i.e., they play a role of pull factors for the tourist attraction (Bec et al., 2019; Young and Markham, 2020).

It has been referred by Bandarin et. al. (2011) that the relationship between culture (and cultural tourism) and development can be very collaborative, hence creating a pragmatic circular path. Based on this and other similar studies like Yang and Wall (2009) it is argued that as culture conservation increases, the wellbeing of the host community increases which in turn brings about development in the region. Due to increased development (and economic empowerment of locals), facilities, more specifically road, transportation, accommodation and food, improve hence attracting more cultural tourists to the region. Most of the tourists who visit these tourist destinations are more interested in the history of the place and its people, they therefore visit historical sites, forts and museums, local handicraft shops, cultural sites, national parks and patronize locally prepared organic food. Such activities encourage residents to commodify their cultural objects to improve their living standards and hence increase their wellbeing. When tourists arrive in these regions, they consume these commodities available at these sites, which will help the community either directly or indirectly to improve their objective wellbeing.

The cultural history of Gilgit-Baltistan is of great interest to visitors since it is located in the far North of Pakistan with mighty mountainous ranges, different National parks, including Khunjerab National park, forts and mosques including Baltit Fort, Altit Fort, an old mosque in Nagar Khaas, fairy meadows, and more importantly the old silk route along the Karakorum Highway. These heritage and other tourist attraction sites must be given an international recognition like their inclusion in the World Heritage Monuments list of the UNESCO. This will give international recognition to these cultural sites as well as to the region, as stated by Ryan and Silvanto (2011).

IMPLICATIONS, LIMITATIONS AND FUTURE RESEARCH DIRECTIONS

This chapter has both theoretical and practical implications. Firstly, a key theoretical implication relates to the need for the context-specific theorization of issues like cultural tourism and wellbeing. The available definitions and constructs have been used to address these topics. However, it is an established fact in the extant literature that cultural perceptions concerning wellbeing tend to differ between different societies, where some aspects of wellbeing may be perceived more important than others (Dalziel et al., 2018). If scholars in the field of cultural tourism incorporate such sensitivities in their theorization, it will be beneficial for the emerging research stream, focusing on cultural tourism in developing countries, where the cultural paradigm is collectivistic. A key practical implication relates to the urgent need for skills development concerning the commodification of local cultural aspects in the Gilgit-Baltistan region. As cultural tourism is an emergent phenomenon in this region, the development of vocational training programs and hands-on guidance to the local entrepreneurs will increase the wellbeing of local residents and contribute to overall social value creation. Social value creation through (vocational) skills development has been highlighted in some recent studies as very important by scholars to ensure the viability of operations in any business field (e.g., Golgeci et al., 2021). Hence, this aspect should be taken seriously by the relevant policymakers to help develop such vocational skills in the region. Specialization in different cultural products like local art, cuisine, music, and heritage can significantly enrich tourists' cultural experiences. So far, there is a lack of such skills development and training initiatives at the policy level in the Gilgit-Baltistan region. Although there has been a visible improvement in infrastructure in this region due to CPEC, due to the mountainous nature of the region, more needs to be done to increase access to culturally attractive sites and valleys as well as ensure all-year-round access, rather than only seasonal access in summer or spring.

It is further important to mention that the ongoing COVID-19 pandemic has significantly influenced every aspect of life, including tourism in mountainous regions, where the economic dependence on tourism is very high (Nepal, 2020). Policymakers in the Gilgit-Baltistan region need to think about incorporating technological elements to enhance the cultural tourism experience in the region by offering interested tourists, different options like virtual tours and developing 3-D tour programs. For this they can benchmark on initiatives undertaken in different western countries since the start of COVID-19. Availability of such options while the world still waits to return to normality will help keep the tourism sector functional and offer new possibilities for commodification of local cultural products.

This study however has several limitations. Firstly, it is based on data collected from selected parts of the Gilgit-Baltistan region, and findings cannot be generalized to other parts of Pakistan, due to significant variation in geography, climate, and cultural dynamics. However, the aim of the paper was to focus on this significantly under-researched region, and the findings provide a bases for future studies to build on this research. In particular, there is a need to undertake in-depth and qualitative research where specificities of different cultural tourism elements should be analyzed in relation to particular types and categories of cultural tourists (e.g., individual or family tourists from Pakistan, international tourists and group tourists). Such in-depth research is expected to offer specific insights concerning the development of tailor-made cultural tourism experiences for different groups, which is expected to strengthen further, local residents' wellbeing engaged with the tourism business. Finally, so far, there is a lack of research on tourists' cultural adjustment strategies when they visit indigenous communities in the region, who may have specific local customs and cultural sensitivities. Future researchers should explore this avenue, as their findings will be beneficial to both visiting tourists in the adjustment process while exploring and immersing themselves in the local culture and local touristic entrepreneurs to understand the process of adjustment of visitors and limitations associated with it.

REFERENCES

Abeyasekera, S. (2005). *Multivariate methods for index construction.* Available online at https://unstats.un.org/unsd/hhsurveys/pdf/Chapter_18.pdf

Ahebwa, W. M., Aporu, J. P., & Nyakaana, J. B. (2016). Bridging community livelihoods and cultural conservation through tourism: A Case study of Kabaka heritage trail in Uganda. *Tourism and Hospitality Research, 16*(2), 103–115. doi:10.1177/1467358415589659

Alatartseva, E., & Barysheva, G. (2015). Well-being: Subjective and objective aspects. *Procedia: Social and Behavioral Sciences, 166*, 36–42. doi:10.1016/j.sbspro.2014.12.479

Allis, T. (2016). Culture and Tourism. In G. Lohmann & A. P. Netto (Eds.), *Tourism theory: Concepts, models and systems*. CABI Press.

Aref, F. (2011). The effects of tourism on quality of life: A case study of Shiraz, Iran. *Life Science Journal, 8*(2), 26–30.

Asmelash, A. G., & Kumar, S. (2019). Assessing progress of tourism sustainability: Developing and validating sustainability indicators. *Tourism Management, 71*, 67–83. doi:10.1016/j.tourman.2018.09.020

Baig, S., Khan, A. A., & Khan, A. A. (2021). A time series analysis of causality between tourist arrivals and climatic effects for nature-based tourism destinations: Evidence from Gilgit-Baltistan, Pakistan. *Environment, Development and Sustainability, 23*(4), 5035–5057. doi:10.100710668-020-00803-0

Baig, S., Qasim, M., Xuemei, L., & Alam, K. M. (2020). Is the China-Pakistan economic corridor an opportunity or a threat for small and micro-entrepreneurs? Empirical evidence from Northern Pakistan. *Sustainability, 12*(5), 1727. doi:10.3390u12051727

Baig, S., & Zehra, S. (2020). China-Pakistan economic corridor, governance, and tourism nexus: Evidence from Gilgit-Baltistan, Pakistan. *Current Issues in Tourism, 23*(23), 2884–2889. doi:10.1080/13683500.2020.1763266

Bandarin, F., Hosagrahar, J., & Sailer Albernaz, F. (2011). Why development needs culture. *Journal of Cultural Heritage Management and Sustainable Development, 1*(1), 15–25. doi:10.1108/20441261111129906

Baron, R. M., & Kenny, D. A. (1986). The moderator-mediator variable distinction in social psychological research: Conceptual, strategic, and statistical considerations. *Journal of Personality and Social Psychology, 51*(6), 1173–1182. doi:10.1037/0022-3514.51.6.1173 PMID:3806354

Bec, A., Moyle, B., Timms, K., Schaffer, V., Skavronskaya, L., & Little, C. (2019). Management of immersive heritage tourism experiences: A conceptual model. *Tourism Management, 72*, 117–120. doi:10.1016/j.tourman.2018.10.033

Besculides, A., Lee, M. E., & McCormick, P. J. (2002). Residents' perceptions of the cultural benefits of tourism. *Annals of Tourism Research, 29*(2), 303–319. doi:10.1016/S0160-7383(01)00066-4

Boniface, P. (2013). *Managing quality cultural tourism*. Routledge. doi:10.4324/9780203424230

Borowiecki, K. J., & Castiglione, C. (2014). Cultural participation and tourism flows: An empirical investigation of Italian provinces. *Tourism Economics, 20*(2), 241-262. doi:10.5367/te.2013.0278

Brohman, J. (1996). New directions in tourism for third world development. *Annals of Tourism Research, 23*(1), 48–70. doi:10.1016/0160-7383(95)00043-7

Brunt, P., & Courtney, P. (1999). Host perceptions of socio-cultural impacts. *Annals of Tourism Research, 26*(3), 493–515. doi:10.1016/S0160-7383(99)00003-1

Cellini, R., & Cuccia, T. (2013). Museum and monument attendance and tourism flow: A time series analysis approach. *Applied Economics, 45*(24), 3473–3482. doi:10.1080/00036846.2012.716150

Cheer, J. M., & Lew, A. A. (Eds.). (2017). *Tourism, resilience and sustainability: Adapting to social, political and economic change*. Routledge. doi:10.4324/9781315464053

Dalziel, P., Saunders, C., & Saunders, J. (2018). *Wellbeing economics: the capabilities approach to prosperity*. Springer Nature. doi:10.1007/978-3-319-93194-4

Dwyer, L., Forsyth, P. & Dwyer, W. (2010). *Tourism economics and policy*. Channel View Publications.

Fang, W. T. (2020). Cultural tourism. In W. T. Fang (Ed.), *Tourism in Emerging Economies* (pp. 75–101). Springer. doi:10.1007/978-981-15-2463-9_4

Fusté-Forné, F. (2019). Seasonality in food tourism: wild foods in peripheral areas. *Tourism Geographies*. doi:10.1080/14616688.2018.1558453

Gölgeci, I., Arslan, A., Khan, Z., & Kontkanen, M. (2021). Foreign firm operations and skills development of local employees in violence-hit countries. *Technological Forecasting and Social Change, 162*, 120376. doi:10.1016/j.techfore.2020.120376

Gursoy, D., & Rutherford, D. G. (2004). Host attitudes toward tourism: An improved structural model. *Annals of Tourism Research, 31*(3), 495–516. doi:10.1016/j.annals.2003.08.008

Kenny, D. A., Kashy, D. A., & Bolger, N. (1998). Data analysis in social psychology. The handbook of social psychology, 1(4), 233-265

Khan, A., Khalid, M., & Ali, S. (2018). *Cultural Tourism under CPEC: A Case Study of Peshawar Valley, KPK*. CPEC. Available online at http://cpec.gov.pk/brain/public/uploads/documents/Magazine2-2.pdf

Kim, K. (2002). *The effects of tourism impacts upon quality of life of residents in the community* (Doctoral dissertation, Virginia Tech). Available online at https://vtechworks.lib.vt.edu/bitstream/handle/10919/29954/Title_and_Text.pdf?sequence=1

Malik, A. R. (2018). The China–Pakistan Economic Corridor (CPEC): a game changer for Pakistan's economy. In B. R. Deepak (Ed.), *China's Global Rebalancing and the New Silk Road* (pp. 69–83). Springer. doi:10.1007/978-981-10-5972-8_7

Mathew, P. V., & Sreejesh, S. (2017). Impact of responsible tourism on destination sustainability and quality of life of community in tourism destinations. *Journal of Hospitality and Tourism Management*, *31*, 83–89. doi:10.1016/j.jhtm.2016.10.001

Nepal, S. K. (2020). Adventure travel and tourism after COVID-19–business as usual or opportunity to reset? *Tourism Geographies*, *22*(3), 646–650. doi:10.1080/14616688.2020.1760926

Nigar, N. (2018). Ecotourism for Sustainable Development in Gilgit-Baltistan. *Strategic Studies*, *38*(3), 72–85.

Olsen, D. H. (2011). Towards a religious view of tourism: Negotiating faith perspectives on tourism. *Tourism, Culture & Communication*, *11*(1), 17–30. doi:10.3727/109830411X13049571092633

Pandey, R. N., Chettri, P., Kunwar, R. R., & Ghimire, G. (1995). *A case study on the effects of tourism on culture and the environment*. UNESCO Principal Regional Office for Asia and the Pacific. Available online at https://resourcecentre.savethechildren.net/sites/default/files/documents/2996.pdf

Perdue, R. R., Tyrrell, T., & Uysal, M. (2010). Understanding the value of tourism: conceptual divergence. In D. Pearce & R. Butler (Eds.), *Tourism research: A 20:20 vision* (pp. 123–134). Goodfellow Publishers.

Preacher, K. J., & Hayes, A. F. (2008). Asymptotic and resampling strategies for assessing and comparing indirect effects in multiple mediator models. *Behavior Research Methods*, *40*(3), 879–891. doi:10.3758/BRM.40.3.879 PMID:18697684

Ratten, V., Costa, C., & Bogers, M. (2019). Artisan, cultural and tourism entrepreneurship. *International Journal of Entrepreneurial Behaviour & Research*, *25*(4), 582–591. doi:10.1108/IJEBR-05-2018-0319

Ratz, T., & Michalko, G. (2011). The contribution of tourism to well-being and welfare: The case of Hungary. *International Journal of Sustainable Development*, *14*(3-4), 332–346. doi:10.1504/IJSD.2011.041968

Richards, G. (Ed.). (2007). *Cultural Tourism: Global and Local Perspectives*. Haworth Press.

Richards, G. (2007). *Cultural tourism: Global and local perspectives*. Psychology Press.

Richards, G. (2018). Cultural tourism: A review of recent research and trends. *Journal of Hospitality and Tourism Management, 36*, 12–21. doi:10.1016/j.jhtm.2018.03.005

Robinson, M., & Picard, D. (2006). *Tourism, Culture and Sustainable Development.* Programme "Culture, tourism, development", Division of cultural policies and intercultural dialogue, Culture Sector, UNESCO. Available online at https://unesdoc.unesco.org/ark:/48223/pf0000147578

Ruhanen, L., & Whitford, M. (2019). Cultural heritage and Indigenous tourism. *Journal of Heritage Tourism, 19*(3), 179–191. doi:10.1080/1743873X.2019.1581788

Ryan, J., & Silvanto, S. (2011). A brand for all the nations: The development of the World Heritage Brand in emerging markets. *Marketing Intelligence & Planning, 29*(3), 305–318. doi:10.1108/02634501111129266

Saarinen, J., & Manwa, H. (2008). Tourism as a socio-cultural encounter: Host-guest relations in tourism development in Botswana. *Botswana Notes and Records, 39*, 43–53.

Shrout, P. E., & Bolger, N. (2002). Mediation in experimental and nonexperimental studies: New procedures and recommendations. *Psychological Methods, 7*(4), 422–445. doi:10.1037/1082-989X.7.4.422 PMID:12530702

Smith, M. K. (2009). *Issues in cultural tourism studies.* Routledge. doi:10.4324/9780203869857

Soontayatron, S. (2013). Thai interpretation of socio-cultural impacts of tourism development in beach resort. *American Journal of Tourism Management, 2*(2), 29–35.

Soutsas, K., Tsantopoulos, G., Arabatzis, G., & Christopoulou, O. (2006). Characteristics of tourism development in mountainous regions using categorical regression: The case of Metsovo (Greece). *International Journal of Sustainable Development and Planning, 1*(1), 32–45. doi:10.2495/SDP-V1-N1-32-45

Suhr, D. D. (2005). Principal component analysis vs. exploratory factor analysis (paper 203-30). In *Proceedings of the thirtieth annual SAS® users group international conference* (Vol. 203, p. 30). Available online at https://support.sas.com/resources/papers/proceedings/proceedings/sugi30/203-30.pdf

UNWTO. (2005). *City Tourism & Culture: The European Experience.* World Tourism Organization and European Travel Commission. Available online at http://81.47.175.201/stodomingo/attachments/article/122/CityTourismCulture.pdf

Wester, P., Mishra, A., Mukherji, A., & Shrestha, A. B. (2019). *The Hindu Kush Himalaya assessment: Mountains, climate change, sustainability and people*. Cham: Springer Nature.

Wood, M. E. (2017). *Sustainable tourism on a finite planet: Environmental, business and policy solutions*. Taylor & Francis. doi:10.4324/9781315439808

Yang, L., & Wall, G. (2009). Ethnic Tourism: A framework and an application. *Tourism Management*, *30*(4), 559–570. doi:10.1016/j.tourman.2008.09.008

Young, M., & Markham, F. (2020). Tourism, capital, and the commodification of place. *Progress in Human Geography*, *44*(2), 276–296. doi:10.1177/0309132519826679

Yurdugul, H. (2008). Minimum sample size for Cronbach's coefficient alpha: a Monte-Carlo study. *Hacettepe Üniversitesi eğitim fakültesi dergisi*. Available online at http://www.efdergi.hacettepe.edu.tr/yonetim/icerik/makaleler/571-published.pdf

Zadel, Z., & Bogdan, S. (2013). Economic impact of cultural tourism. *UTMS Journal of Economics (Skopje)*, *4*(3), 355–366.

Zieba, M. (2016). Tourism Flows and the Demand for Regional and City Theatres in Austria. *Journal of Cultural Economics*, *40*(2), 191–221. doi:10.100710824-015-9250-9

Chapter 7
Crowdsourcing Translation for Community-Based Tourism Empowerment in Thailand

Narongdej Phanthaphoommee
Mahidol University, Thailand

Singhanat Nomnian
Mahidol University, Thailand

ABSTRACT

English has not only become a lingua franca for global communication, but it also carries a symbolic power that influences local people and stakeholders in community-based tourism (CBT) in Thailand. Thai villagers encounter communicative challenges due to their limited English proficiency. Thai-English speaking tour guides are valued, as they provide translation services on behalf of CBT members whose interactions with foreign tourists are minimal and marginalized. Losses in translation regarding local wisdom and heritage are inevitably unavoidable. This chapter addresses a conceptual model of crowdsourcing translation for CBT. The study hopes that the sociolinguistic and cultural gaps can be bridged for the enhancement of other CBT communities. Crowdsourcing in the form of collaborative translation and volunteering translators' efforts can potentially provide an alternative lens for possible translation scenarios. Tourism operators and stakeholders can apply these practices to assist the local people to promote lesser-known tourist attractions in remote areas in Thailand.

DOI: 10.4018/978-1-7998-7335-8.ch007

INTRODUCTION

Community-based tourism (CBT) has played a significant role in Southeast Asia in recent years. According to ASEAN Secretariat (2016), CBT is initiated by local people in a community in order to create a mutual engagement with relevant stakeholders and reciprocal partnership with tourism enterprises to support community projects and businesses for income generation and the improvement of the socio-economic well-being of communities, cultural preservation, and environmental sustainability. Tourism activities and enterprises are owned and managed by the community whose income generations and benefits will then return to the community for its socio-economic growth and sustainable development (Walia, 2020). Local community engagement is key to CBT as residents are involved in providing information based on their local knowledge-creating opportunities and local enterprises to accommodate tourists whose visits can potentially promote jobs and generate income for the community (Lo & Janta, 2020). CBT can thus be considered as the social empowerment underpinned by interactions and support between stakeholders (Dolezal & Novelli, 2020). In contrast to mass tourism, CBT is an alternative tourism approach that has grown reputation with public and private sectors as the strategic planning and policies for natural resource conservation, sustainability, and development in Thailand (Nitikasetsoontorn, 2015; Nomnian et al., 2020). CBT-related government policies, agencies, and businesses are, therefore, vital to strengthen the collaboration between the tourism industry and local community leading to shared benefits among stakeholders (Kattiyapornpong & Chuntamara, 2020).

The Thai government's Pracharat policy plays a significant role in strengthening and creating employment, social welfare, security, local heritage, healthcare, and key necessities for local people through work experience, training, and folk wisdom to be utilized as a part of human capital development (Inphoo & Nomnian, 2019; Nuchdang, 2019). Responding to such a policy, Thai AirAsia Co. Ltd. has bridged the gaps between the public and private sectors and their contribution to society by uplifting local communities' quality of life and standard of living through CBT under the project titled "Journey D".

Since 2017, Journey D has enhanced CBT by working closely with local communities on improving their tourism standards and forming a network of CBT experts and stakeholders to engage in the development of the following three key areas:

1. *Community-based Tourism Standards Enhancement and Development.* Thai AirAsia worked with expert organizations to hold workshops for community tourism leaders to improve their skills in areas such as homestay standards, waste management, tourism site management, communication, pricing and

capital management, and activity design. Their key partner was *Local Alike*, which designed a program for enhancing community-based tourism leaders and their areas to sustainably welcome visitors in search of environmental and cultural tourism.

2. *Community Products Development.* To create new revenue streams for communities, the program supported the development of locally made products, emphasizing each area's distinction and uniqueness. Products were chosen based on the capabilities of each community and help was provided in areas such as quality development, packaging, and pricing. Marketing channels were also found to ensure that revenue would return to the communities. This effort was handled by *Local Alike, Big Trees Group,* and *Thai AirAsia.*
3. *English Knowledge Enhancement through "English on Air".* Working with a university faculty, Journey D designed an English learning program specifically for community tourism operators, with volunteers from within *AirAsia All Stars* being responsible for delivering the English lessons.

Thai AirAsia and its partner agencies made sure to choose communities ready for assistance and tourism, selecting areas based on their natural resources, community stability, unity, and openness to development. The program took place in 2017 in the four communities of Pha Mhee in Chiang Rai province, Baan Kok Muang in Buriram province, Koh Klang in Krabi, and Phrom Lok in Nakhon Si Thammarat province. All four communities were able to establish their own Journey D community tourism commissions comprising committees handling tourism administration, visitor coordination, homestays, food, special activities, and transport. The commission also considered the unique aspects of community products and helped to develop and design the quality and packaging so that they could be put to market. The communities were chosen based on the readiness of the community to become a CBT tourist destination, along with their readiness to welcome tourists and the need for English skills. "*English on Air*" targeted homestay operators, minor CBT-related operators, local guides, and youth.

Thai AirAsia recognized the importance of promoting and developing CBT, which has become increasingly popular among both Thai and foreign tourists. The "*English on Air*" initiative was developed to improve the ability of communities to use English in order to expand CBT. Activities involved teaching English to various communities in collaboration with academic staff who designed the English language syllabus for the CBT course. Participants mainly included Thai AirAsia representatives and volunteers from all departments who offered their time to teach English to the community locals.

Although CBT affects local people's ways of life, standard of living, businesses, and socio-cultural transformations, it may cause challenges and adverse effects on community disempowerment and minoritization of ethnic groups (Dolezal et al., 2020). According to Nomnian et al. (2020) and Charoenkongka and Nomnian (2019), the limited English language proficiency of community leaders and members potentially hinders CBT in terms of host-guest communication that requires interdependency among bilingual tour guides. In addition, the communicative English topics in the tourism and hospitality industry are related to community life, local attractions, local food, and occupations (Jhaiyanuntana & Nomnian, 2020).

In this chapter, the CBT community that serves as the basis for discussion is Ban Phrom Lok, which is a subdistrict located in Phrom Khiri District in Nakhon Si Thammarat province in southern Thailand and is the hometown of the second author. Based on Tourism Authority of Thailand (2011), this remote village is surrounded by Khao Luang National Park with the highest peak of 1,835 meters above the sea level with the highest peak in the South and the third highest in Thailand. According to Journey D (2021), tourists can choose to stay at homestays and learn from the locals who will welcome them with authentic southern dishes and take them around the village. Phrom Lok CBT focuses on training local guides, the development of tourism routes, and tourism activities that enable tourists to experience the way of life in the community. Attractions and activities in the area include Phrom Lok Waterfall, cycling, hiking up Phromkhiri Mountain, making batik, and toys from local materials, shopping at the local market, paying homage to King Taksin at Khao Khunphanom Temple, and visiting Tak Fa cave and Maha-ut Monastery. For foreign tourists, an English-speaking tour guide is provided with a service fee of US$90 per day for 8 people, which is included in a tour price if required (Journey D, 2021).

Drawing upon Nomnian et al. (2020), the translation service is offered by English-speaking guides from outside the community such as urban-based tour guides and cultural brokers, who have become liaisons and translators on behalf of the locals. As a result, cultural interpretation risks being misleading or invalid. Although foreign tourists would like to interact with these local people, they usually find communication impossible without the help of tour guides, and this demonstrates the unequal power relations in Thailand's CBT. English is, therefore, not only an international communicative tool, but it also carries power exerted by language users such as tour guides, interpreters, and translators whose linguistic capital is greatly valued and relied upon by CBT communities.

The development of communicative English skills and ICT initiatives offer positive opportunities for lesser-known tourism attractions and remote areas in remote tourist attractions in Thailand (Phumpho & Nomnian, 2019). Unlike digitalized urban zones, remote communities in Thailand are deprived of ICT infrastructure and digital networks for local and global outreach. In international tourism interaction, English is needed by communities whose actual ethnic identity risks becoming shrouded or misinterpreted in translation. Due to the dominant *lingua franca* status of English,

ICT infrastructure and machine translation devices should be made available for the community's language use, learning, and education in order to empower and sustain CBT communities.

This chapter, therefore, proposes the use of crowdsourcing translation, or online collaboration translation by volunteer translators, to buttress and facilitate international communication between community members and foreign tourists in the hope that the sociolinguistic and cultural gaps can be bridged for the enhancement of CBT in remote areas in Thailand.

TOURISM TEXTS AS PROMOTIONAL TOOLS

In promoting CBT, detailed information about the localities is deemed essential in communicating with and attracting tourists around the world. For the tourism industry, there are various mediums for distributing information about destinations and activities such as blogs, websites, tourism guides, brochures, and signage. Tourist texts on these mediums typically feature a creative language. Although they mostly concentrate on a wide range of leisure pursuits, some text types are created specifically to convey information among professionals and are often filled with more technical terms than texts aimed at simply gaining the attention of tourists. Basically, tourist texts can be divided into three types: (a) *description* (of destiny, vacation, or tour); (b) *information* (on facilities, services, terms, and conditions); and (c) *professional communication* (between professionals, price lists, reservation systems) (Skibitska, 2015). To translate these text types for marketing purposes requires a huge effort in terms of lexical selection to match the style of each type. It is necessary for translators to pay close attention before deciding whether to retain the level of creativity or reduce the lexical density of the source text in the target text.

Many cultures and languages share similar techniques in producing tourist texts, but the extent to which they are employed depends largely on the creators and the commissioners. According to Sulaiman and Wilson (2019, p. 25-30), there are seven techniques of language often used in tourism promotional materials. The first one is *ego-targeting* referring to the direct address to the readers and potential consumers to single them out from the crowd and make them feel special and privileged. This technique is expected to create a bond between the would-be visitors and the promoted destination. Second, *keying* is concerned with using keywords to invoke an emotional response to the texts, such as 'escape from everyday life', 'holiday retreat', or 'indulgence in nature'. The third technique is *contrasting*. Lexical oppositions, both implicit and explicit, are emphasized in the prospectus, thereby contrasting the readers' mundane life with the environment of the destination being promoted. The fourth one is *exoticizing,* which means using foreign words to create a stylistic effect and strangerhood of the destination in the reader's mind. By contrast, the fifth technique, *comparing*, tries to downplay the strangeness and familiarize the readers with the destination's image. Next is the use of *poetic devices* such as alliteration, assonance, puns, or metaphor to invoke positive feelings. The last technique is *humour*, which adds amusing, surprising elements to capture the reader's attention. It is plausible that the translations of these techniques used by people in the local community without a complete understanding of industry terminologies could hinder the delivery of the intended message for foreign tourists or even cause misunderstandings among CBT members and stakeholders. However, in order to understand the full impact of the different uses of a language to gain the attention of prospective foreign tourists of another culture, it is necessary to investigate studies on translation for tourism from around the world, especially those related to the CBT context in Thailand.

Research on cross-language tourism and hospitality tends to reflect the invisibility of translators and the translation decision-making process (Suebwongsuwan & Nomnian, 2020). Such studies as using image destination for non-visitors (Cherifi et al., 2014), online consumer reviews of dining experiences (Huang, 2017), or the motivation and service expectation of Chinese tourists to Western Europe (Prayag et al., 2015) do not give enough credit to the translation process that took place before the final translated product was made available for customers. Many studies on cross-cultural tourism do not even mention translation as the medium of communication between the locals and the tourists (cf. Koc, 2020; Su et al., 2016; Wu & Pearce, 2014; Zhang et al., 2018). Some mention the role of editor (Shani & Uriely, 2012) or the post-editing process by consulting online resources (Afdal, 2016; Mkono, 2012), but usually without clear implications as to how these resources may be used in different tourist destinations.

Although there are some concerns over cross-cultural awareness in the translation of tourist texts (cf. Argoni, 2012; Jun, 2019; Xu, 2019), misunderstanding about the role of the translator/editor continues to prevail among tourism translation commissions. In the study of tourism promotional brochures, Sulaiman (2016) maintains that the commissioners typically ignore the option of a professional translator because of their disapproving view towards the translator as merely someone who carries words and sentences across languages rather than a cultural mediator. This is in line with studies on the translation commissioning process in the Indonesian tourist industry where the quality of the translation that was commissioned was rather inadequate, thereby undermining the text's value as tourism promotional material (Napu, 2018; 2019).

It is quite surprising that studies on the issue of tourism and translation in Thailand are somewhat limited in number despite it being one of the most popular tourist destinations in the world. A number of studies about tourist texts in Thailand, mostly written in Thai, seem to be preoccupied with prescriptive translation methods for achieving equivalency effect rather than the process or networking among translation agents. Such studies include the translation of Thai culture in museums (Vanijdee, 2015), translating services for Chinese (Pavasiriporn & Juntaronanont, 2017), signs in multiple languages (Sarod & Kraisame, 2019), and the management of cultural tourism in less attractive destinations (Phomnimitkul, Putjorn, & Vechpong, 2020). The lack of research on translation and tourism at a regional level or even within Thailand makes it an urgent matter to develop a body of knowledge, particularly translation networking that serves multiple tourism sectors to effectively attract more visitors, as well as to constructively yield an approach for testing further research.

MACHINE TRANSLATION (MT) TOOLS AS COLLABORATIVE EFFORT

Thanks to the advent of translation technology, trained translators can find an assistant in machine translation (MT) tools (e.g., Bing translator or Google Translate) to generate automatic translations. Professional or in-house translators can equally use computer-assisted translation (CAT) tools that feature source text and target text alignment, concordance for term searching and extraction, and translation memory tools to create databases of previous translation (Hartley, 2009). Despite overlapping concepts, MT tools are gradually being accepted as one of the CAT tools by which trained translators can interact with one another. These features increase translation speed, facilitate workflow (e.g., pre-and post-editing), and ensure consistency in different texts by different translators (Bowker, 2015).

Machine translation (MT) is an automated process, by which computer software is used to render a text from one natural language to another. There are two main types of MT: statistical and rule-based (SYSTRAN, n.d.). Based on probabilistic mathematics, the statistical type works well when involving large and qualified sets of corpus data, but the translation produced by this mechanism is unpredictable and sometimes inconsistent, and its corpus creation can cost a fortune to run the system; it requires hardware to build and operate the huge translation models. The rule-based type, on the other hand, provides better management and quality especially in terms of dictionary-based customization. It can ensure predictability and compliance with already-existing terminology and is considered a more standardized mechanism. However, no matter which type the translator prefers, it can handle the translation on their behalf with a quick-access document. The MT technology basically allows translators to deposit their translation memories that can be retrieved whenever they choose. The MT tools also provide the translators with the use of glossaries and grammar checkers (Pym, 2011). This can be of benefit to those translators who work in concert and collaborate across space and time.

For a functionalist approach (cf. Nord, 2018), the translation process typically focuses on collaboration between translators, commissioners or any one of these agents to bring about in a certain way a translated product. This kind of collaborative project opens the possibility of editing and revising in a single text by several agents for the precise purposes of translation for/by the community, volunteer translation, fansubbing, crowdsourcing, or online collaborative translations (Jiménez-Crespo, 2017, pp. 17-18). Studies on collaborations among translation studies scholars span between two sets: the traditional method of collaboration and the phenomena of technology-aided collaborative translation. The first set features such works as Cordingley and Manning (2016) and Brown (2018), primarily discussing analog system and the evidence that translation is always a collaborative effort. The second set largely deals with interactivity in digital platforms used by a large group of (un) identified translators, such as those of O'Hagan (2009) and Jiménez-Crespo (2017) and the newly coined 'translaboration' by Alfer and Zwischenberger (2020). In fact, the latter seems to prove the argument of the first in that self-managed, online collaboration (e.g., translations for TED Talk, non-profit organisations) is the unsolicited activity confirmative of working environment wherein the joint action among free-will individuals with the help of technology reigns supreme.

O'Brien (2011, p. 17) explains that collaborative translation can be considered in two ways. In the narrow sense, the actual collaboration by agents involved to produce the translation becomes realised with the help of technology facilitating members of an online community in initiating and translating particular documents. By using online applications, such as OneDrive or other computer-assisted translation tools, this can be arranged by professionals with a relevant background or novice translators who are willing to invest their time in the network. Collaboration between translators and editors for revision and editing by various agents can be included. In a broader sense, collaborative translation means a group of translating and non-translating agents can participate in a translation project. Group members might not be translators, but they take part in a broad range of connections and interplays that happen prior to and after the project. The collaborative translation is often geared up to and assisted by MT tools, normally executed by non-professional translators networking within the virtual world. These collaborative phenomena basically rely on digital platforms and state-of-the-art technologies that reflect the routine interactions and unofficial relations between agents to create far more effective translated texts.

This leads to another unsolicited form of translation: volunteer translation. Since the members of collaborating groups comprise mainly non-professional translators who dedicate themselves to commissioned or non-commissioned tasks, volunteerism in translation can be illustrated by collaborations, especially among non-profit organisations such as Translators without Borders, The Rosetta Foundation, and PerMondo. However, quite often, there is no editor or proofreading process involved in this voluntary set-up, as demonstrated by a parallel corpus (English, Italian, and Spanish) study by Gigliotti (2017). Analysing texts translated by volunteer translators for non-profit organisations, Gigliotti found that this group of volunteers for NPOs failed to maintain a good quality of the translations, although there were some decent works.

Another relevant concept is crowdsourcing, which refers to the recruitment of any group of people engaged in a specific task, originally practised in online platforms such as Facebook and Wikipedia (Howe, 2006). The most well-defined typology for crowdsource is perhaps that of Estellés-Arolas and González-Ladrón-De-Guevara (2012). They categorize the concept according to activities ranging from *crowdcasting* (online contest for an innovation), *crowdstorming* or *crowdsupport* (massive online brainstorming to help others with problems), and *crowdfunding* (funding request in exchange for a reward) to *crowdopinion* (requesting opinions from the crowds for specific issues). Online collaborative translation falls into the category of *crowdcontent* or *crowdproduction* where the crowd puts its efforts and knowledge into creating diverse types of content, and the participants are responsible for the production of content but not a competition among peers. In this sense, the *crowdproduction* or online translation is typically associated with volunteerism, as O'Brien (2011, p. 18) posits, often prompted by a project leader to ensure the best translations on any selected platform. Crowdsource as collaborative translation, along with the voluntary efforts by individual translators, offers an alternative perspective on a range of possible translation scenarios, and tourism practitioners can apply these merging practices in assisting locals to promote their tourist destinations.

THE USE OF CROWDSOURCING APPROACH FOR CBT EMPOWERMENT

To empower communities for CBT and make effective the translation of these texts, the present chapter proposes that online collaborative translation should serve as an enabler for the benefit of local people to promote tourism, particularly with the aid of MT tools in crowdsourcing translation initiated by members of an online translator community. Although a recent survey on perceptions toward technology-enhanced tools reveals a generally conservative attitude among professional translators (Wang & Ping, 2020), MT and CAT tools are still of benefit by assisting a leader of a crowdsourcing group to take on such translation tasks for CBT. Translators should refrain from the notion that MT or CAT tools fully replace human translation, precisely because many specific translations require human skills in dealing with socio-cultural connotations or semantic problems that a computer cannot solve (Pym, 2013). Technology in translation should go hand-in-hand with and assist humans, as the term 'computer-assisted translation' suggests, by integrating it into the whole translation training process.

By linking to the available technology, such as a term-based translation platform, crowdsourcing from outsiders can facilitate the voluntary translation of content (e.g., local wisdom presentations, culture-specific items) or useful information (e.g., dangerous routes, availability of assistance) for tourist attractions in Thai CBT. The objective of this proposal is to introduce outside-community crowdsourcing translation to the local people, so as to empower those with less proficiency in English and stimulate tourist spending that ultimately generates increased income for the villagers.

Applying crowdsourcing translation to CBT, the stakeholders can gain benefits in two ways. First, the online translation project is cost-effective. It can function as one of the AirAsia community development schemes, complementary to teaching English in tourist attraction communities, by recruiting those with English proficiency to volunteer their services without having to commute to the sites. Second, tourist texts can be a perfect set of materials for translation training. Online collaborative platforms can serve as a pedagogical tool for students of translation studies. As asserted by Marczak (2018, p. 11), it is necessary to equip students with knowledge of how to put online technology into maximum use by "implementing student-focused work modes through which students will be able to learn how to independently seek, validate and update knowledge far beyond the point of graduation." By volunteering in this kind of crowdsourcing project, students can, therefore, practice their translating skills in the transition from a classroom-based environment to a real-world experience. This teaching method can groom them for future specialized careers and automation of the translation process.

There are three practical models already in place for and in relation to the translation of tourist texts: translation in thematic analysis (Esfehani & Walters, 2018), cultural-conceptual translation (2019), and Kató community (Translator without Borders, 2020). This present study details advantages and addresses their drawbacks in order to propose a suitable translation model for CBT.

Based primarily on Attride-Stirling's (2001) and Braun and Clarke's (2006) studies of thematic interpretation, Esfehani and Walters (2018) propose a model for micro-level analysis with the help of translation in tourism and hospitality research by expanding frameworks for properly conducting thematic analysis. Their model adds the translation to the 6-phase thematic analysis to help bilingual researchers to overcome language differences when dealing with a network of related themes in the interpretation of tourism promotional materials. According to Esfehani and Walters (ibid., p. 3162-3167), Phase 1 is the first reading of texts to gain a holistic understanding of the content and intuitive sense (e.g., initial thoughts and points). Phase 2 involves the repeated readings to decode texts (e.g., valuable insights, missing elements, contradictions, nuances). Phase 3 is the development of basic themes, manually or digitally, by categorizing all similar codes. Phase 4 is the consolidation of organizing themes by checking for consistency and re-examining all data extracted to confirm the coherent pattern. Phase 5 is the analysis of global themes to disclose overarching networks that capture the essence of the texts by checking for coherence to ensure a common relationship within the entire thematic network. Phase 6 is the description of networks by using visual tools to obtain an empirical, transparent, and easy-to-understand presentation of material.

Translation plays its part in different phases: Pre-Phase, Point A at the beginning of data analysis, Point B at the end, and Point C between Phase 2 and 3. Verbatim transcription of empirical material is carried out in the Pre-Phase, while literal translation is recommended for Point A before a deep analysis of the material by 'qualified translators' to share the workload with the researcher, research workload, and careful consideration of the trustworthiness of translation. Point B corresponds to Phase 6 in the thematic analysis, which is the translation of the results, and final report or paper. Translation Point C squeezes in between Phases 2 and 3 – an internal procedure in which the researcher's familiarity with materials and codes derived from Phases 1 and 2 is applied.

Another model for translation of tourism promotional materials that put the importance of cultural differences at its heart is the so-called "cultural-conceptual translation" (CCT) by Sulaiman and Wilson (2019). The CCT model focuses on push and pull factors conditioning translation techniques: translation purpose and cultural conceptualization to fulfil the purpose.

The first technique is the understanding that the purpose of translation of tourist promotional material is to present 'destination image' in the tourist's mind and persuade them to purchase a holiday or travel package. This technique is significant and instrumental to promote complex tourist attractions. The second techniques, drawn upon Sharifian's (2011) thesis, concerns itself with cultural conceptualization brought into life by the interaction among diverse cultural groups. In order to conjure up a desirous image, pull factors (e.g., destination attributes, featured experiences) are customized in accordance with push factors (tourists' demands/tastes). This process is driven by both culture and marketing campaigns in which translation is at play in the advertising to generate expectations in potential tourists from different cultural backgrounds. Different cultures conceptualize reality and construct meaning differently. The tourist's home-grown cultural values governing how he/she perceives the world are, therefore, crucial in driving one's demands.

The heart of the CCT model is to conceptualize diverse values to match those of prospective tourists and ultimately familiarize them with images of the local destination by effectively rendering tourist promotional materials into different tongues. To achieve this, the translator is required to thoroughly comprehend the culture of the destination through understanding in the context, which can be gained from "living the culture rather than merely learning about it" (Hall, 1976 as cited in Sulaiman & Wilson, 2019, p. 61). When successful, the project destination image could persuade readers of the materials to become visitors. What the translator needs to acquire, in a nutshell, is adequate knowledge of the pull factors of tourist destinations, specific attractions, and ranges of activity. The translator should also keep close contact with the commissioners who have the final say in the target texts, for they could provide the translator with essential information during the translation process, including any change in policy pivotal to the effectiveness of the final product (Sulaiman & Wilson, 2019, pp. 62-63).

This model recognizes the purpose of the source text and the translator's competence in the translation of tourist promotional materials, but it seems insufficient to simply offer a holistic view of the translation process. Other crucial factors are not mentioned either (namely, agents involved, detailed analysis of the original text, problems occurring during execution of the task), nor is there a discussion of the unique cultural environment that is available at each tourist destination. Also, the idea of "living the culture" seems to limit the number of potential translators since they always have different degrees of familiarization with the source culture. The question is how can one be sure how much experience is enough; it seems that the translator who knows best and "lives the culture" has to be one who has been living in the source culture for a long time and not simply a translator residing in that place.

The last model, which takes volunteerism and the translator's actions into account, is Kató community (Translators without Borders, n.d.), originally called Trommons platform (a short form of Translation Commons) (Du et al., 2015). Developed by the Localisation Research Centre at the University of Limerick, Trommons is an online platform that used to be maintained by The Rosetta Foundation to cooperate between non-profit translation projects for various organisations. In 2017, The Rosetta Foundation merged with Translators without Borders (TWB) to function as the Kató translation platform with state-of-the-art open-source translation technology. This global crowdsourcing community is the fruit of the merger that unites these two spearheading charitable organizations aiming to improve access to information in different languages (Translators without Borders, 2017).

On its web page, a number of tasks are created for volunteers to claim after registering their accounts and searching for ones that suit them. On each task page, a project name is indicated on the header row, including source and target languages, timeline and deadline of the selected task and its current status. The staff at Kató normally assume the role of project manager and divide the task into small sections convenient for translators to work with (Joanna, 2020a). Based on a free and open-source online CAT tool called MateCat (n.d.), Kató has created its own translation memory to store glossaries associated with their project to provide volunteers with suggestions on appropriate and key terms in any specific language (Manuel, 2019). There is also a channel for inquiry during the translation process, including a chat room and blog with TWB staff.

Once a task is finished and submitted to the platform, quality assurance ensues to check that it is consistent with other combined individual sections. A peer review system by at least two more translators is, therefore, established to offer constructive criticism and recommend any revisions of the translations (Joanna, 2020b). Both neophytes and trained translators can also enrol in online courses designed to provide them with a better understanding of the rationale behind humanitarian translation as one of the objectives of TWB and the translation process at the platform.

Partially adopting the procedure for non-profit collaboration workflow available on the Kató platform (Translator without Borders, 2020) and the CCT model (Sulaiman & Wilson, 2019), the present chapter puts forward a cross-culture and crowdsourcing workflow for translation for CBT by compiling the advantages of the above-mentioned proposed models and presented as a new conceptual model that would benefit the Thai tourism industry, as illustrated in **Figure 1**.

The workflow comprises three lines of procedure: community-initiated, translator-initiated, and customer-initiated. Each procedure has steps for members in each line to follow. The first step is the community-initiated procedure. The tasks are created by a volunteer project manager who is familiar with the tourist sites chosen for translation. This could be a local, who approaches any known translators or makes an online project announcement to prospective managers and translators. After the initial team is set up (project manager, translators, IT technician), the manager selects the suitable platform to work on and makes the second announcement about the project. A registration of volunteer translators across the country would be required to find the exact number of translators and their respective expertise in order to facilitate and divide the tasks. The online working platform can be basic and friendly-used, such as an online web board, Facebook group or Google Doc, to keep track of discussions and the translation process. They could also be more of a sophisticated platform (e.g., MateCat, as in the Kató's case) if the project manager or any professional translator in the team suggests so. The translation brief is later provided to explain the purpose of the project, overall strategy, terminology, deadline, and core messages that the community wants to disseminate.

The second step is the translator-initiated procedure, beginning when the tasks are divided and distributed by the manager, who uploads the source texts (both segmented or the entire piece) onto the selected platform and sets the deadline. The assignment can be rendered into as many languages as necessary to serve a diverse group of prospective tourists. After uploading the translation project and tasks divided into sections, the volunteers can work on the sections or the whole of the text if they prefer. Any *gratis* machine translation program (e.g. Google Translate,

Figure 1. A proposed model of crowdsourcing translation for CBT

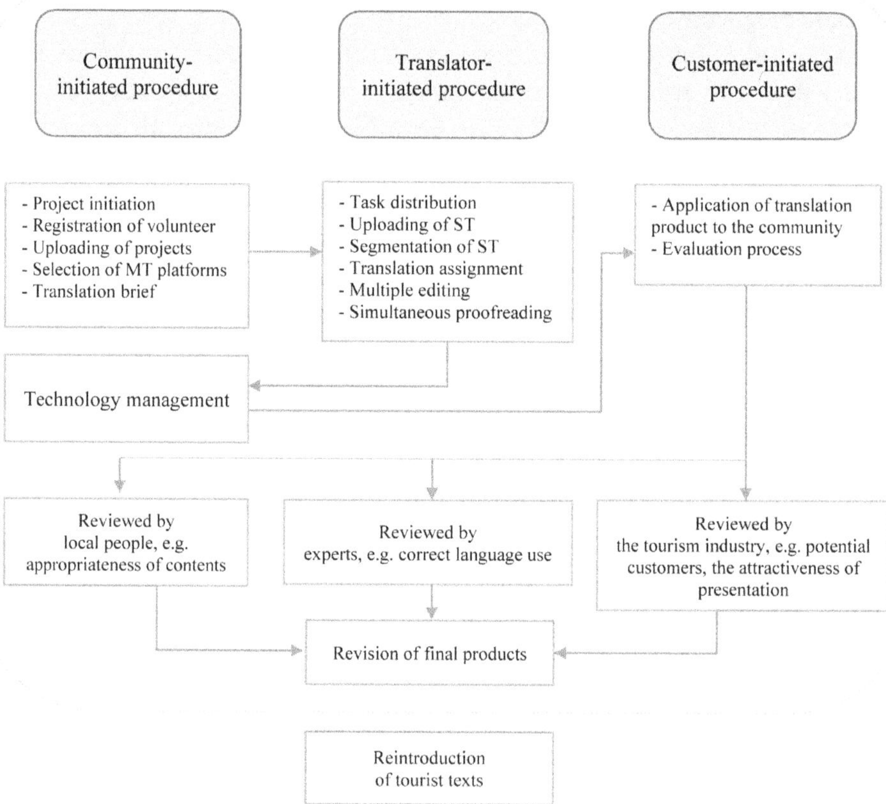

SYSTRAN, DayTranslations) to help facilitate the project can be chosen for the team, or individual translators may choose these that suit them. They can use a basic online dictionary or consult glossaries already uploaded onto other sophisticated platforms (e.g., MateCat, Memsource, Smartcat) maintained by the manager for the project at hand. Once the translated sections are completed, the co-editing process commences with at least two translators (hence, *crowdopinion*), based on the expertise or preferences specified in the registration. Simultaneous proofreading is the final stage, and this can be carried out by volunteers or the manager him/herself after circulation of a final draft. Experts in English (and other languages) or translation studies take part in the review of the proofread texts. If there is any feedback, it can be returned to the team for revision according to comments from the experts.

The third step is the community-initiated procedure. After the translations are combined and finalized, the volunteers or IT technicians based in the community can help upload both original and approved translations onto the selected online social media platform or distribute them in QR code format to test their validity. It is then reviewed by the local people to gauge their reception and confirm the correctness of the translation as well as the effectiveness of the digital distribution (e.g., email list, social media accounts).

The fourth step is the post-distribution process in which a number of tourists are asked to read and appraise the promotional information. The feedback from the would-be visitors entails an evaluation of the translated information by returning it to the local community. The fifth step is a simultaneous review, but this time by outer-circle stakeholders, such as the tourism industry, translation experts, and the local villagers' perspective. By reviewing the translated contents during this stage, the local villagers can also improve their English skills thanks to the translators' choices of lexes and sentences. The sixth step sees the volunteer team revise the translated texts in response to the comments and opinions received, before finally returning the translation to the community. In addition, for the purposes of ethics with respect to the project and all parties, a code of conduct is drafted and agreed upon prior to and during the process. The basic human right that no violence be conducted in the course of gathering content for the project must be guaranteed.[1]

METHODOLOGY

To empirically test the model, a pilot investigation into the effectiveness of, at least, the translation part in the proposed model was set up with a small group of translation studies student volunteers at a Thai university. The participants in this project remained anonymous. This qualitative method started when the task of translating tourist signages was given to the *ad hoc* translation team. Its five team members were assigned roles in the project, namely, project manager and translators. Available in the public domain, the samples of data were randomly selected from two Ban Phrom Lok tourist promotional websites: nakhononline.com and 108kaset.com.

After the roles in the translation had been assigned on 4 November 2020, the volunteers were asked to follow the procedures of the model proposed in the above section and translate the samples on an online platform and translation tool of their choice for two weeks. In order to discuss its effectiveness with the proposed model, the record of communication between the team was kept for analysis after the project ended.

Thereafter, the five participants were also interviewed to gain insightful information during the translation process and management among volunteers. The interviews were recorded and transcribed for the purpose of the discussion with the proposed model and existing literature on crowdsourcing translation and CBT. This online translation project was coordinated and managed by Thai university students who volunteer their time. The authors of this paper only acted as observers to the whole process and took part in the discussion of its results.

RESULTS

The platform the team used to distribute and compile the tasks was OneDrive, on which simultaneous editing and commenting can be managed. Another online platform for informal communication between members is LINE – one of Thailand's most popular phone-based text exchange applications. **Figures 2** and **3** below are examples of signages and online collaboration among volunteer translators.

Figure 2. The sign for a pond at one of the Buddhist temples in Nakhon Si Thammarat Province (adapted from https://www.nakhononline.com/2488)

For the first task, since the text of the signage is relatively short and concise, the translators can work on the whole text together. After discussions among the team members on the selected platform, the final translation approved by the group is as follows.

Source text:

บ่อน้ำยา

เป็นบ่อที่พ่อท่านคล้ายได้ตั้งชื่อให้จากความเชื่อว่าน้ำในบ่มีสรรพคุณเป็นยารักษาโรค ปวดเมื่อย ผิวหนัง ใช้อาบน้ำหรือทาบริเวณที่เจ็บป่วย

Translated text:

Healing pond

This pond was given a name by Phra Kru Pisit Auttakarn (Khlai), former abbot of Suan Kwan temple Nakornsrithammarat province [emphasis added]. There's a belief that the water in this pond has a magical/medical ability to heal diseases including lassitude, dermatitis by bathing in this pond or applying it where needed.

The history of communication between translators demonstrates that discussions in the group mostly were about lexical choices and how to re-present the Thai cultural uniqueness in English. For instance, they brainstormed on ways of rendering the title in English; the choices boiled down to whether 'Healing pond' or 'Magical pond' would best suit the context, and they agreed on the first. Culture-specific items (CSIs) play an important role in this example. The term พ่อท่านคล้าย (*Pho Than Klai*) is the epithet for the likely head monk residing in the temple where the sign is located. On the LINE chat group, one translator raised the question of whether they should employ the domestication strategy and search for an equivalent of Christian clergy ranking to that of a Buddhist monk. The team settled on the translation procedure of explicitation and addition of information by using the phrase 'former abbot of Suan Kwan temple Nakornsrithammarat province' to clarify who *Pho Than Klai* is. They also made him more formal in the translated text by using the official way of addressing the monk: *Phra Kru* followed by his actual name *Pisit Auttakarn* with his nickname, *Khlai*, in parentheses.

Culture-specific items (CSI) remain a problem in the second example as shown in **Figure 3**.

To deal with CSIs, the team managed to translate ลายพระหัตถ์ (*lai phra had*), literally meaning 'the royal handwriting', by explaining and adding more information in the target text. They also gave more background knowledge as to whose handwriting it belonged to: 'His Majesty King Rama the 9th's handwriting'.

The most frequently used translation procedure for CSIs was transliteration. All four levels of a waterfall (หนานบ่อน้ำวน, หนานวังไม้ปัก, หนานวังหัวบัว, หนานวังอ้ายแล) were rendered as *Nan Bhor Nam Von*, *Nan Wang Mai Pak*, *Nan Wang Hua Bua,* and Nan Wang Ai Lae, respectively. The team chose to shun an explanation as to what the referential meanings of those terms are. A possible reason for this is that the team treated these terms as common proper nouns for which the translator need not divulge their real meanings. In doing so, the translated text loses the flavour of the local culture, since cultural naming conventions are often based on indigenous narratives, i.e. *Ai Lae* (Brother Lae), and so have historical significance among villagers.

Figure 3. The map of Phrom Lok waterfall in Nakhon Si Thammarat Province (adapted from https://108kaset.com/farm/2018/08/09)

One interesting opinion that proves a good use of online resources and collaboration comes with the term คลอง (*klong*), literally meaning 'canal'. By referring to an explication from an online encyclopaedia, one translator proposed instead the English term 'river', which is more correct according to the target culture perception. The term 'canal' implies an artificial watercourse, while 'river' is natural.[2]

DISCUSSION

Crowdsourcing Translation

After applying this initial collaboration method, a group of translators has made several observations pertaining to the workflow. First, using crowdsourcing translation is helpful to the team and makes it easy to collaborate among the team members. This is because working on visual space allows for the possibility of distance communication between translators located in different places, especially

Table 1. Source text and translation choices for the map of Phrom Lok waterfall

Source text	Translation choices
Along the blue line (the river)	
ชั้นที่ 4 หนานวังอ้ายแล	Fourth floor: Nan Wang Ai Lae
ชั้นที่ 3 หนานวังหัวบัว	Third floor: Nan Wang Hua Bua
ชั้นที่ 2 หนานวังไม้ปัก	Second floor: Nan Wang Mai Pak
ลายพระหัตถ์	His Majesty King Rama the 9th's handwriting
บ่อน้ำวน	Whirlpool
ชั้นที่ 1 หนานบ่อน้ำวน	First floor: Nan Bhor Nam Von
บ่อน้ำวน	Whirlpool
ศาลาแปดเหลี่ยม	Hexagonal pavilion (paed liam pavillion)
ร้านค้าสวัสดิการ	Convenience store
Along the small orange line (the trekking rail)	
ทางเดินเท้า	Footpath
ป้อม	Fortress
ศาลา / ท่านยืนอยู่ตรงนี้	Pavilion / You are here
ลานจอดรถ	Parking lot
สุขาชาย / สุขาหญิง	Men's restroom / women's restroom
ศาลาตาหลวง	-
ที่ทำการหน่วยพิทักษ์	Ranger Unit
ด่านเก็บค่าธรรมเนียม	Tollgate
On the bottom left-hand corner	
เครื่องหมาย	Symbols
ถนน	Road
ทางเดิเแท้า	Footpath
คลอง	River
ชั้นน้ำตก	Waterfall tier

during the COVID-19 pandemic when physical commuting is not recommended. However, if the project scope is larger and involves a lot more complicated texts, there needs to be more efficient management of each assignment. Another advantage of crowdsourcing is the ability to brainstorm, whereby members of the team can voice their individual opinions. It is cost-effective and serves as a peer-checking process by which translators with different expertise and skills can contribute to the group without having to wait until the editing stage. However, some translators may not fully understand the premise of the translated text (as mentioned in Sulaiman &

Wilson's (2019) model), so it may be necessary to recruit more volunteers in order to propose a wider range of possible equivalents, particularly when the team comes across the culture-specific terms.

The second observation concerns the platform selected for this pilot study. Since it is a small project, the team does not rely on any specific MT tools. OneDrive, the selected collaborative platform, has no spontaneous notification; the translators need to regularly check their email or keep the page on their computer screen active to see whether there are any new comments or suggestions from their colleagues. Also, the division of text is not clear, which results in confusing communication on the platform. This limitation may be due to mismanagement among members, rather than a flaw in the workflow of the model *per se*. For a future test or the actual project itself, the *ad hoc* team suggests Facebook group as an alternative where each post can be a new piece of section for brainstorming, and its comment box is limited to that particular post. This platform has already been found to be beneficial for teaching English among Thai university students (Inpeng & Nomnian, 2020).

Last but not least, this pilot investigation embodies the new concept of 'translaboration' advanced by Alfer and Zwischenberger (2020). It also demonstrates the applicability of the Thai CBT translation model in comparison to the Kató and cultural-conceptual translation methods (cf. Sulaiman & Wilson, 2019). This study broadens their well-informed advantages to include the practice of CBT in Thailand. However, the information gleaned from the interview reveals that during the actual translation process the volunteers did not take into account the thematic approach (Esfehani & Walters, 2018) because it is too detailed and the pilot project's time is limited.

Empowering Villagers of CBT

As the result of the pilot study informs, the local communities can be put at the centre as one of the main features of CBT. By doing so, crowdsourcing translation as a form of social empowerment can strengthen relationships and support among stakeholders, as suggested by Dolezal & Novelli (2020). This concurs with the ideas advanced by Phumpho and Nomnian (2019) and Nomnian et al. (2020). Apart from relying on a bilingual tour guide, the local villagers of CBT with limited English proficiency can also learn from this CBT translation project by participating in reviewing process (particularly the third and fifth step in the proposed model) to improve their English communication skills, especially in relation to topics such as community life and local diet.

This chapter asserts that as English is the prevalent *lingua franca*, ICT services and machine translation technologies should be made accessible for group language use, learning, and schooling in order to empower and support CBT communities.

Therefore, the present project can serve as a foundation for CBT-related government policies, agencies, and businesses to strengthen the collaboration between the tourism industry and local community leading to shared benefits among stakeholders, as suggested by Kattiyapornpong & Chuntamara (2020). This crowdsource translation initiative, in addition to the Thai AirAsia project of teaching English in CBT communities, will undoubtedly benefit the villagers of Phrom Lok District and other state-supported CBT areas around the world.

All in all, although it is a partial application of the model—the tourism community has not yet been involved and there are some limitations (e.g., inadequate number of volunteers)—the reflections from the translators are positive, particularly regarding how the project was run. What is apparently needed is feedback from tourism practitioners.

CONCLUSION

This chapter discusses the advantages and disadvantages of three key models regarding the translation of tourist texts: translation in thematic analysis (Esfehani & Walters, 2018), cultural-conceptual translation (2019), and Kató community (Translator without Borders, 2020) in order to make recommendations for a conceptual model of crowdsourcing translation for CBT. This model, along with many tourist promotional materials in Thailand, is a practical source of teaching materials for translation in a group, as suggested by Marczak (2018, p. 3), who posits that the digital pedagogical method should be encouraged for translation studies students to learn and work out solutions to the translation tasks they embark upon. This accords with suggestions by Pym (2011, 2013), who argues that educators should take advantage of digital technologies at any appropriate stage of learning. Students of translation studies are also encouraged to use technologies of their choice in the most profitable way and in as many settings as they possibly could.

The technology with which they familiarize themselves during the training program affords students the chance to gain work experience and a professional profile, especially through collaboration among translators, which is rapidly becoming the norm in the translation industry. Online communication through preferred platforms, without predetermined sets of translation tools or times, can free students of the burden of performing their tasks in only one locale. Even while travelling, they could be learning by this translation-*cum*-volunteer project for CBT – a reciprocal benefit for all stakeholders, as long as quality assurance is present, which the proposed model offers. Thus, this conceptual model is expected to yield fruitful results, corresponding with the aims of the Thai AirAsia project and the government's policy of promoting and sustaining CBT in Thailand.

This chapter aims to shed some light on the use of crowdsourcing translation to boost CBT in tourist areas where there seem to be language barriers between the hosts and the foreign tourists. This online collaborative translation will not only ensure that the true meaning of the tourism texts is not missing but will also empower community members and volunteer translators who can facilitate international communication in the hope that the sociolinguistic and cultural gaps can be bridged for the enhancement of CBT in remote areas in Thailand.

REFERENCES

Afdal, T. (2016). Consulting external resources during post-editing a machine translated tourism text by an Indonesian translator. *Translation Journal, 19*(3). https://translationjournal.net/July-2016/consulting-external-resources-during-post-editing-machine-translated-tourism-text-by-an-indonesian-translator.html

Agorni, M. (2012). Tourism communication: the translator's responsibility in the translation of cultural difference. *Pasos. Revista de turismo y patrimonio cultural, 10*(4), 5-11. https://www.redalyc.org/pdf/881/88123115002.pdf

Alfer, A., & Zwischenberger, C. (Eds.). (2020). Translaboration. Exploring collaboration in translation and translation in collaboration. Special issue of Target, Volume 32, Issue 2. John Benjamins.

ASEAN Secretariat, . (2016). *ASEAN community based tourism standard*. ASEAN Secretariat.

Attride-Stirling, J. (2001). Thematic networks: An analytic tool for qualitative research. *Qualitative Research, 1*(3), 385–405. doi:10.1177/146879410100100307

Bowker, L. (2015). Computer-aided translation: Translator training. In S.-W. Chan (Ed.), *The Routledge encyclopedia of translation technology* (pp. 88–104). Routledge.

Braun, V., & Clarke, V. (2006). Using thematic analysis in psychology. *Qualitative Research in Psychology, 3*(2), 77–101. doi:10.1191/1478088706qp063oa

Brown, H. (2018). Rethinking agency and creativity: Translation, collaboration and gender in early modern Germany. *Translation Studies, 11*(1), 84–102. doi:10.1080/14781700.2017.1300103

Charoenkongka, A., & Nomnian, S. (2019). Needs analysis of communicative English through community-based tourism: A case study of Koh Klang community in Krabi province. *Journal of International Students, 9*(2), 211–232.

Cherifi, B., Smith, A., Maitland, R., & Stevenson, N. (2014). Destination images of non-visitors. *Annals of Tourism Research, 49*, 190–202. doi:10.1016/j.annals.2014.09.008

Cordingley, A., & Manning, C. F. (Eds.). (2016). *Collaborative translation: From the Renaissance to the digital age*. Bloomsbury.

Dolezal, C., & Novelli, M. (2020). Power in community-based tourism: Empowerment and partnership in Bali. *Journal of Sustainable Tourism*, 1–19. Advance online publication. doi:10.1080/09669582.2020.1838527

Dolezal, C., Trupp, A., & Bui, H. T. (Eds.). (2020). *Tourism and development in Southeast Asia*. Routledge. doi:10.4324/9780429264191

Du, J., Way, A., Qiu, Z., Wasla, A., & Schaler, R. (2015). Domain adaptation for social localisation-based SMT: A case study using Trommons platform. In S. O'Brien, & M. Simard (Eds.), *Proceedings of the 4th Workshop on Post-Editing Technology and Practice* (pp. 57–75). Retrieved October 18, 2020, from https://amtaweb.org/wpcontent/uploads/2015/10/MTSummitXV_WPTP4Proceedings.pdf

Esfehani, M. H., & Walters, T. (2018). Lost in translation? Cross-language thematic analysis in tourism and hospitality research. *International Journal of Contemporary Hospitality Management, 30*(11), 3158–3174. doi:10.1108/IJCHM-10-2017-0701

Estellés-Arolas, E., & González-Ladrón-De-Guevara, F. (2012). Towards an integrated crowdsourcing definition. *Journal of Information Science, 38*(2), 189–200. doi:10.1177/0165551512437638

Hartley, A. (2009). Technology and translation. In J. Munday (Ed.), *The Routledge companion to translation studies* (pp. 106–127). Routledge.

Howe, J. (2006, July 6). The rise of crowdsourcing. *Wired*. Retrieved October 18, 2020, from https://www.wired.com/2006/06/crowds

Huang, J. (2017). The dining experience of Beijing roast duck: A comparative study of Chinese and English online consumer reviews. *International Journal of Hospitality Management, 66*, 117–129. doi:10.1016/j.ijhm.2017.07.003

Inpeng, S., & Nomnian, S. (2020). The use of Facebook in a TEFL program based on the TPACK model. *LEARN Journal: Language Education and Acquisition Research Network Journal, 13*(2), 369–393.

Inphoo, P., & Nomnian, S. (2019). Dramatizing a Thai northeastern folklore to lessen high school students' communication anxiety. *PASAA, 57*(1), 33–66.

Jhaiyanuntana, A., & Nomnian, S. (2020). Intercultural communication challenges and strategies for the Thai undergraduate hotel interns. *PASAA*, *59*(1), 204–235.

Jiménez-Crespo, M. A. (2017). *Crowdsourcing and online collaborative translations: Expanding the limits of translation studies*. John Benjamins. doi:10.1075/btl.131

Joanna. (2020a, November 10). All about revision tasks. *Translators without Borders*. https://community.translatorswb.org/t/all-about-revision-tasks/25147

Joanna. (2020b, November 10). What are chunks? *Translators without Borders*. https://community.translatorswb.org/t/what-are-chunks/25146

Journey, D. (2021). *Promlok...Best Shade of Khao Luang 2D1N*. http://journey-d.com/trip/phrom-lok-nakhon-si-2d1n?lang=en

Jun, W. (2019). The Application of cross-cultural awareness in English translation of tourist attractions. *International Conference on Humanities, Cultures, Arts and Design (ICHCAD 2019)*. https://www.webofproceedings.org/proceedings_series/ART2L/ICHCAD%202019/AD051.pdf

Kattiyapornpong, U., & Chuntamara, C. (2020). Assessing sustainable community-based tourism development in Thailand. In S. K. Walia (Ed.), *The Routledge Handbook of Community-Based Tourism Management* (pp. 397–414). Routledge. doi:10.4324/9780429274664-37

Koc, E. (2020). *Cross-cultural aspects of tourism and hospitality: A services marketing and management perspective*. Routledge. doi:10.4324/9781003018193

Lo, Y.-C., & Janta, P. (2020). Resident's perspective on developing community-based tourism – A qualitative study of Muen Ngoen Kong Community, Chiang Mai, Thailand. *Frontiers in Psychology*, *11*, 1493. doi:10.3389/fpsyg.2020.01493 PMID:32848976

Manuel. (2019, October 1). Glossaries in Kató TM. *Translators without Borders*. https://community.translatorswb.org/t/glossaries-in-kato-tm/10955

Marczak, M. (2018). Translation Pedagogy in the Digital Age. How digital technologies have been altering translator education. *Angles. New Perspectives on the Anglophone World*, (7). doi:10.4000/angles.895

MateCat. (n.d.). *How to use glossary terms while translating*. https://site.matecat.com/support/translating-projects/use-glossary-terms-translating

Mkono, M. (2012). A netnographic examination of constructive authenticity in Victoria Falls tourist (restaurant) experiences. *International Journal of Hospitality Management, 31*(2), 387–394. doi:10.1016/j.ijhm.2011.06.013

Napu, N. (2018). Does commissioning process matter? Understanding translation quality with a case of tourism promotional texts. *Journal of Language and Linguistic Studies, 14*(4), 352–363.

Napu, N. (2019). Tourism promotional materials: Translation problems and implications on the text's effectiveness for tourism promotion. *The 1st International Conference on Education, Social Sciences and Humanities (ICESSHum 2019).* 10.2991/icesshum-19.2019.71

Nitikasetsoontorn, S. (2015). The success factors of community-based tourism in Thailand. *NIDA Development Journal, 5*(2), 24–58.

Nomnian, S., Trupp, A., Niyomthong, W., Tangcharoensathaporn, P., & Charoenkonka, A. (2020). Language and community-based tourism: Use, needs, dependency, and limitations. *Austrian Journal of South-East Asian Studies, 13*(1), 57–79. doi:10.14764/10.ASEAS-0029

Nord, C. (2018). *Translation as a purposeful activity: Functionalist approaches explained* (2nd ed.). St Jerome.

Nuchdang, Y. (2019). Thailand development according to Pracharat's policy along with people's participation in Chachoengsao province. *NRRU Community Research Journal, 13*(1), 13–25.

O'Brien, S. (2011). Collaborative translation. In Y. Gambier & L. Van Doorslaer (Eds.), Handbook of Translation Studies Volume 2 (pp. 17-20). John Benjamins. doi:10.1075/hts.2.col1

O'Hagan, M. (2009). Evolution of user-generated translation: Fansubs, translation hacking and crowdsourcing. *The Journal of Internationalization and Localization, 1*(1), 94–121. doi:10.1075/jial.1.04hag

Pavasiriporn, R., & Juntaronanont, P. (2017). A study on Chinese services related to tourism in Bangkok inner Rattanakosin Island. *Journal of Chinese Language and Culture, Huachiew Chalermprakiet University, 4*(1), 119-128. https://so02.tci-thaijo.org/indcx.php/clcjn/article/view/194066

Phomnimitkul, K., Putjorn, T., & Vechpong, T. (2020). The culture management of the second-tier provinces to support local experience-based tourism under the LINK project. *Journal of Liberal Arts, Prince of Songkla University, 12*(1), 200-226. https://so03.tci-thaijo.org/index.php/journal-la/article/view/244540/165633

Phumpho, R., & Nomnian, S. (2019). Challenges of Thai businesspeople using English in ASEAN. *Kasetsart Journal of Social Sciences, 40*(3), 743–750.

Prayag, G., Cohen, S. A., & Yan, H. (2015). Potential Chinese travellers to western Europe: Segmenting motivations and service expectations. *Current Issues in Tourism, 18*(8), 725–743. doi:10.1080/13683500.2013.868413

Pym, A. (2011). What technology does to translating. *Translation and Interpreting, 3*(1), 1-9. http://www.trans-int.org/index.php/transint/article/view/121/81

Pym, A. (2013). Translation skill-sets in a machine translation age. *Meta, 58*(3), 487–503. doi:10.7202/1025047ar

Ribeiro, M. A., Pinto, P., Silva, J. A., & Woosnam, K. M. (2017). Residents' attitudes and the adoption of pro-tourism behaviours: The case of developing island countries. *Tourism Management, 61*, 523–537. doi:10.1016/j.tourman.2017.03.004

Sarod, T., & Kraisame, S. (2019). Urban multilingualism: A case study of the relationship between linguistic landscape and business in North Nana. *Journal of Language and Culture, 38*(1), 24–40. https://so03.tci-thaijo.org/index.php/JLC/article/view/241369/163900

Shani, A., & Uriely, N. (2012). VFR tourism: The host experience. *Annals of Tourism Research, 39*(1), 421–440. doi:10.1016/j.annals.2011.07.003

Sharifian, F. (2011). *Cultural conceptualisations and language: Theoretical framework and applications*. John Benjamins. doi:10.1075/clscc.1

Skibitska, O. (2015). The language of tourism: Translating terms in tourist texts. *Translation Journal, 18*(4). https://www.translationjournal.net/October-2015/the-language-of-tourism-translating-terms-in-tourist-texts.html

Su, M. M., Long, Y., Wall, G., & Jin, M. (2016). Tourist–community interactions in ethnic tourism: Tuva villages, Kanas Scenic Area, China. *Journal of Tourism and Cultural Change, 14*(1), 1–26. doi:10.1080/14766825.2014.976228

Suebwongsuwan, W., & Nomnian, S. (2020). Thai hotel undergraduate interns' awareness and attitudes towards English as a lingua franca. *Indonesian Journal of Applied Linguistics, 9*(3), 704–716. doi:10.17509/ijal.v9i3.23221

Sulaiman, M. Z. (2016). The misunderstood concept of translation in tourism promotion. *Translation & Interpreting*, *8*(1), 53–68. doi:10.12807/ti.108201.2016.a04

Sulaiman, M. Z., & Wilson, R. (2019). *Translation and tourism: Strategies for effective cross-cultural promotion*. Springer. doi:10.1007/978-981-13-6343-6

SYSTRAN. (n.d.). *What is machine translation? Rule-based machine translation vs. statistical machine translation*. https://www.systransoft.com/systran/translation-technology/what-is-machine-translation

Tourism Authority of Thailand. (2011). *Nakhon Si Thammarat*. Retrieved from https://www.turismotailandes.com/wpcontent/uploads/2015/07/Nakhon%20Si%20Thammarat.pdf

Translators without Borders. (2017, June 15). *Translators without Borders and The Rosetta Foundation are merging*. https://translatorswithoutborders.org/translators-without-borders-and-the-rosetta-foundation-are-merging

Translators without Borders. (2020). *TWB Kató*. https://kato.translatorswb.org

Vanijdee, A. (2015). Translating Thai cultural information into language learning activities: the case of English for local museum personnel. *ASEAN Journal of Education, 1*(1), 53-68. https://so01.tci-thaijo.org/index.php/AJE/article/view/180800

Wang, B., & Ping, Y. (2020). Perceptions of machine translation and computer-aided translation by professionals and the general public: A survey study based on articles in professional journals and in the media. *International Journal of Translation, Interpretation, and Applied Linguistics*, *2*(2), 1–14. doi:10.4018/IJTIAL.20200701.oa1

Wu, M. Y., & Pearce, P. L. (2014). Tourism blogging motivations: Why do Chinese tourists create little 'lonely planets'? *Journal of Travel Research*, *55*(4), 537–549. doi:10.1177/0047287514553057

Xu, C. (2019). Cross-culture in tourism English translation. *Education Research Frontier*, *9*(3), 106–109.

Zhang, H., Cho, T., Wang, H., & Ge, Q. (2018). The influence of cross-cultural awareness and tourist experience on authenticity, tourist satisfaction and acculturation in World Cultural Heritage Sites of Korea. *Sustainability*, *10*(4), 927. doi:10.3390u10040927

ENDNOTES

1 This can be instantiated by the Code of Conduct for volunteer translators on Translators without Border's website (https://translatorswithoutborders.org/wp-content/uploads/2020/07/Plain-language-Code-of-Conduct-for-Translators-July-2020.pdf).

2 https://www.worldatlas.com/articles/what-are-the-differences-between-a-river-and-a-canal.html

Chapter 8
Application of Geoinformation Technology to the Management of Community-Based Natural Resources for Tourism Development in Northern Tanzania

Norbert John Ngowi
https://orcid.org/0000-0003-0885-7597
Mzumbe University, Tanzania

ABSTRACT

The approaches to natural resources management have evolved. Disparities in their adoption are likely to produce a long-lasting negative impact on the resources and the livelihood security of the community depending on them. The use of geoinformation by the local community is a critical measure to the sustainability of its resources. Nonetheless, the application of geoinformation technologies to the community-based natural resources for the tourism industry is highly unknown. This chapter reviewed the application of geoinformation technology to the management of community-based natural resources in the Pangani District of Northern Tanzania. It considers how geoinformation technology is used in the management of tourism activities for community development. Specifically, the chapter discusses community developments resulting from that as well as challenges associated with the use of geographical information systems and remote sensing technologies. The chapter concludes with key recommendations for improving those challenges.

DOI: 10.4018/978-1-7998-7335-8.ch008

BACKGROUND

The application of geoinformation technology for socio-economic development and management of natural resources in particular has been on the increase in most parts of the developed world. However, it remains exceptionally challenging in developing countries. Studies by Azzari et al. (2013), Pena-Regueiro et al. (2020), Sebastiá-Frasquet et al. (2014), and Sumari et al. (2018) show the relevance of geoinformation technology, such as Geographic Information System (GIS) and Remote Sensing (RS) in geodata collection, analysis and planning, fostering economic development, improving health, and in mitigating conflicts in various parts of the world. Studies by Kumar et al. (2015), Quan et al. (2001), Rahman (2010), Turk & Gumusay (2004), and Verka & Angelina (2008) show that RS and GIS have been used in many areas of the tourism sector including - planning, digital mapping, showing data trends, and logistics, among different uses. In Tanzania, the technology has been used in several sectors of the economy such as in the population census by NBS (2013), tourism management and in the land revenue collection. In relation to the management of natural resources for instance, a study by Kironde (2009) shows that GIS and RS have been used in land surveys, registration, transfer of land rights and/or transaction, and land dispute resolution mechanisms. However, despite the benefits, little is known about the application of geoinformation technologies in the management of community-based natural resources for tourism and community development at the village level. This chapter relies mainly on the literature to investigate how geoinformation technology is used for the management of community-based natural resources for tourism development. Specifically, it discusses community developments resulting from that as well as challenges associated with the use of GIS and RS technologies. The review is within the context of rapid advances in geoinformation technology. The chapter concludes with recommendations for improving the utility of GIS and RS technologies for tourism development in Pangani district.

Community-Based Natural Resources Management (CBNRM) is the governance initiative of integrating local community into the management of the environment with the aim of meeting cultural, social, and ecological goals at the local level (Agrawal & Gibson, 1999; Yeatman, et al. 2003). In the early stages, CBNRM was highly pushed by development partners through a neo-liberal approach of reducing state control and influence on natural resources. Worldwide, since 1990s, a number of global, regional and local efforts aiming to promote CBNRM in forests, water, fisheries, and land have been recognised with policy changes. Holden et al. (2014) and Kellert et al. (2000) insist that CBNRM have to balance ecology, economy and cultural goals. For instance, in Nepal, community Water Users Group (WUG) and Forest Users Group (FUG) have been used over 30 years for managing irrigation systems and forests, respectively (Pant et al., 2005). In Malawi, co-management is used in Community Fisheries Management (CFM) and Community-Based Wildlife Management (CBWM) (Zulu, 2012).

The use of CBNRM in Tanzania is not an entirely new idea. Decentralising power to local user groups was adopted from the 1990's when the governement opted for devolution strategies. This focuses on reforming the land tenure system to increase the capacity of local authorities to support communities to manage their resources (Duvail et al., 2006; Heffernan, 2020; URT, 2007). Today, different names have been ascribed to CBNRM for the different types of natural resources. In fisheries, the term Beach Management Units (BMUs) is used while Participatory Forest Management (PFM) is used in the forest sector, in both Community Based Forest Management (CBFM) and Joint Forest Management (JFM). The wildlife sector is manifested in Community Based Conservation (CBC) or Wildlife Management Areas (WMAs) through community land ownership which has been existing since 2002 (Pailler, 2015).

Pailler et al. (2015) show that the CBNRM approach in Tanzania has provided the local community with the opportunity to protect natural resources. It is established that through Wildlife Management Area (WMA), encroachment along the buffer zones of the protected areas in Selous, Mikumi, and Udzungwa parks have declined; hence, conservation has been enhanced through communities participation in the management of wildlife resources. Other community benefits included improvement of the community livelihoods through Income Generating Activities (IGAs) in rural area such as selling of tourism – related goods and services, revenue collected in the wildlife sector, and enhanced people's livelihood. The benefits were recorded by Wilfred (2010) in Uyumbu and Ipole WMAs, and USAID (2020) in Makame WMA where 5% of the revenue collected amounting to US$ 43,000 was dedicated to elephant protection. However, grazing land attraction has increased livestock populations causing human-wildlife conflicts over resource use (Wilfred, 2010). Akunaay et al. (2003) show that Community-Based Tourism (CBT) is significant in reducing poverty among the rural population, improving rural economy, and increasing diversification of the tourism sector in Tanzania.

Although CBNRM is an area that has attracted a considerable number of research over the past few years, the implication of the application of the geoinformation technologies with involvement of community for tourism development remains largely unknown. Therefore, this chapter aims to answer the question: *how has geoinformation technology been applied to the community-based natural resources for tourism development in northern Tanzania?.*

LITERATURE

Tourism includes the activities in which a person or a group of persons travel and stay in places outside of their usual environment for not more than one consecutive year either for leisure, business, education or other purposes (URT, 1999). Camilleri (2018), identifies two types of tourism: domestic tourism and international tourism. In recent times, tourism has become a significant driver of the national economy and youth employment (Christie et al., 2012). In many parts of Africa, for instance, tourism has increasingly become one of significant potential sources of revenue for national economies as well as for addressing poverty among the rural poor (Akunaay et al., 2003; Kinyondo & Pelizzo, 2020; Kyara et al., 2021; Lwoga, 2017 & 2013; Steyn & Spencer, 2013; Tonya & Matilda 2019). Community Based Tourism abbreviated hereto as CBT is defined as the type of tourism that takes take into consideration, the environment, social systems, community ownership, community culture, and the benefits from tourism activities for sustainable development (see Asker et al., 2010; Nelson, 2003 & Suansri, 2003). Community Based Tourism (CBT) requires partnership and/or joint agreement with central government institutions (Nelson, 2003). CBT activities are supervised and owned by the community members for their benefits with the overall goal of raising environmental awareness for increased income through visitors/tourists (Suansri, 2003). Yanes et al. (2019) report that CBT is based on management paradigms such as self-reliance and empowerment of local community to participate in tourism activities on a sustainable basis. Nguyen (2007), on the other hand, shows that ecological tourism is the fastest growing segment of the tourism industry. It differs from traditional tourism as it benefits the community around, the tourist attraction and protects environment.

Tourism in Tanzania is the second most important sector after the manufacturing industry, contributing about 10.7% of national Gross Domestic Product (GDP) which is equivalent to about US$ 6,600 million (Kyara et al., 2021). For instance, the north-eastern Tanzania zone plays an important role in the national economy because the main tourism attractions are found there. These include Mount Kilimanjaro, Ngorongoro Crater, Manyara, Tarangire and Serengeti National. However, tourism in northern Tanzania remains primarily focused on conventional destinations within state-managed national parks due to increased encroachment and overutilisation of natural resources. In tourist areas and in the six World Heritage Sites of Serengeti, Ngorongoro, Selous, Kilimanjaro, Zanzibar Stone, Kilwa Kisiwani and Songo Mnara Ruins and Zanzibar Archipelago, tourists are attracted by natural resources like water, landscapes, coastlines and biodiversity (URT, 2001). In Tanzania, cultural tourism contributes more to the socio - economic wellbeing of rural communities because many tourists are searching for broader experiences that are not found in wildlife resources. The growth of this sector therefore enhances the economy of the local communities through viable tourism activities at campsites, traditional catering establishments and historical sites. Maasai communities are the national pioneers in cultural tourism as they were the first to initiate it in the Ngorongoro area by producing art, jewelry, beaded handicrafts, and dancing to traditional songs to attract tourist to the Ngorongoro District (Carr & Melubo, 2018). Other examples of active CBT activities in Tanzania are found in the Ngorongoro, Monduli, Simanjiro, Babati, Mbulu, and Kiteto Districts. For instance Nelson (2003) reports that the Ololosokwan Village CBT in Ngorongoro District earns annually over US$ 55,000 through ecotourism. Also, Carr & Melubo (2018) reports that tourism activities in the Ngorongoro district contribute about 4% of the district revenue. Part of the income is given back to the community to be used in schools, healthcare centres, village offices, constructions, payment of health insurances and school fees for children from poor families in Loliondo and Ngorongoro districts. The same is used for paying the salaries of village game scouts who conduct surveillance against poaching activities (Mgonja et al., 2015). CBT also promotes national cultures for more than 120 ethinic groups in Tanzania. However, the growth of CBT in the country has been hampered by a number of factors. The problems include legitimacy of some of activities and/or initiatives. Other challenges are poor road networks, insufficient communication networks, unreliable electricity and inadequate clean water supply (Mgonja et al., 2015). Geoinformation technology particularly, the GIS and RS technologies have become important tools for the management of natural resources such as land, wildlife, and forests (Eslami & Azad, 2016). Mcconchie & Mckinnon (2002) and Oisebe (2012) report that GIS and RS are used mainly at the mapping of natural resources, making inventory and updating the natural resources data. This is done by production of land resource maps which guide the development of management plans as the decision making tools. Further, Quan et al. (2001) highlight that community participation during information storing enhances the production of comprehensive maps. However, Mwape (2010) argues that GIS is not only useful in showing the community locations of interest to tourism destinations but also it provides information on poaching and the wildlife movement (Mambo & Makunga, 2017). On the other hand, Chen (2014), Eslami & Azad (2016), and Sonti (2015), show that GIS can indicate locations and sizes of different plants, rivers and lakes within the given ecosystem, and can as well monitor environmental trends. In addition, GIS and RS are used in ecotourism activities (especially those concerning undisturbed natural areas) for education, and nature exploration (Diamantis, 1999). They also organise ecotourism information and characteristics of customers for planners and developers (Rahman, 2010).

Worldwide, GIS and RS have been widely used in monitoring and mapping of natural resources (See Kumar et al., 2015; Quan et al., 2001; Rahman, 2010;Turk & Gumusay, 2004; Verka & Angelina, 2008). The aim is to design a management plan for natural resources conservation. Due to the increase in pressure on natural resources, GIS and RS have emerged as potential tools for uncovering natural resource changes (Kumar, et al., 2015). In the first palce, GIS and RS have been used in land use suitability analysis (Ayhan et al., 2020; Store & Kangas, 2001) and in tourism planning projects that intergrate GIS software with photograph, free text, video and sound (Bahaire & Elliott-White, 1999). Also, they are used in mapping the risk assessment (Coppock, 1994; Mwape, 2010). Moreover, GIS and RS help in the design of early disaster warning systems, planning evacuation routes, and posting disaster risk assessment for reconstruction (Toharudin et al., 2021).

GIS and RS play a major role in tourism management through tourism research, planning, marketing and tourism development. Turk & Gumusay (2004) and Sipkes (2002) explain that GIS is used in tourism development and wildlife which is a key aspect of tourism. According to Herricks & Fisher (2003), it is easy to manage and assess a wildlife when their locations are identified and their numbers are evaluated. Thus, GIS and RS also support in examining wildlife habitat requirements; ranging populations of wildlife, establishing disease levels and reserving historical and present population of wildlife database.

In the Mafia Island Archipelago of Tanzania, flipper tags (a component of GIS) are mounted on the tails of sea turtles to monitor their seasonal behavior. Through this, local tour guides in Juani village know the location of sea turtles and direct tourists to the perfect destination/area where many turtles are available. Eventually, tourism revenue has significantly increased for the Juani village through the CBT involving marine species. This ecotourism initiative has resulted in improvements in school and health services delivery. The youth have also benefited from employment opportunities such as local tour guiding activities (Muir & Sense, 2005). In Zanzibar and in many parts of mainland Tanzania, the mobile GIS App and GIS T App are used by tour guides and tourists in accessing services such as local bars and restaurants, boutiques, guesthouses, and cultural heritage sites (Buberwa & Msusa, 2019). Elsewhere, in Ethiopia, GIS has been used to monitor and protect Ethiopian wolves which serve as tourist attractions (ESRI, 2010). Similarly, GIS Apps are used in Indonesia by Muslim tourists for easy location of mosques within local communities (Afnarius et al., 2020). Other scholars (Kumar et al., 2015; Quan et al., 2001; Rahman, 2010; Turk & Gumusay, 2004; and Verka & Angelina, 2008) have also used GIS and RIS to study the various components of tourism as further discussed in this chapter.

METHODS

Description of Study Area

The United Republic of Tanzania (URT) is a country within the African Great Lakes region. It is a union of two independent states which are Tanzania Mainland (formerly known as Tanzanyika) and Zanzibar, covering an area of 947,303 square kilometers. The 2012 Population and Housing Census reported that there were more than 45 million persons (NBS, 2013). However, recently the World Bank (2018) estimated 56.32 million people. Land on the mainland Tanzania (which is the focus of this chapter), is estimated to be nearly 88 Million Hectares (Ha) - covering water and other forms of landscapes (Kilahama, 2008). The land sector in the country has been contributing significantly to the national economy and livelihoods of the rural poor through mining, agriculture, fishing, tourism activities among others. While, land use planning is the driving force to the sustainability of land resources in the country, URT (1997) affirms that land use planning in many districts has been very slow. Kironde (2009) partly attributes this slow spead of land use planning to guidelines which define land management and other associated challenges including: legal pluralism, degradation of land resources, insecurity of the land tenure, state control over land, land disputes, and gendered land rights, among others.

Pangani District (which is the case study) is a gateway to the interior tourism attraction sites along the coast of the Indian Ocean of North Eastern Tanzania. The district is located between $5^0$15' and 6^0 S, and between 38^0 35' and 39^0 E) (See Figure 1). The altitude of the area ranges from 0 to 95 m above sea level [masl] (Pangani District Council, 2003, 2018). Pangani district covers an area of about 1,830.8 square kilometers with an estimated population of 54,025 persons in 2012, with 26,870 (49.7%) males and 27,155 (50.3%) females (NBS, 2013). The average household size was about 4 persons and the population density was about twenty four (24) persons per sq. km.

The district experiences both hot and humid climate in the three main rainy seasons differing in intensities with more rain in the higher interior areas. The main rain occurs between March and June with an average annual rainfall of 1000 mm and a variation between 600 mm and 1400 mm. The short rains occur between October and December while occasional showers extend between June and September.

The average temperature is 29^0C with a variation from 24^oC to 33^oC experienced in the hottest months i.e. December and February, while May to June are the coolest months. The average range of diurnal humidity is between 81% to 68% in the morning and evening. Soil types are predominantly sandy along the seashores and sandy loams in the higher areas. Clay loams, and clays are found in depressions in the landscape but are prone to water logging during the rains; hence, create seasonal

Figure 1. Map of Pangani district
Source: Pangani District Council, 2003

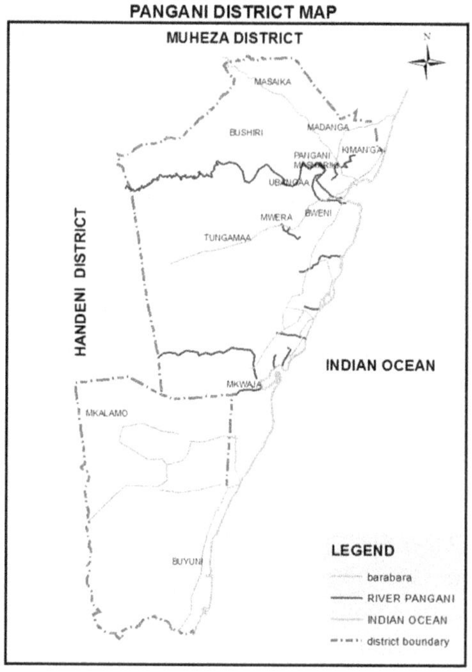

swamps. On the other hand, the district is endowed with potential land consisting of coconut and cashew nut and the sisal-maize agro-ecological zones. The land use falls in the three main categories of ownership, namely village land (60.3%), reserved land (34.6%), and the general land (5.1%) (Table 1).

The district natural resources are highly diverse, including coastal and mangrove forests, national parks, agriculturally fertile land for sisal and plantation forests, lime mining, consequently high land use demand for tourism and economic development (Pangani District Council, 2018).

Data Collection and Analysis

A review of the literature was the main method used to determine the application of GIS and RS technology in the management of community-based natural resources for tourism development in the study area (Snyder, 2019). About 69 field related literature were reviewed: 50 journal articles, 5 books, 6 reports, 6 policy documents, and 2 web pages. Moreover, the detailed procedure for systematic review (SR) and meta-analysis (MA) framework as reported by Tawfik et al. (2019) was followed. To conclude the findings with confidence, the review method was combined with

Table 1. Land use in the district

Land use	Hectares	(%)
Settlements	3,843	2.1
Sisal estates	14,287	7.8
Fruit trees and mixed cropping	10,511	5.7
Continual and rotational annual crops	12,444	6.8
Rice	1,098	0.6
Grazing land	23,790	13.0
Msubugwe Forest Reserve	9,337.1	5.1
Bush land	94,959.9	51.9
Mangrove forests	3,660	2.0
Wetlands	9,150	5.0
Total	183,080	100

Source: Pangani District Council (2003)

information gathered through personal communication with the natural resources department in Pangani District. This method was preferred due to the dynamic nature of the tourism industry.

Five steps were used in undertaking the systematic review process. The first step was corroboration of the study question to get a deeper insight of contribution of geoinformation technology and factors affecting CBNRM for CBT activities in the country. The second step was exploration of databases for the relevant sources of information. This step involved developing key words (geoinformation technology, CBT in Tanzania, GIS and tourism, RIS and tourism, among others. The key words were later used to search from Google Scholar enough paper articles for conducting the analysis. The inclusion criteria were: all up-to-date paper articles that contain information answering the research question or the chapter area of focus, target audience, enough data or the comparison between two studied interventions. But the most important criterion was for papers freely downloadable in pdf format. To reduce bias resulting from missing literature, manual searching of the policy documents at specific shops selling Government books was conducted. The third step involved cleaning of data. Publications available in texts or abstracts only or lacking wider coverage of the chapter were excluded in the analysis. The fourth step was about checking different versions of data abstracts developed in the previous step for a comprehensive representation of the information. The author checked if all cited literatures appear appropriately to the list of references; this included verifying: name of the author, year and title of publication, volume, pages, publisher, proper capturing of the type of the literature (paper, book, website, etc.). Step five was the final full-text screen; the manual was defined and applied in the interpretation of the full data set of the study findings.

Limitation of the Methods Used

Although the combination of the methods used in the systematic review and analysis is recommended, comprehensive search of data was strained with time, resources, and internet connectivity.

RESULTS AND DISCUSSION

The Status of Community-Based Natural Resources for Tourism Development

Community-Based Natural Resources Management as a model for tourism development has been promoted worldwide. In low-icome and countries with transition economies in particular, it is considered to have started in the late 1970s (Stone & Rogerson, 2011). Akunaay et al. (2003), Anderson (2011), Kinyondo & Pelizzo (2020), Kyara et al. (2021), Steyn & Spencer (2013), Tonya & Matilda (2019), and URT (1999), highlighted that about 10.7% of the GDP come from tourism. Although total exports from tourism has grown steadily in some selected coutries in Africa such as Angola, Ethiopia, Equatorial Guinea, Ivory coast and Togo, the situation in Tanzania is different (Kinyondo & Pelizzo, 2020). This is due to some challenges hindering the country from fully taping its tourism potentials. They include: low level of community engagement in tourism planning and policy implementation (Steyn & Spencer, 2013); lack of diversification beyond scenic and wildlife based attractions, high annual deforestation rates of about 900,000 hectares, expansion of human activities close to protected areas, and degradation of natural resources (Akunaay et al., 2003). Anderson (2011) pointed out that lack of tourism products diversity, for instance beach tourism in few areas like Zanzibar, and accessibility of tourist attraction sites among others, accounts for the low share of tourism when compared to competitors like Kenya, and South Africa. Whereas the literature reviewed has established among others, the significance of geoinformation in the promotion of tourism activities in many parts of the world, very few have discussed local community capacity in using GIS and RIS technologies for planning and managing community tourism activities. This chapter aims to contribute to this knowledge gap because of its significance in enhancing sustainability of natural resources for tourism.

Pangani District which is the focus of this chapter, like other areas in the country (Akunaay et al., 2003; Tonya & Matilda, 2019) is endowed with different tourism attraction sites. These include the coastal environment, forests, sandy beaches, coral reefs, highland marine reserves, historical sites, National Parks along its large game populations, as well as the old Pangani town. These sites offer great tourism opportunities including snorkelling, jogging, beach football, diving, and fishing, as well as watching green tortoises and dolphins *(Plates A and B)*. Others are boating and sightseeing of historical sites such as German monuments along the Mauya valley *(Plate C)*. Also, watching wildlife bathing in the Indian Ocean *(Plate D)*. These tourism activities do not only generate revenue but also they are culturally and ecologically friendly (Figure 2).

Figure 2. Community tourism sites (A) fishing (B) sandy beach (C) mangrove forest (D) wildlife

Community Forest Reserve is a type of forest category where its entire system of management is under the community themselves. All decisions are made through village assembly and implemented by the community. The forest is used for different purposes including timber production as well as non-timber products like medicinal plants, thatch, water, and wildlife (URT, 2007). Under community forest management, Pangani district has a number of forests which harbour different types of birds and tree species including *Acacia* and mangroves (Table 2).

Through Joint Forest Management [JFM] (URT, 2007), the local community is involved in the management of natural resources in collaboration with the central government authorities. The output is the Joint Management Agreement (JMA) that spells out how the costs and benefits of managing the forest resources should be shared between the forest owner and the manager. There are a number of forests that fall under JFM arrangement in the district (Table 3).

Table 2. Community forests

Forest name	Village	Hectares	Status	Remarks
Mawata	Mkwaja	4,953.34	Surveyed	Not gazzetted
Kwatango	Mtango	1,092.43	Surveyed	Not gazzetted
Misakazi	Mtango	1,242.54	Surveyed	Not gazzetted
Beho	Mseko	3,577.91	Surveyed	Not gazzetted
Kwasinge	Kigurusimba	331.37	Surveyed	Not gazzetted
Kibubu	Mivumoni	3266	Surveyed	Not gazzetted
Mtonga	Mtonga	1,187.53	Surveyed	Not gazzetted
Kwavinonde	Msaraza	82.57	Surveyed	Not gazzetted
Tengwe	Mkalamo	Not known	Not surveyed	Not gazzetted
Kwezichacha	Mtango	Not known	Not surveyed	Not gazzetted
Mbulizaga	Mbulizaga	Not known	Not surveyed	Not gazzetted
Buyuni	Buyuni	Not known	Not surveyed	Not gazzetted

Source: Pangani District Council (2018)

Table 3. Forests under joint arrangement in the district

Forest name	Village	Forest type	Hectares	Status
Msubugwe	Mtango, Mtonga and Mseko	Coastal	4,410	Assessed
Garafuno	Mtango	Coastal	195	Assessed
Kumba	Pangani Magharibi	Mangrove	193.94	Assessed
Mwembeni	Mwembeni	Mangrove	219.4	Assessed
Bweni	Bweni	Mangrove	137.2	Assessed
KISA	Kipumbwi and Sange	Mangrove	422	Assessed
Mkwajuni	Mkwajuni	Mangrove	827.19	Assessed
Ushongo Mabaoni	Ushongo	Mangrove	291.12	Assessed
Ushongo Mtoni	Ushongo	Mangrove	32.69	Assessed
Mkwaja	Mkwaja	Mangrove	466.9	Assessed
Buyuni	Buyuni	Mangrove	466.6	Assessed

Source: Pangani District Council (2018)

Use of Geoinformation Technologies in the Management of Natural Resources

Geoinformation for Tourism Development

The application of geoinformation technologies in Natural Resources Management for tourism development has been studied in many parts of the world (Kumar et al., 2015; Quan et al., 2001; Rahman, 2010; Turk & Gumusay, 2004; and Verka & Angelina, 2008). Kumar et al. (2015) established that multi-temporal, spectral and spatial resolutions of the GIS and RIS do offer great opportunity for management and monitoring of forests and wildlife resources in Andaman, India. For instance, they established that RIS and GIS are useful for the generation of cover, human activities, and wildlife suitability data for some wildlife specific species in the context of changing human population. Verka & Angelina (2008) reported that various GIS components like ArCGIS have facilitated the development of Cajetina, Nova Varos, as well as Zlatibor and Zlatar areas potentially known for tourism in Serbia. Also, RIS and GIS have contributed to the growth and expansion of tourism infrastructure to include: integration of tourist attractions in planning, service delivery and transportation. This was possible due to the ability of GIS to provide and identify data trends using digital maps, internet and mobile mapping through digital files. Quan et al. (2001) showed that GIS has been an important tool in stakeholder participation, planning, data collection, and integration of social and natural science issues using spatial database in Brazil, Ghana, Uganda and Tanzania repectively.

On the other hand, Rahman (2010) reported that GIS has been used as a decision making tool for sustainable ecotourism development in different areas worldwide. In Bangladesh, the study shows that there has been little application of GIS technology in tourism research although it seem to be have increased in the field of leisure because of the GIS ability to manipulate both spatial and attribute data. In the US, UK, and Canada, GIS has been used for the planning of tourism and ecotourism activities, location identification (particularly in the dense forests), facility management, destination monitoring, services management, tourism resources inventory, marketing, and research. The results from a study conducted in Eminonu District, in Turkey, by Turk & Gamusay (2004), showed that querying in GIS design, has helped in determining suitable places for tourist attractions including areas suitable for locating hotels, suitable places for sightseeing, and the shortest route to these areas.

In 1999, participatory mapping of village boundaries with the objective of identifying individual land parcels was piloted in Mbozi district, Tanzania. After mapping, villages were able to apply for land certificates for various uses. Subsequently, gathering data for unplanned areas in the Local Government Authorities for issuance of residential licenses was piloted in the Dar es Salaam Region between 2004 and

2005 (URT, 1999). Likewise, in Pangani district, this exercise was introduced in 2005. Land surveyors from nearby Handeni district and the Ministry of Lands were involved. Introduction of new procedures for data management such as survey and mapping, collection using GIS recording property rights facilitate land transactions. Currently, over 90% of villages own land certificates. With the help of GIS and RS technologies, the land property register can be produced within the district.

Dispute Resolution among Land Users

Conflicts arising from land uses (ownership and boundaries) are a major problem in many parts of Tanzania (Kironde, 2009). The Ministry of Land and Human Settlement Development reports on daily basis the use of GIS technology in data collection, storage, analysis and reporting which has helped to resolve conflicts in many parts of the country. Adoption of GIS technology has helped in the establishment of land tribunals. Although these tribunals were in place in some villages and wards at Pangani district, members were not trained to execute their duties and responsibilities. Due to this weakness existing in the land, governance institutions at the district, villages and wards, barely offer their services including resolution of land disputes leaving many complaints unsolved.

In Pangani, as with most districts in the mainland Tanzania, land conflicts are mainly between different land users; agriculturalists and livestock keepers, villages and reserve land, as well as over boundaries and individual plots ownership (Pangani District Council, 2003). In some villages like Msaraza, Kigurusimba, Mikinguni, Stahabu, Mtonga, Mtango, Mbulizaga, Mkalamo and Mkwaja, conflicts between farmers and pastoralists have dominated for many years with no permanent solution. Even villages with land use plans like Kigurusimba, stahabu, Msaraza, Mtonga, Mtango, Mbulizaga and Mkalamo have not been able to resolve land use conflicts due to lack of enforcement.

Challenges of GIS and RS Application to the Management of Tourism Activities

Although GIS and RIS technologies are rapidly growing tools for managing tourism activities in many countries, a number of factors are reported to have limited the utility of these tools. Whereas Dunn et al. (1997) for instance, emphasise the significance of the availability of personnel with GIS technical expertise in all organisations, a comparison of studies conducted elsewhere show that GIS has ability of handling huge data sets and integrating various discplines of natural resources in a more holistic way (Kumar et al., 2015). Turk & Gumsay (2004) also show that Arc Info Software has been instrumental in solving problems caused by the lack of standards across the data especially in the process of tranfering spatial data to GIS. In Bangladesh, Rahman (2010) found that application of GIS tools has aided planning of tourism activities which are complex and more dynamic. However, despite the fact that Pangani district has enormous tourism potentials, it faces a number of major challenges not only related to RIS and GIS. For instance, Anderson (2011) show that main challenges for the growth of tourism in Tanzania include lack of tourism products diversity, poor marketing strategy, low linkages within local economy, insufficient institutional and technical capabilities, and shortage of appropriate skills. The challenges in the district which make it fully adopt geoinformation include lack of skilled personnel in the tourism industry. As a result, district tourism guidelines to guide tourism activities in the district are not respected. For instance, available statistics in the district show that out of 13 personnel required for the full functioning of the land department, which is responsible for natural resources management, there is a deficit of eight (8) staff equivalent to 61.5% in total while there is no single personnel with GIS and RS speciality in the cartography profession (Table 4).

Table 4. Human resources capacity

Section	Available	Deficit	Required (Total)
	Number and %	Number and %	Number and %
Land section	2 (40%)	3 (60%)	5 (100%)
Town and village planning	1 (50%)	1 (50%)	2 (100%)
Land survey	2 (50%)	2 (50%)	4 (100%)
Cartography	0 (0%)	2 (100%)	2 (100%)
Total	5 (38.5%)	8 (61.5%)	13 (100%)

Source: Pangani District Council (2018)

Lack of awareness, non-involvement of local communities in the participatory planning and decision-making in tourism activities have made local communities reluctant in setting aside areas for tourism investment. Furthermore, inadequate transport infrastructure and lack of clean and safe water have hampered efforts to promote tourism in the district. In addition, shortage of GIS and RS knowledge/skilled personnel as shown in Table 4 have led to inefficiencies and ineffective application of GIS and RS technologies in people's day-to-day activities. The lack of GIS and RS skills and knowledge has led to the inability to obtain relevant information for the production of different reports required for community-based natural resources management including tourism related development activities.

Dunn et al. (1997) report that GIS operations require a certain minimum of hardware. However, it is noted that apart from the computerised system for revenue collection from land as well as tourism, the Pangani district does not have sufficient hardware and software (computers printer, plotter, digitizer or scanner) and reliable electricity to enable GIS and RS technologies to work efficiently as far as natural resources management is concerned. Similarly, there is the problem of inadequate internet infrastructure in the district has resulted in limited access to data, leading to delays in obtaining information on tourism services. Inadequate funding is another bottleneck facing the natural resources department in the district. For instance, between 2018/2019 and 2019/2020, only about 8 mil. Tshs and 5 mil. Tshs (equivalent to US$ 3478 and 2174) respectively were allocated to the department. A similar trend was also observed in the budget allocated for tourism activities. The annual budget was meager in comparison to the needs. It was not sufficient to execute the main categories of natural resources management of land services of surveying and registration, transfer of rights, and conflict/dispute resolution. Although GIS infrastructure was reported as a limiting factor in Pangani, Quan et al. (2001) show that factors such as the scale used, compartibility of dataset formats, accurancy of recording in the observation points, and accuracy in entering the attribute data need to be controlled for quality GIS data.

CONCLUSION AND RECOMMENDATION

This chapter analyses the application of GIS and RS technology to selected community-based natural resources for tourism development in northern Tanzania using a documentary review methodology complemented with personal communication. It was found that the application of geoinformation technology was low. The effect of low adoption affected accrued benefits from natural resources including land registration, improved security of tenure, access to credits from financial institutions and control over natural resources for community based activities.

The chapter recommends that to overcome the main challenges of applying GIS and RS technology to community-based natural resources management for tourism development in the Pangani area, some strategies should be employed. These are recruitment of personnel who are skilled in geoinformation, development of internet infrastructure for easy access of some of the GIS software freely available online, and improving existing internet network infrastructure to boost weak internet connection to fully support GIS technology. Moreover, encouragement of various stakeholders with resources and expertise to invest in GIS and RS technology for community based tourism development in the district is recommended.

REFERENCES

Afnarius, S., Akbar, F., & Yuliani, F. (2020). Developing web-based and mobile-based GIS for places of worship information to support halal tourism: A case study in Bukittinggi, Indonesia. *ISPRS International Journal of Geo-Information*, *9*(1), 1–18. doi:10.3390/ijgi9010052

Agrawal, A., & Gibson, C. (1999). Enchantment and disenchantment: The role of community in natural resource conservation. *World Development*, *27*(4), 629–649. doi:10.1016/S0305-750X(98)00161-2

Akunaay, M., Nelson, F., & Singleton, E. (2003, December). Community based tourism in Tanzania: Potential and perils in practice. In *Second Peace Through Tourism Conference* (pp. 7-12). Academic Press.

Anderson, W. (2011). Challenges of tourism development in the developing countries: The case of Tanzania. *Journal of Tourism*, *12*(2), 17–36.

Asker, S., Louise, B., Carrard, N., & Paddon, M. (2010). *Effective Community Based Tourism : (Issue June)*. Sustainable Tourism Cooperative Research Centre.

Ayhan, Ç., Cengi̇z Taşlı, T., Özkök, F., & Tatlı, H. (2020). Land use suitability analysis of rural tourism activities: Yenice, Turkey. *Tourism Management, 76*(July). doi:10.1016/j.tourman.2019.07.003

Azzari, M., Zamperlin, P., & Landi, F. (2013). GIS in geography teaching. *Journal of Research and Didactics in Geography, 2*(2), 27-42.

Bahaire, T., & Elliott-White, M. (1999). The application of geographical information systems (GIS) in sustainable tourism planning: A review. *Journal of Sustainable Tourism*, *7*(2), 159–174. doi:10.1080/09669589908667333

Buberwa, R. F., & Msusa, A. (2019). Developing a mobile gis tour guide app for dar-es-salaam city, Tanzania. *International Journal of Scientific and Technology Research*, *8*(4), 237–242.

Camilleri, M. A. (2018). The Tourism Industry. *An Overview*, (October), 3–27. Advance online publication. doi:10.1007/978-3-319-49849-2_1

Carr, A., & Melubo, K. (2018). *Developing indigenous tourism in the bomas: Critiquing issues from within the Maasai community in Tanzania*. Academic Press.

Chen, D. (2014)... *Application Of Gis in Environmental Impact Assessment., 994*, 4855–4860. doi:10.4028/www.scientific.net/AMR.989-994.4855

Christie, I., Fernandes, E., Messerli, H., & Twining-Ward, L. (2012). *Tourism in Africa: The World Bank harnessing tourism for growth and improved livelihoods.* https://www.worldbank.org/content/dam/Worldbank/document/Africa/Report/africa-tourism-report-2013-overview.pdf

Coppock, T. (1994). *GIS and Natural Hazards: An overview from a GIS perspectives.* Academic Press.

Diamantis, D. (1999). The concept of ecotourism: Evolution and trends. *Current Issues in Tourism*, *2*(2–3), 93–122. doi:10.1080/13683509908667847

Dieke, P. U. (2000). Developing tourism in Africa: Issues for policy consideration. *Development Policy Management in Sub-Saharan Africa*, *7*(1), 19–26.

Dunn, C. E., Atkins, P. J., & Townsend, J. G. (1997). GIS for development: A contradiction in terms? *Area*, *29*(2), 151–159. doi:10.1111/j.1475-4762.1997.tb00017.x

Duvail, S., Hamerlynck, O., Nandi, R. X., Mwambeso, P., & Elibariki, R. (2006). Participatory Mapping for Local Management of Natural Resources in Villages of the Rufiji District, Tanzania. *The Electronic Journal on Information Systems in Developing Countries*, *25*(1), 1–6. doi:10.1002/j.1681-4835.2006.tb00167.x

Eslami, A., & Azad, I. (2016). *Implementation of GIS in Natural Resources.* Academic Press.

ESRI. (2010, Oct.). GIS for Wildlife Management. *GIS Best Practices,* 40. https://www.esri.com/library/bestpractices/wildlife-management.pdf

Heffernan, A. (2020). *Community-Based Natural Resource Management and Global Climate Change in Namibia.* Academic Press.

Herricks, E., & Fisher, J. (2003). Application of GIS Technologies to Wildlife Management at DFW Airport. *2003 Bird Strike.* http://digitalcommons.unl.edu/cgi/viewcontent.cgi?article=1005&context=birdstrike2003

Holden, E., Linnerud, K., & Banister, D. (2014). Sustainable development: Our common future revisited. *Global Environmental Change*, *26*, 130–139. doi:10.1016/j.gloenvcha.2014.04.006

Kellert, S. R., Mehta, J. N., Ebbin, S. A., & Lichtenfeld, L. L. (2000). Community natural resource management: Promise, rhetoric, and reality. *Society & Natural Resources*, *13*(8), 705–715. doi:10.1080/089419200750035575

Kilahama, F. (2008). Impact of increased charcoal consumption to forests and woodlands in Tanzania. *Forest Ecology and Management*, *308*, 45–55.

Kinyondo, A., & Pelizzo, R. (2020). Socioeconomic Impact of Tourism: The Case of Tanzania. *African Journal of Economic Review*, *8*(2), 239–257.

Kironde, J. L. (2009). Improving land sector governance in Africa: the case of Tanzania. Proceedings of "Land governance in support of the MDGs: Responding to New Challenges.

Kolios, S., Vorobev, A. V., Vorobeva, G., & Stylios, C. (2017). *GIS and environmental monitoring. In Applications in the marine, atmospheric and geomagnetic fields*. Springer International Publishing AG.

Kumar, N., Yamaç, S. S., & Velmurugan, A. (2015). Applications of remote sensing and GIS in natural resource management. *Journal of the Andaman Science Association*, *20*(1), 1–6.

Kyara, V. C., Rahman, M. M., & Khanam, R. (2021). Tourism expansion and economic growth in Tanzania: A causality analysis. *Heliyon*, *7*(5), e06966. doi:10.1016/j.heliyon.2021.e06966 PMID:34027171

Lwoga, M. N. B. (2013). Tourism development in Tanzania before and after Independence: Sustainability perspectives. *The Eastern African Journal of Hospitality, Leisure and Tourism*, *1*(2), 1–26.

Lwoga, N. B. (2017). Local collaboration network in management of built heritage. *Journal of Cultural Heritage Management and Sustainable Development*, *7*(3), 226–243. doi:10.1108/JCHMSD-05-2016-0034

Mambo, P. W., & Makunga, J. E. (2017). Application of Remote Sensing and Gis for Assessing Land Cover Resources Variability in the the Selous Game Reserves, Tanzania. *European Journal of Technology*, *1*(2), 74. doi:10.47672/ejt.227

Mcconchie, J., & Mckinnon, J. (2002). Using GIS to Produce Community-Based. *Asean Biodiversity*, 27–34.

Mgonja, J. T., Sirima, A., Backman, K. F., & Backman, S. J. (2015). Cultural community-based tourism in Tanzania: Lessons learned and way forward. *Development Southern Africa*, *32*(3), 377–391. doi:10.1080/0376835X.2015.1010710

Muir, C., & Sense, S. (2005). *The status of marine turtles in the United Republic of Tanzania, East Africa. Sea sense*. Tanzania Turtle & Dugong Conservation Programme.

Mwape, A. (2010). *Geographical Information Systems and Natural Resource Management in Zambia*. http://mro.massey.ac.nz/bitstream/handle/10179/1657/02_whole.pdf?sequence=1

National Bureau of Statistics (NBS). (2013). *2012 Population and housing census distribution by administrative areas*. Ministry of Finance.

Nelson, F. (2003). *Community-based tourism in Northern Tanzania: Increasing opportunities, escalating conflicts and an uncertain future*. Rozenberg Publishers.

Nguyen, D. M. (2007). *Building a GIS database for ecotourism development in Ba Vi District*. Ha Tay Province.

Oisebe, P. R. (2012). *GIS and natural resource management*. GIS Lounge.

Pailler, S., Naidoo, R., Burgess, N. D., Freeman, O. E., & Fisher, B. (2015). Impacts of community-based natural resource management on wealth, food security and child health in Tanzania. *PLoS One*, *10*(7), e0133252. doi:10.1371/journal.pone.0133252 PMID:26186210

Pangani District Council. (2003). *Investment Profile*. Unpublished.

Pangani District Council. (2003). *Land Annual Report*. Unpublished.

Pangani District Council. (2018). *Annual Report*. Unpublished.

Pant, D., Thapa, S., Singh, A., Bhattarai, M., & Molden, D. J. (2005). *Integrated management of water, forest and land resources in Nepal: opportunities for improved livelihood* (No. 618-2016-41137). Academic Press.

Pena-Regueiro, J., Sebastiá-Frasquet, M. T., Estornell, J., & Aguilar-Maldonado, J. A. (2020). Sentinel-2 Application to the Surface Characterization of Small Water Bodies in Wetlands. *Water (Basel)*, *12*(5), 1487. doi:10.3390/w12051487

Quan, J., Oudwater, N., Pender, J., & Martin, A. (2001). GIS and Participatory Approaches in Natural Resources Research. *Socio-Economic Methodologies for Natural Resources Research. Best Practice Guidelines.*, 1–36.

Rahman, M. (2010). Application of GIS in Ecotourism Development: a case study in sundarbans. Academic Press.

Sebastiá-Frasquet, M. T., Altur, V., & Sanchis, J. A. (2014). Wetland planning: Current problems and environmental management proposals at supra-municipal scale (Spanish Mediterranean Coast). *Water (Basel)*, *6*(3), 620–641. doi:10.3390/w6030620

Sipkes, J. (2002). Use of GIS and remote sensing in East Africa. *GIM International, 16*(3), 6–9.

Snyder, H. (2019). Literature review as a research methodology: An overview and guidelines. *Journal of Business Research, 104*, 333–339. doi:10.1016/j.jbusres.2019.07.039

Sonti, S. (2015). Application of Geographic Information System (GIS) in Forest Management. *Journal of Geography & Natural Disasters, 5*(3). Advance online publication. doi:10.4172/2167-0587.1000145

Steyn, J. N., & Spencer, J. P. (2013). Are sustainable tourism policies and strategies working in Tanzania?: Tourism and hospitality. *African Journal for Physical Health Education, Recreation and Dance, 19*(2), 299–312.

Stone, M. T., & Rogerson, C. M. (2011). Community-based natural resource management and tourism: Nata bird sanctuary, Botswana. *Tourism Review International, 15*(1-2), 159–169. doi:10.3727/154427211X13139345020570

Store, R., & Kangas, J. (2001). Integrating spatial multi-criteria evaluation and expert knowledge for GIS-based habitat suitability modelling. *Landscape and Urban Planning, 55*(2), 79–93. doi:10.1016/S0169-2046(01)00120-7

Suansri, P. (2003). *Community based tourism handbook*. Responsible Ecological Social Tour-REST.

Sumari, N. S., Shao, Z., & Kira, E. (2018). Challenges and opportunities for the advancement of GIS education in Tanzania. *Journal of Education and Practice, 8*(28), 67–75.

Tawfik, G. M., Dila, K. A. S., Mohamed, M. Y. F., Tam, D. N. H., Kien, N. D., Ahmed, A. M., & Huy, N. T. (2019). A step by step guide for conducting a systematic review and meta-analysis with simulation data. *Tropical Medicine and Health, 47*(1), 1–9. doi:10.118641182-019-0165-6 PMID:31388330

Toharudin, T., Suprijadi, J., Caraka, R. E., Pontoh, R. S., Chen, R. C., Lee, Y., & Pardamean, B. (2021). Social Vulnerability and How It Matters: A Bibliometric Analysis. *International Journal of Criminology and Sociology, 10*, 610–619. doi:10.6000/1929-4409.2021.10.71

Tonya, E. M., & Matilda, L. (2019, April). Unexplored Tourist Sites in Tanzania: Experience of Mbeya and Songwe Regions. In *International Conference on the Future of Tourism (ICFT)* (pp. 1-17). OUT.

Turk, T., & Gumusay, M. U. (2004, July). GIS design and application for tourism. In *XXth ISPRS Congress* (pp. 12-23). Academic Press.

United Republic of Tanzania (URT). (1997). *National Land Policy*. Ministry of Land and Human Settlement Development.

United Republic of Tanzania (URT). (1999). *National Tourism Policy*. Ministry of Natural Resources and Tourism.

United Republic of Tanzania (URT). (1999). *Land Act*. Government Printers.

United Republic of Tanzania (URT). (2001). *Overview of the Tourism Sector 2001*(May). URT.

United Republic of Tanzania (URT). (2007). *Community Based Forest Management Guidelines; for the establishment of village land forest reserves and community forest reserves. Ministry of Natural Resources and Tourism*. Forest and Beekeeping Division.

United Republic of Tanzania (URT). (2007). *The Wildlife Policy of Tanzania*. Government Printer.

USAID. (2020). *Making Wildlife Management Areas Deliver for Conservation and Communities*. USAID.

Verka, J., & Angelina, N. (2008). The application of GIS and its components in tourism. *Yugoslav Journal of Operations Research*, *2*(18), 261–272.

Wilfred, P. (2010). Towards sustainable wildlife management areas in Tanzania. *Tropical Conservation Science*, *3*(1), 103–116. doi:10.1177/194008291000300102

World Bank. (2018). *The World Bank Annual Report 2018*. The World Bank.

Yanes, A., Zielinski, S., Diaz Cano, M., & Kim, S. I. (2019). Community-based tourism in developing countries: A framework for policy evaluation. *Sustainability*, *11*(9), 2506. doi:10.3390u11092506

Yeatman, L., Du Toit, M., & Andreasen, L. (2003). *Developing approaches for improved natural woodland resource use and management: Lessons learnt in the communal areas of Bushbuckridge (Part 2)*. DWAF.

Zulu, L. C. (2012). Neoliberalization, decentralization and community-based natural resources management in Malawi: The first sixteen years and looking ahead. *Progress in Development Studies*, *12*(2-3), 193–212. doi:10.1177/146499341101200307

KEY TERMS AND DEFINITIONS

Community-Based Natural Resource Management: Is the management of natural resources by the community user group on village land or in co-management with the Central Government on state land or open areas, and applies equally to forestry, fisheries, wetlands and wildlife.

Community-Based Tourism: Refers to the tourism activities planned and implemented on the land owned by the villages in partnership with the local community.

Decentralization by Devolution: Describes recent local government reforms to empower administration at the local governance level.

Decentralized Natural Resource Management: A process of devolving management to the local government of natural resources = CBNRM under D*D.

District Council: A district council comprises a body of the elected, political representation of the ward, who govern the district and its development activities.

Joint Management: A general term to describe how a community can collaborate with central or local government in the management of gazetted natural resource areas (ie forest, fish, wetlands, marine or wildlife) reserves. A state of "co-management" prevails, based on Joint Management Agreements (JMA), Joint Management Plans (JMP), etc.

Participatory Forest Management: Is a term in Community Based Natural Resource Management of forest resources.

Village Assembly: This is the meeting of all adult members of the village held at least four times a year, and described in the Local Government (District Authorities) Act, 1982 as 'the supreme authority in the village'.

Wildlife Management Area: Means the whole or part of village land which has been placed under the management authority of a Village Council (and or CBO, an Authorized Association) for purposes of conservation and sustainable utilization of natural resources (ie forest, fish, wetlands or wildlife).

NOTE

The author's translation is supported with the definitions adopted from the literature (United Republic of Tanzania, 2012 and Akunaay et al., 2003).

Conclusion

Implications of Current Developments for Sustainable Community-Based Tourism Development

Tourism as a business and industry has been largely driven by economic interests. It is the economic prospects such as revenue, employment, and infrastructure development associated with tourism development that makes the industry attractive to most destinations and investors. Although the economic prospects are imperative, it has skewed attention from other equally important issues connected with the tourism development process. One critical perspective which this book seeks to project is the host community's involvement in tourism development. Host communities cannot be taken out of any discussion regarding tourism development. This is because they play very critical roles in the tourism development process. Murphy (1985) identifies host communities as the nucleus of the tourism product. He opines that since host communities are the most affected by tourism development, involving them in the development process will result in a win-win situation for both communities and the tourism industry. The extant literature has extensively highlighted the instrumental role played by host communities in the tourism development process.

The host community is considered very crucial in shaping visitors' choice of destinations and their experiences (Murphy, 1985; Nunkoo & Ramkissoon, 2011). Furthermore, the overall success and sustainability of tourism and its related initiatives (Martin, de los Salmones Sanchez & Herrero, 2018) are contingent on the host community's support. The host community's participation in tourism development helps to control the pace of the development process, properly integrate tourism into the local economy and deliver a more fulfilling tourist experience (Murphy, 1965). To attain the needed support from host communities, it is important to understand tourism and its related issues from their perspective. Yet, studies have shown that host communities have not been adequately involved in the tourism development process (Tosun, 2000; Bello, Lovelock & Carr, 2017).

This brings to the fore the subject of community-based tourism (CBT). CBT is a model of tourism development that emphasizes the importance of the host community.

It has also over the years proven to be a pathway to achieving environmental conservation, poverty reduction, improved livelihoods, and community development. However, the actual implementation of CBT is not without challenges. The efficacy of CBT in delivering improved livelihoods, poverty alleviation and environmental conservation in multiple contexts have been questioned as many CBT initiatives have failed to deliver on their objectives. In the light of increasing advocacy for tourism to be developed more sustainably and more community involvement in the tourism development process, the discourse on the prospects and effectiveness of CBT will continue to be relevant. The next section reflects on the different perspectives reflected in the various chapters of the book. This is followed by the future outlook for CBT and the implications for sustainable CBT development.

PERSPECTIVES FROM THIS BOOK

This book highlights the intricacies of community-based tourism. The cases presented in the various chapters do not encompass all the issues regarding CBT. Nevertheless, they provide clear examples of major themes that require the attention of all stakeholders. Chapters 1 and 2 set the tone for the book by providing a review of the extant literature, discussing the benefits, challenges, and prospects of CBT which is the central theme of the book. In all the chapters of this book, there is an acknowledgement of the prospects of CBT with possible lessons and best practices that could be learnt and replicated in other contexts (chapter 3). But there are still gaps between what CBT stands for and what is actually delivers. The cases presented in chapters 4 and 5 touches on the issue of communities' involvement in tourism development. The issue of active host community participation in tourism development remains one of the cardinal arguments of CBT. Yet in many instances, it appears to be more of a rhetoric than reality. The same was observed in the cases presented in chapters 4 and 5 where the communities were passively engaged in the decision-making processes regarding tourism development in their spaces. The presence and control of external forces such as government agencies over the CBT process can reduce the benefits that accrue to the host community, affecting their support for the project and undermining the sustainability of the entire initiative. An important theme that emerged in chapter 4 is the role of indigenous knowledge and values in the tourism development process. The indigenous knowledge possessed by host communities could be harnessed for product development and environmental conservation.

Fairness and equity in the sharing of tourism benefits are of much concern to host communities. Chapter 6 provides an understanding of the issue by discussing the literature on distributive justice in CBT development. The justice theme is equally reflected in the cases discussed in chapters 4, 5, 7. To gain host communities' support for CBT, there should be fairness in how the benefits derived from tourism are shared. In the absence of that sense of equity or perceived justice in the sharing of the associated benefits, host communities will be apathetic towards tourism development. Technology should be embraced and utilized by CBT projects. Chapters 8 and 9 have indicated how this could be possible with the aid of Geoinformation technology and crowdsourcing translation technologies. Technology could be used in solving operational problems and enhancing visitors' experiences. The contributions in this book confirm the challenges and prospects of employing CBT for livelihood enhancement, environmental conservation, and enhancement of visitor experiences. It also highlights how technological innovations and the application of distributive justice could help tackle the challenges confronting CBT. The cases presented in this book provide valuable lessons for CBT initiatives in other jurisdictions.

THE FUTURE OUTLOOK

Though there have been several studies on CBT, there are still many research gaps that need to be addressed. The various perspectives expressed in this book point to the fact that CBT is still evolving and, in the future, it would be further shaped by forces such as changing demographics, technological advances, growing demand for nature-based tourism, dwindling external funding as well as health and safety. The ongoing COVID-19 pandemic and its impact on the tourism industry has raised questions about the adaptive capacity and resilience of host communities and CBT initiatives to crises.

Changing Demographics

A noticeable feature of the contemporary world is population growth. A total of 82 million people were added to the global population to reach 7.7 billion in 2019 (UNCTAD, 2020). Much of the population growth is however occurring in the developing world. More than 90% of the global population growth occurred in developing economies, mainly Asia and Oceania (UNCTAD, 2020). Parts of the developing world that have experienced population growth have equally experienced explosive growth in travel. China especially has become a dominant economic force and a major source market for travel. It is expected that all other things being equal, the growth in population coupled with improved economic growth will lead to an overall increase in travel in general and travel to natural areas in particular. However, such unprecedented growths in population and tourist arrivals could result in the overuse of natural attractions thereby negatively affecting natural resources.

According to Mckee et al. (2004), population growth coupled with more people having social and economic opportunities to travel can lead to the irreversible and inevitable destruction of the earth's natural resources which is perceived as one of the greatest threats to species biodiversity and ecosystem function.

Also, most of the CBT projects are located in rural communities in developing countries. In such destinations, where there are limited economic opportunities and the people largely subsist on natural resources. Hence, population growth will put pressure on the natural resources used for CBT. Changing demographics also manifest in changes in household composition. In the major tourism source markets in Europe and North America, households are getting smaller. Sole-parent and single-person households are becoming more prevalent in those countries. This has been attributed to higher divorce rates and low fertility rates. Household composition affects tourists' choices of nature-based tourism activities (Tangeland and Aas, 2011).

Another demographic change that has implications for travel to natural environments, is an ageing population. Improved healthcare and advances in medicine have resulted in longer life expectancy among people in developed countries. This, coupled with reduced fertility rates has accounted for an ageing population in Europe, North America and Japan which are the main countries of origin of international tourists (Buckley et al., 2015). Older people tend to travel to natural environments. A report on travel trends by the Association of British Travel Agents [ABTA] (2015) has indicated that older tourists seek more active holidays, participate in sporting holidays and travel to the mountains and lakes (ABTA, 2015). They value 'sight-doing' as opposed to sightseeing and this means being active, discovering, learning, and finding excitement in new experiences (Amadeus, 2014).

Growing Demand for CBT

The increasing demand for authentic travel experiences has been identified as one of the major global travel trends today, as an increasing number of tourists seek to immerse themselves in local cultures and environments and to generate a greater understanding of the cultures of destinations (World Bank Group, 2018). The fact is there has been a growing demand for travel to natural areas and this has provided opportunities for CBT development. A report by Euronews Travel (2020) on tourism trends after 2020 identifies wilderness tourism, ecotourism and authentic tourism as forms of tourism that will be in high demand. For the main European markets such as Germany, France, Netherlands, Belgium, Poland, and the Czech Republic, visiting nature has been identified as the primary motive for selecting tourist destinations by tourists (Centre for the Promotion of Imports, 2020).

The desire to be away from crowded places and to be in secluded and beautiful natural environments especially after the pandemic is a major reason for this. As a result of the COVID-19 pandemic, people will try to avoid mass tourism destinations in order to reduce the risk of contracting the virus. This will probably be the norm for some tourists even in the post-COVID era. People will also seek authentic cultural experiences travellers also want to leave positive footprints by giving something back to communities, whether through direct payments or volunteering (Euronews travel, 2020).

Technological Advances

Advances in technology will continue to shape the future of the tourism industry including CBT. Technology is and will continue to be a significant force driving change in the tourism industry. Social media has become an important tool for e-marketing which also facilitates electronic word-of-mouth and helps project destination image. Also, mobile technology and the internet have removed boundaries to communication and opened up remote and unknown areas for travel. This means that even the remotest and most rural communities will have to adopt technology in order to reach out and cater for tourists. The use of advanced technology will also enable tourist destinations to improve products and services rendered to their visitors and reduce their operating costs.

The growing influence of millennials who make up a significant proportion of the populations of source markets like the USA is one of the reasons. Millennials were born between 1980 and 1995 and are avid consumers of ICT products. For destinations to appeal to especially millennials, they have to deploy technology, especially ICT. Fortunately, millennials also have a penchant for travelling to communities to learn new things, develop themselves and support the local communities they visit.

Dwindling External Funding and Revenues

Many CBT projects in developing countries were established through donor funding. Donor agencies like the European Union (EU), United Kingdom Department of International Development (DFID), Netherlands Development Organisation (SNV), United States Agency for International Development (USAID) have been instrumental in the development of CBT projects across the globe. Thus, CBT projects in rural communities in the developing world have largely depended on donor funding. However, the devastating impact of COVID-19 is likely to result in a reduction in donor funding. Based on previous crises, it has been predicted that there will be a decrease in donor funding for conservation in Africa for about two years as a result of weakened economies and changing priorities of governments (Lindsey et al., 2020). Thus COVID-19 has resulted in a drastic decline in tourism revenue. This will not only affect the livelihoods of communities that depend on tourism but will also reduce the funds available for conservation. This is against the backdrop that travel restrictions have already led to declines in revenues from CBT and other services within the tourism value chain. According to Spenceley (2021), severe travel restrictions and park closures in the first half of 2020 led to a dramatic decline in the numbers of visitors to protected areas, outdoor spaces and World Heritage sites across the world, with many of these areas closed due to COVID-19. Saunders (2020). In South Africa for instance, the average future bookings for lodges and protected areas in May 2020 were 70% lower than in 2019; 61% of lodges were fully closed while 33% were only partially operational (Saunders, 2020). As the pandemic lingers on with an even more devastating third wave fueled by the delta variant, this situation will persist for a while and the recovery even after the pandemic has subsided, may take some time. This means that managers of CBT have to be more innovative in order to survive.

Growing Importance of Sanitation, Health and Wellness

The COVID-19 pandemic has drawn attention to the need for people to prioritize their health and wellness. In a COVID-19 and post-COVID eras, tourists will travel to places where their health and safety is guaranteed and will avoid crowded, polluted and dirty places. Some scholars have mentioned the healing properties of the sea, trees, and fossil-based treatments at destinations (Dryglas and Salamaga, 2018). People perceive that travelling to natural and therapeutic environments such as health resorts, national parks, beaches, rivers, and waterfalls will help improve their health and well-being. Nature-based attractions and activities have been found to satisfy the wellness-seeking motives of tourists (Woo et al., 2015). Thus, travel to natural environments helps promote the well-being of tourists. In fact, studies have shown that participating in nature-based tourism, leisure and recreation has positive effects on tourists' wellbeing and quality of life (Buckley, 2020; Campon-Cerro et al., 2020; Hanna et al., 2019). The demand for health and wellness travel will increase in a post-COVID-19 world as people seek to replenish their bodies and minds, destress and build immunity. CBT destinations could capitalize on their natural environments and tap into this growing market. However, tourists will need some assurance that destinations are safe to travel to. CBT destinations therefore need to take safety precautions to minimize the risk of people contracting diseases.

SUSTAINABLE COMMUNITY-BASED-TOURISM: IMPLICATIONS FOR POLICY AND PRACTICE

Sustainability of CBT

The prospects and challenges of CBT development, coupled with the future trends, present both opportunities and threats to the attainment of sustainable CBT.. Undeniably, the sustainability concept has become an important development model in the 21st century. Sustainable tourism has been defined as 'all forms of activities, management and development of tourism that preserve natural, economic, and social integrity and guarantee the maintenance of natural and cultural resources' (Niedziółka, 2014:160). Sustainable tourism emphasizes the long-term harmonious development among tourism and society, environment, resources, and local economies, and promotes the effective and sustainable use of resources, balancing the relationship between different stakeholders and ensuring fairness in intergenerational development opportunities (Guo et al., 2019).

Sustainable tourism means tourism must not be developed at the expense of environmental conservation, cultural preservation and the general well-being of society as a whole. According to the United Nations Environment Programme (UNEP) and United Nations World Tourism Organization [UNWTO] (2005), the bottom line of sustainable tourism is socio-cultural, economic, and environmental sustainability. Specifically, sustainable tourism development must conserve the cultural heritage and traditional values of host communities and enhance intercultural understanding. It must ensure the long-term economic viability of host communities and provide socio-economic benefits such as stable employment, income-earning opportunities, and poverty alleviation to all stakeholders in a fair manner. Furthermore, it should promote the optimal and wise use of natural resources, maintain essential ecological processes, and help to conserve natural heritage and biodiversity (United Nations Environment Programme and World Tourism Organization, 2005).

CBT has been described as a type of sustainable tourism (Yanes et al., 2019). This is because CBT is founded on the principles of sustainable tourism development, the triple-bottom-line of people, profits and planet. CBT must provide economic opportunities that enhance the quality of life of residents and protect the natural environment (Lee, 2013; Ohe & Kurihara, 2013). CBT is considered successful only when it has positive impacts on the social, economic and environmental circumstances of a community (Asker, et al., 2010; Tasci, Semrad, & Yilmaz, 2013). In effect, CBT should provide socio-economic opportunities for host communities without compromising the quality of the natural environment and the cultures of the communities. At the heart of sustainable CBT is the provision of life satisfaction sustainability realms such as material well-being, community well-being, emotional well-being, and health and safety well-being to community members (Woo et al., 2015). To enhance the wellbeing of communities, maximize the benefits to be derived by host communities, and in view of the future outlook, a truly sustainable CBT can be achieved through partnerships and collaborations, empowerment of local communities, adoption of technology and building resilience.

Partnerships and Collaboration

One of the reasons why some community-based projects have failed to live up to expectations is the varied and conflicting interests of different stakeholders. Stakeholders such as host communities, NGOs, private sector businesses, government and tourists tend to have their expectations which sometimes do not coincide with the objectives of the project. There is also the problem of local elites projecting their interests above the interests of marginalized groups (Knight & Cottrell, 2016). Also in many communities, there is government control of tourism development. However, the backing and collaboration of all stakeholders are necessary for sustainable CBT to be achieved (Okazaki, 2008). According to Stone (2015), collaboration is a key principle in community-based tourism approaches in developing countries. Hall (1996) argues that collaborative partnerships methodologies in tourism planning have the potential to take in a wide set of stakeholders, escalate political participation and social equality, and contribute to more sustainable forms of tourism. Through a participatory approach, effective collaboration towards CBT development could be achieved. To ensure that the views of all stakeholders are incorporated into the decision-making process, a local tourism development committee should be formed with membership drawn from various stakeholder groups and organizations. This according to Jamal & Getz (1995) is a recommended step for achieving local management and empowerment, though it could also pose a challenge to the distribution of widespread benefits for the entire community. To address the problem of governmental and elite control of tourism development in communities, governments should develop policy frameworks that guide the implementation and management of CBT activities through the active participation of all relevant stakeholders (Moswete, Thapa, & Child, 2012). A typical example is the collaborative management approach adopted by governments in some destinations like Ghana known as Community Resource Management Areas (CREMAs). Through this, local stakeholders are involved in the management of a particular resource or conservation area through some form of a negotiated agreement.

Community Involvement, Empowerment and Control

Most of the barriers to sustainable CBT result from unequal power relations within communities and between communities and other stakeholders. First, there is usually the problem of elite capture and control of CBT projects. Then there is also the issue of external influence and control by funding agencies, NGOs, governments and private enterprises. According to Wearing, McDonald, & Ponting, (2005), the problem of power relations is at the heart of the equitable and sustainable development of CBT. When power is concentrated in a central government authority it usually results in bureaucracy which delays decision-making. It also creates a system of dependency whereby local communities have to rely on external agencies. This is because current CBT models are 'overly reliant on Western "experts" and development agencies and that far too little attention has been paid to local and non-Western perspectives and knowledge' (Le et al., 2012, p. 362). In actual fact, the CBT refers to a form of tourism that is managed and controlled by the community (Leksakundilok & Hirsch, 2008). Giving local people control over tourism development enables them to maximize the benefits to be derived from tourism because their needs, concerns and problems are factored into the tourism project. Thus, communities must be empowered so that they can take full control over tourism development. Empowerment refers to a shift in the balance between the powerful and powerless, and the dominant and dependent (Sofield, 2003). As a result of empowerment, community members can determine their own priorities and actions for development (Cole, 2006). Han et al. (2014) posit that increased power is reflected in financial resources, skills and technical expertise, political influence, control over land and resources, community cohesion, and involvement in local planning and management. Empowerment of local communities will enhance their participation in tourism development and enable them to maximize the benefits to be derived from tourism. It will result in an improvement in their quality of life, the conservation of natural and cultural resources and the quality of the experience to be gained by tourists.

Adoption and Deployment of ICT

Technology has become one of the major forces driving change in the tourism industry. ICT has become a way of life for many people, including modern travellers who are always connected to the internet (Gretzel, Fesenmaier, & O'Leary, 2006). ICT is, therefore, an appropriate medium for engaging with tourists; marketing destinations, maintaining relationships with tourists, exchanging information and executing sales transactions. The proliferation of ICT, especially the internet, provides new opportunities for the marketing of CBT. ICT can help destinations to minimize costs and improve tourists' experiences. The adoption of ICT by destinations is particularly imperative in a COVID-19 era. COVID-19 has provided destinations with opportunities to employ ICT for marketing and operations. In a very competitive tourism industry, ICTs enable CBT destinations to maintain a global presence (Spencer et al., 2012). ICT has made every part of the world, including even the most remote and isolated tourism operations cheaper and more widely accessible (di Castri & & Balaji, 2002). Even the remotest destinations could become visible and accessible through well-developed websites. Advances in ICT has given birth to the concept of e-Tourism. This is the use of information and communication technology to increase the efficiency of tourism operations, provide various tourism services to customers, and make the implementation of tourism marketing more accessible in the form of Telematics (Condratov, 2013). In the same vein, ICTs are being deployed in rural communities for CBT operations, and is commonly referred to as e-CBT. According to di Castri & Balaji (2002), e-CBT has the potential for achieving the objectives of pro-poor tourism for two important reasons: First, it reverses the current pattern whereby much of the tourism industry is controlled by external financial interests. Secondly, it promotes micro-enterprise tourism which acts as a catalyst for the promotion of other local economic activities like community fisheries, traditional agriculture, handicraft production, and conservation practices, as well as helping to enhance the quality of natural and cultural resource utilisation. Also, e-CBT can disintermediate and deconstruct the tourism value chain and thereby drive income closer towards the local communities which are the actual providers of tourism experiences (UNCTAD, 2001).

Another significant trend resulting from advances in ICT that has a direct bearing on CBT is virtual tours which allow people to experience attractions in the comfort of their homes. Virtual tours of protected areas and natural environments have gained popularity globally in recent times. WildEarth, for instance, had a fivefold increase in viewers of its virtual safaris in the first two weeks of March 2020 (Cocks, and Tassiem, 2020). Though the increase in popularity of virtual tours could be attributed to the COVID-19 pandemic which has resulted in a number of travel restrictions, this trend has become part of the new normal and will continue into the post-COVID era. Destinations must seize this opportunity and leverage ICT for the realization of sustainable CBT. For local communities to adopt and employ ICT, they will require digital access capability not only in terms of material access but also product usage skills and content among other things (Minghetti & Buhalis, 2010). However, rural economies often experience a digital divide (Minghetti & Buhalis, 2010). CBT projects are usually located in rural communities in developing countries where the internet connection is usually unstable. This is obviously a challenge to the adoption of ICT for sustainable CBT development but governments must work towards bridging this digital divide.

Facilitating Recovery and Building Resilience

The COVID-19 pandemic has dealt a devastating blow to the tourism industry and reversed its fortunes. Travel restrictions, flight cancellations and other protocols occasioned by the outbreak of the pandemic have resulted in a plummet in tourist arrivals, declines in tourism revenues and job losses. CBT destinations have been seriously impacted because they usually depend on inbound tourists from the main generating markets in Europe and North America. The development of vaccines has offered a ray of hope but it appears the road to recovery could be long as new variants of the pandemic have emerged. Historically, the tourism industry has always been resilient to such crises. CBT destinations must therefore institute some measures towards recovery and build resilience. Governments and Destination Management Organizations (DMOs) need to rethink and reshape tourism policy towards recovery and building resilience. The G20 Rome guidelines for the future of tourism are based on the seven interrelated policy areas namely;

- Safe mobility: restoring and maintaining confidence in travel;
- Crisis management: minimising the impact of future crises affecting tourism;
- Resilience: securing a robust and stable tourism sector in uncertain times;
- Inclusiveness: widening community engagement and benefits from tourism;
- Green transformation: managing tourism to sustain global and local environments;
- Digital transition: enabling all tourism stakeholders to benefit fully from digital opportunities; and
- Investment and infrastructure: focusing resources on a sustainable future for tourism. (OECD, 2020).

Many destinations are turning to their domestic markets to drive recovery. To tap into the domestic market, destinations must provide packages that appeal to the domestic market and provide incentives to stimulate their interest in travel to natural environments. This includes discounts, holiday vouchers, sales promotions, and reduced rates. Such packages should be promoted through mass media and social media channels. In South Africa for instance, some nature-based tourism operators are creating new tourism products with lower rates for domestic visitors (Southern African Development Community (SADC) Transfrontier Conservation Area (TFCA) Network (2020).

Due to the multisectoral nature of the tourism industry and the presence of diverse stakeholders, it is important for CBT destinations to adopt a collaborative approach towards recovery. Public-private partnership is crucial for recovery from the adverse impacts of COVID-19. Communities must partner with government, private sector businesses and other stakeholders to develop innovative products and services which will help accelerate recovery. Ultimately, government support for the tourism industry will determine the pace of recovery. Government can drive recovery through stimulus packages for CBT projects and other tourism businesses along the value chain to help revive tourism. These could be in the form of interest-free or soft loans, tax rebates, tax holidays, subsidies, employment support and other credit facilities.

CONCLUSION

CBT remains a good model for achieving community development, improved livelihoods and environmental conservation. However, the operationalization of the CBT model is fraught with challenges. The various chapters in this book have added to the debate on the potency of the CBT concept for achieving socio-economic development and environmental conservation in communities. A common thread through most of the discussions is that CBT projects have not succeeded in adequately involving local communities in the tourism development process. This represents the most significant challenge to achieving sustainable CBT development as it prevents local communities from reaping the full benefits of tourism development and breeds apathy. This and other challenges have to be addressed by communities, governments and other stakeholders to ensure that communities achieve sustainable CBT development. The attainment of sustainable CBT development also depends on aligning CBT policies and practices with future trends. Sustainable CBT could be achieved through partnerships and collaborations, empowerment of local communities, adoption of technology and building resilience.

Ishmael Mensah
University of Cape Coast, Ghana

Ewoenam Afenyo-Agbe
University of Cape Coast, Ghana

REFERENCES

Amadeus. (2014). *Trending with seniors. Understanding the active senior consumer traveler*. Retrieved from: https://extranets.us.amadeus.com/pdf/amadeus_active_seniors_report.pdf

Asker, S., Boronyak, L., Carrard, N., & Paddon, P. (2010). *Effective Community Based Tourism: Best Practice Manual*. Sustainable Tourism Cooperative Research Center.

Association of British Travel Agents. (2015). *Travel Trends 2016*. Available at: https://abta.com/assets/uploads/general/ABTA_Travel_Trends_2016_210116.pdf

Bello, F. G., Lovelock, B., & Carr, N. (2017). Constraints of community participation in protected area-based tourism planning: The case of Malawi. *Journal of Ecotourism*, *16*(2), 131–151. doi:10.1080/14724049.2016.1251444

Buckley, R. (2020). Nature tourism and mental health: Parks, happiness, and causation. *Journal of Sustainable Tourism*, *28*(9), 1409–1424. doi:10.1080/09669582.2020.1742725

Buckley, R., Gretzel, U., Scott, D., Weaver, D., & Becken, S. (2015). Tourism megatrends. *Tourism Recreation Research*, *40*(1), 59–70. doi:10.1080/02508281.2015.1005942

Campón-Cerro, A. M., Di-Clemente, E., Hernández-Mogollón, J. M., & Folgado-Fernández, J. A. (2020). Healthy water-based tourism experiences: Their contribution to quality of life, satisfaction and loyalty. *International Journal of Environmental Research and Public Health*, *17*(6), 1961. doi:10.3390/ijerph17061961 PMID:32192098

Centre for the Promotion of Imports (2020). *The European market potential for nature and ecotourism*. Author.

Cocks, T., & Tassiem, S. (2020). South Africa "virtual safaris" liven up lockdown with jackals and leopard cubs. *Reuters*. Available at: https://www.eventbrite.com/e/become-a-certified-virtual-tour-guide-tickets-100932710454?utm-medium=discovery&utm-campaign=social&utm-content=attendeeshare&aff=esfb&utm-source=fb&utm-term=listing#

Cole, S. (2006). Information and empowerment: The keys to achieving sustainable tourism. *Journal of Sustainable Tourism*, *14*(6), 629–644. doi:10.2167/jost607.0

Condratov, I. (2013). E-tourism: Concept and evolution. *Ecoforum*, *2*(1), 58–61.

di Castri, F., & Balaji, V. (Eds.). (2002). *Tourism, Biodiversity and Information*. Backhuys. doi:10.1016/S0160-7383(01)00046-9

Dryglas, D., & Salamaga, M. (2018). Segmentation by push motives in health tourism destinations: A case study of Polish spa resorts. *Journal of Destination Marketing & Management*, *9*, 234–246. doi:10.1016/j.jdmm.2018.01.008

Euronews Travel. (2020). *Travel Trend Report October 2020*. Available at: http://www.switch-asia.eu/resource/travel-after-2020-what-will-tourism-look-like-in-our-new-reality/

Gretzel, U., Fesenmaier, D. R., & O'leary, J. T. (2006). The transformation of consumer behaviour. *Tourism Business Frontiers: Consumers. Products and Industry*, *9*, 18.

Guo, Y., Jiang, J., & Li, S. (2019). A sustainable tourism policy research review. *Sustainability*, *11*(11), 3187. doi:10.3390u11113187

Hall, C. M. (1996). *Introduction to Tourism in Australia: Impacts, Planning and Development*. Addison, Wesley and Longman.

Han, G., Wu, P., Huang, Y., & Yang, Z. (2014). Tourism development and the disempowerment of host residents: Types and formative mechanisms. *Tourism Geographies*, *16*(5), 717–740. doi:10.1080/14616688.2014.957718

Hanna, P., Wijesinghe, S., Paliatsos, I., Walker, C., Adams, M., & Kimbu, A. (2019). Active engagement with nature: Outdoor adventure tourism, sustainability and wellbeing. *Journal of Sustainable Tourism*, *27*(9), 1355–1373. doi:10.1080/09669582.2019.1621883

Jamal, T. B., & Getz, D. (1995). Collaboration theory and community tourism planning. *Annals of Tourism Research*, *22*(1), 186–204. doi:10.1016/0160-7383(94)00067-3

Kim, H., Lee, S., Uysal, M., Kim, J., & Ahn, K. (2015). Nature-based tourism: Motivation and subjective well-being. *Journal of Travel & Tourism Marketing*, *32*(sup1), S76-S96

Knight, D. W., & Cottrell, S. P. (2016). Evaluating tourism-linked empowerment in cuzco, Peru. *Annals of Tourism Research*, *56*, 32–47. doi:10.1016/j.annals.2015.11.007

Le, T.-A., Weaver, D., & Lawton, L. (2012). Factors influencing the performance of community- based tourism in the semi-periphery of Viet Nam: A resident perspective from Ta Van Village. In Book of proceedings: The new golden age of tourism and hospitality (pp. 361–365). CAUTHE – Council for Australian University Tourism and Hospitality Education, Melbourne Convention and Exhibition Centre, Book 2.

Lee, T. H. (2013). Influence analysis of community resident support for sustainable tourism development. *Tourism Management*, *34*, 37–46. doi:10.1016/j.tourman.2012.03.007

Leksakundilok, A., & Hirsch, P. (2008). Community-based tourism in Thailand. In J. Connell & B. Rugendyke (Eds.), *Tourism at the grassroot. Villagers and Visitors in the Asia-Pacific* (pp. 214–235). Routledge.

Lindsey, P., Allan, J., Brehony, P., Dickman, A., Robson, A., Begg, C., & Tyrrell, P. (2020). Conserving Africa's wildlife and wildlands through the COVID-19 crisis and beyond. *Nature Ecology & Evolution, 4*(10), 1300–1310. doi:10.103841559-020-1275-6 PMID:32728187

Martín, H. S., de los Salmones Sanchez, M. M. G., & Herrero, Á. (2018). Residents' attitudes and behavioural support for tourism in host communities. *Journal of Travel & Tourism Marketing, 35*(2), 231–243. doi:10.1080/10548408.2017.1357518

McKee, J. K., Sciulli, P. W., Fooce, C. D., & Waite, T. A. (2004). Forecasting global biodiversity threats associated with human population growth. *Biological Conservation, 115*(1), 161–164. doi:10.1016/S0006-3207(03)00099-5

Moswete, N., Thapa, B., & Child, B. (2012). Attitudes and opinions of local and national public sector stakeholders towards kgalagadi transfrontier park, Botswana. *International Journal of Sustainable Development and World Ecology, 19*(1), 67–80. doi:10.1080/13504509.2011.592551

Murphy, P. (1985). *Tourism: A Community Approach*. Methuen.

Niedziółka, I. (2014). Sustainable tourism development. *Regional Formation and Development Studies, 8*(3), 157–166.

Nunkoo, R., & Ramkissoon, H. (2011). Developing a community support model for tourism. *Annals of Tourism Research, 38*(3), 964–988. doi:10.1016/j.annals.2011.01.017

Ohe, Y., & Kurihara, S. (2013). Evaluating the complementary relationship between local brand farm products and rural tourism: Evidence from Japan. *Tourism Management, 35*, 278–283. doi:10.1016/j.tourman.2012.07.003

Okazaki, E. (2008). A community-based tourism model: Its conception and use. *Journal of Sustainable Tourism, 16*(5), 511–529. doi:10.1080/09669580802159594

Organization for Economic Cooperation and Development. (2020). *G20 Rome guidelines for the future of tourism*. OECD Report to G20 Tourism Working Group. Available at: https://www.g20.org/wp-content/uploads/2021/05/G20-Rome-guidelines-for-the-future-of-tourism_OECD-report-to-G20-TWG_CLEAN-COVER.pdf

Saunders, G. (2020). *South Africa's safari and game lodge industry*. Unpublished report.

Sofield, T. H. (Ed.). (2003). *Empowerment For Sustainable Tourism Development*. Emerald Group Publishing.

Southern African Development Community, Transfrontier Conservation Area Network. (2020). *Building a resilient tourism sector in SADC TFCAs.* Week 7. Webinar session 13. 20 August 2020. Available at: https://www. youtube.com/watch?v=9FFKP61zS08&feature=youtu.be

Spenceley, A. (2021). *Tourism and Visitation in Protected Areas Amid COVID-19: Guidance for Protected Area Authorities and Managers.* European Commission.

Spencer, A. J., Buhalis, D., & Moital, M. (2012). A Hierarchical Model of Technology Adoption for Small Owner-Managed Travel Firms: An Organisational Decision-Making and Leadership Perspective. *Tourism Management, 33*(5), 1195–1208. doi:10.1016/j.tourman.2011.11.011

Stone, M. T. (2015). Community-based ecotourism: A collaborative partnerships perspective. *Journal of Ecotourism, 14*(2-3), 166–184. doi:10.1080/14724049.2015.1023309

Tangeland, T., & Aas, Ø. (2011). Household composition and the importance of experience attributes of nature based tourism activity products – A Norwegian case study of outdoor recreationists. *Tourism Management, 32*(4), 822–832. doi:10.1016/j.tourman.2010.07.005

Tasci, A. D., Semrad, K. J., & Yilmaz, S. S. (2013). *Community Based Tourism Finding the Equilibrium in the COMCEC Context Setting the Pathway for the Future.* COMCEC Coordination Office.

Tosun, C. (2000). Limits to community participation in the tourism development process in developing countries. *Tourism Management, 21*(6), 613–633. doi:10.1016/S0261-5177(00)00009-1

UNCTAD. (2001). *High-level meeting on Tourism and Development in the Least Developed Countries.* Maspalomas Convention and Exhibition Centre, Las Palmas. http://r0.unctad.org/conference/preconf/tourism.en.htm

United Nations Conference on Trade and Development. (2020). *Handbook of Statistics.* UNCTAD.

United Nations Environment Programme and World Tourism Organization. (2005). *Making Tourism More Sustainable A Guide For Policy Makers.* United Nations Environment Programme.

Wearing, S., McDonald, M., & Ponting, J. (2005). Building a decommodified research paradigm in tourism: The contri- bution of NGOs. *Journal of Sustainable Tourism, 13*(5), 424–437. doi:10.1080/09669580508668571

Woo, E., Kim, H., & Uysal, M. (2015). Life satisfaction and support for tourism development. *Annals of Tourism Research*, *50*, 84–97. doi:10.1016/j.annals.2014.11.001

World Bank Group. (2018). *Demand analysis for tourism in African local communities*. Available at: https://openknowledge.worldbank.org/bitstream/handle/10986/31071/IFC-Community-Based-Tourism-Document-4-12-18-HR-PAGES.pdf?sequence=1&isAllowed=y

Yanes, A., Zielinski, S., Diaz Cano, M., & Kim, S. I. (2019). Community-based tourism in developing countries: A framework for policy evaluation. *Sustainability*, *11*(9), 2506. doi:10.3390u11092506

Compilation of References

Abane, A. M., Awusabo-Asare, K., & Kissi, A. K. (1999). In who's Interest? Individual and Societal needs in the Creation of Forest Reserves: The Case of Kakum in Ghana. *Bulletin of the Ghana Geographical Association*, *2*, 21–30.

Abdullah, A. (2020, August 14). Covid-19 Pierces "Heartbeat" of Ghana's Tourism Hard. *Ghana News Agency*. https://www.gna.org.gh/1.18646544

Abeyasekera, S. (2005). *Multivariate methods for index construction*. Available online at https://unstats.un.org/unsd/hhsurveys/pdf/Chapter_18.pdf

Abukari, H., & Mwalyosi, R. B. (2020). Local communities' perceptions about the impact of protected areas on livelihoods and community development. *Global Ecology and Conservation*, *22*, 22. doi:10.1016/j.gecco.2020.e00909

Abukhalifeh, A. N., & Wondirad, A. (2019). Contributions of community-based tourism to the socio-economic well-being of local communities: The case of Pulau Redang Island, Malaysia. *International Journal of Tourism Sciences*, *19*(2), 80–97. doi:10.1080/15980634.2019.1621521

Ackuayi, E. D., Kotoku, G., Dzeto, G. K., & Bonsu-Owu, A. (2014). Contribution of the Hospitality and Tourism Sector to the Development of Local Communities in the Hohoe Municipality, Ghana. *Journal of Economics and Sustainable Development*, *5*(23), 209–219.

Adetola, B. O. (2015). Ecotourism impacts on support zone communities around Cross River National Park, Nigeria. *Journal of Sustainable Development in Africa*, *17*(5), 153–163.

Adiyia, B., & Vanneste, D. (2018). Local tourism value chain linkages as pro-poor tools for regional development in Western Uganda. *Development Southern Africa*, *35*(2), 210–224. doi:10.1080/0376835X.2018.1428529

Afdal, T. (2016). Consulting external resources during post-editing a machine translated tourism text by an Indonesian translator. *Translation Journal, 19*(3). https://translationjournal.net/July-2016/consulting-external-resources-during-post-editing-machine-translated-tourism-text-by-an-indonesian-translator.html

Afenyo, E. A., & Amuquandoh, F. E. (2014). Who benefits from community-based ecotourism development? Insights from Tafi Atome, Ghana. *Tourism Planning & Development, 11*(2), 179–190. doi:10.1080/21568316.2013.864994

Afify, A. M. N. (2011). Challenges and advantages of community participation as an approach for sustainable urban development in Egypt. *Journal of Sustainable Development, 4*(1), 23–36.

Afnarius, S., Akbar, F., & Yuliani, F. (2020). Developing web-based and mobile-based GIS for places of worship information to support halal tourism: A case study in Bukittinggi, Indonesia. *ISPRS International Journal of Geo-Information, 9*(1), 1–18. doi:10.3390/ijgi9010052

Agbo, V. M. (2019). *From conflict to collaboration: Atewa Forest governance* (Master's thesis). University of Waterloo.

Agorni, M. (2012). Tourism communication: the translator's responsibility in the translation of cultural difference. *Pasos. Revista de turismo y patrimonio cultural, 10*(4), 5-11. https://www.redalyc.org/pdf/881/88123115002.pdf

Agrawal, A., & Gibson, C. (1999). Enchantment and disenchantment: The role of community in natural resource conservation. *World Development, 27*(4), 629–649. doi:10.1016/S0305-750X(98)00161-2

Agyeman, Y. B., Aboagye, O. K., & Ashie, E. (2019). Visitor Satisfaction at Kakum National Park in Ghana. *Tourism Recreation Research, 44*(2), 178–189. doi:10.1080/02508281.2019.1566048

Ahebwa, W. M., Aporu, J. P., & Nyakaana, J. B. (2016). Bridging community livelihoods and cultural conservation through tourism: A Case study of Kabaka heritage trail in Uganda. *Tourism and Hospitality Research, 16*(2), 103–115. doi:10.1177/1467358415589659

Ahmadu, N. O. (2011). Hospitality and tourism in Nigerian polytechnics and colleges of Technology: Challenges and prospects. *HATMAN Journal of Hospitality and Tourism, 1*(1), 65–70.

Aitchison, C. (1999). *New cultural geographies: The spatiality of leisure, and evolution.* Academic Press.

Akama, J. (2011). Efficacy of Tourism as a Tool for Local Community Development. A case Study of Mombasa Kenya. *Journal of Social Sciences, 1*(1), 1–16.

Akama, J. S. (1996). Western environmental values and nature-based tourism in Kenya. *Tourism Management, 17*(8), 567–574. doi:10.1016/S0261-5177(96)00077-5

Akankwasah, B., Mukadasi, B., Kansiime, F., & Basuta-Isabirye, G. (2010). The Effects of Crop Raiding on Household Food Security in the Albertine Rift: A Case Study of Queen Elizabeth National Park, Western Uganda. *Human Dimensions of Wildlife, 15*(1), 45–54. doi:10.1080/10871200903115070

Akarsua, T. N., Foroudib, P., & Melewarb, T. (2020). What makes Airbnb likeable? Exploring the nexus between service attractiveness, country image, perceived authenticity and experience from a social exchange theory perspective within an emerging economy context. *International Journal of Hospitality Management*, 91.

Akunaay, M., Nelson, F., & Singleton, E. (2003, December). Community based tourism in Tanzania: Potential and perils in practice. In *Second Peace Through Tourism Conference* (pp. 7-12). Academic Press.

Akyeampong, O. A. (2011). Pro-poor Tourism: Residents' Expectations, Experiences and Perceptions in the Kakum National Park Area of Ghana. *Journal of Sustainable Tourism, 19*(2), 197–213. doi:10.1080/09669582.2010.509508

Alatartseva, E., & Barysheva, G. (2015). Well-being: Subjective and objective aspects. *Procedia: Social and Behavioral Sciences, 166*, 36–42. doi:10.1016/j.sbspro.2014.12.479

Alfer, A., & Zwischenberger, C. (Eds.). (2020). Translaboration. Exploring collaboration in translation and translation in collaboration. Special issue of Target, Volume 32, Issue 2. John Benjamins.

Allis, T. (2016). Culture and Tourism. In G. Lohmann & A. P. Netto (Eds.), *Tourism theory: Concepts, models and systems*. CABI Press.

Amalu, T. E., Otop, O. O., Duluora, E. I., Omeje, V. U., & Emeana, S. K. (2017). Socio-economic impacts of ecotourism attractions in Enugu State, Nigeria. *Geojournal Spatially Integrated Social Sciences and Humanities*. doi:10.1007/s10708-017-9830-7

Amalu, T. E., & Ajake, A. O. (2018). Developing natural lakes for socio-economic development: The case of Nike Lake, Enugu State, Nigeria. *GeoJournal, 78*(4), 156–172.

Amalu, T. E., Ugwu, O., Idam, N. S., & Oko-Isu, P. E. (2019). Impacts of seasonality and patronage of lakes on the economy of host communities (a case of Nike Lake, Enugu Nigeria). *Journal of Travel. Tourism and Recreation, 1*(1), 7 13.

Amerta, I. M. S. (2017). Community based tourism development. *International Journal of Social Science and Humanity, 1*(3), 97–107.

Amoah, M., & Wiafe, E. D. (2012). Livelihoods of Fringe Communities and the Impacts on the Management of Conservation Area: The Case of Kakum National Park in Ghana. *International Forestry Review, 14*(2), 131–144. doi:10.1505/146554812800923381

Amoako-Atta, E., Dayour, F., & Bonye, S. Z. (2020). Community Participation in the Management of Wechiau Community Hippo Sanctuary, Ghana. *Ghana Journal of Development Studies, 17*(1), 1–24. doi:10.4314/gjds.v17i1.1

Amuquandoh, F. E. (2017). Tourists' Motivation for Visiting Kakum National Park, Ghana. *Ghana Journal of Geography, 9*(1), 152–168.

Anderson, W. (2011). Challenges of tourism development in the developing countries: The case of Tanzania. *Journal of Tourism*, *12*(2), 17–36.

Andrade, G. S., & Rhodes, J. R. (2012). Protected areas and local communities: An inevitable partnership toward successful conservation strategies? *Ecology and Society*, *17*(4), art14. doi:10.5751/ES-05216-170414

Antwi, L. S. (2020, March 21). Covid-19: Kakum National Park Closes Down. *Graphic Showbiz*. https://www.graphic.com.gh/entertainment/covid-19-kakum-national-park-closes-down.html

Anuar, A. N. A., & Sood, N. A. A. M. (2017). Community Based Tourism: Understanding, Benefits and Challenges. *Journal of Tourism & Hospitality (Los Angeles, Calif.)*, *6*(01), 1. doi:10.4172/2167-0269.1000263

Ap, J. (1992). Residents' perceptions on tourism impacts. *Annals of Tourism Research*, *19*(4), 665–690. doi:10.1016/0160-7383(92)90060-3

Appiah-Opoku, S. (2011). Using Protected Areas as a Tool for Biodiversity Conservation and Ecotourism: A Case Study of Kakum National Park in Ghana. *Society & Natural Resources*, *24*(5), 500–510. doi:10.1080/08941920.2010.495108

Archabald, K., & Naughton-Treves, L. (2001). Tourism Revenue-sharing around National Parks in Western Uganda; Early Efforts to Identify and Reward Local Communities. *Environmental Conservation*, *28*(2), 135–149. doi:10.1017/S0376892901000145

Aref, F. (2011). The effects of tourism on quality of life: A case study of Shiraz, Iran. *Life Science Journal*, *8*(2), 26–30.

Aref, F., Gill, S. S., & Aref, F. (2010). Tourism development in local communities: As a community development approach. *The Journal of American Science*, *6*(2), 155–161.

Arnstein, S. R. (1971). A Ladder of Citizen Participation in the USA. London: Royal Town Planning Institute.

Arnstein, S. R. (1969). A ladder of citizen participation. *Journal of the American Institute of Planners*, *35*(4), 216–224. doi:10.1080/01944366908977225

Arntzen, J. W., Molokomme, D. L., Terry, E. M., Moleele, N., Tshosa, O., & Mazambani, D. (2003). *Final report of the review of community-based natural resource review in Botswana: Report no. 1.* Gaborone: Centre for Applied Research for the National CBNRM Forum.

Arora, N. K. (2018). Biodiversity conservation for sustainable future. *Environmental Sustainability*, *1*(2), 109–111. doi:10.100742398-018-0023-1

ASEAN Secretariat, . (2016). *ASEAN community based tourism standard.* ASEAN Secretariat.

ASEAN. (2016). *ASEAN community based tourism standard.* Jakarta: Association of Southeast Asian Nations. Retrieved from https://www.asean.org/storage/2012/05/ASEAN-Community-Based-Tourism-Standard.pdf

Ashley, C. (2000). *The Impacts of Tourism on Rural Livelihoods: Namibia's Experience*. Overseas Development Institute.

Ashley, C., Roe, D., & Goodwin, H. (2001). *Pro-poor tourism strategies: Making tourism work for the poor: A review of experience*. Overseas Development Institute.

Asker, S., Louise, B., Carrard, N., & Paddon, M. (2010). *Effective Community Based Tourism : (Issue June)*. Sustainable Tourism Cooperative Research Centre.

Asmelash, A. G., & Kumar, S. (2019). Assessing progress of tourism sustainability: Developing and validating sustainability indicators. *Tourism Management, 71*, 67–83. doi:10.1016/j.tourman.2018.09.020

Asuk, S. A., Ebu, V. T., & Ifebueme, N. M. (2018). Assessment of community-based ecotourism prospects in southern Nigeria: Case study of Iko Esai community. *International Journal of Agricultural Policy and Research, 6*(7), 111-121. https://doi.org/10.15739/IJAPR.18.013

Ateljevic, I., Morgan, N., & Pritchard, A. (Eds.). (2013). *The critical turn in tourism studies: Creating an academy of hope* (Vol. 22). Routledge. doi:10.4324/9780203806586

Atkinson, J. M. (1990). Ethnomethodological approaches to socio-legal studies. *Ethnomethodological Sociology*, 450-472.

Attride-Stirling, J. (2001). Thematic networks: An analytic tool for qualitative research. *Qualitative Research, 1*(3), 385–405. doi:10.1177/146879410100100307

Austin, Z., McVittie, A., McCracken, D., Moxey, A., Moran, D., & White, P. C. (2016). The co-benefits of biodiversity conservation programmes on wider ecosystem services. *Ecosystem Services, 20*, 37–43. doi:10.1016/j.ecoser.2016.06.002

Ayhan, Ç., Cengíz Taşlı, T., Özkök, F., & Tatlı, H. (2020). Land use suitability analysis of rural tourism activities: Yenice, Turkey. *Tourism Management, 76*(July). doi:10.1016/j.tourman.2019.07.003

Ayivor, J. S., Nyametso, J. K., & Ayivor, S. (2020). Protected Area Governance and Its Influence on Local Perceptions, Attitudes and Collaboration. *Land (Basel), 9*(9), 310. doi:10.3390/land9090310

Ayorekire, J., Obua, J., & Byaruhanga, M. B. (2019). Broadening Uganda's tourism product base through indigenous cultural resource utilization. In M. Mkono (Ed.), *Positive tourism in Africa* (pp. 79–90). Routledge. doi:10.4324/9780429428685-7

Azzari, M., Zamperlin, P., & Landi, F. (2013). GIS in geography teaching. *Journal of Research and Didactics in Geography, 2*(2), 27-42.

Bagus, S. I., Imade, S. U., Nyoman, S. I. A., & Putu, W. S. N. (2019). Community Based Tourism as Sustainable Tourism Support. *RJOAS, 10*(94), 70–78.

Bahaire, T., & Elliott-White, M. (1999). The application of geographical information systems (GIS) in sustainable tourism planning: A review. *Journal of Sustainable Tourism*, *7*(2), 159–174. doi:10.1080/09669589908667333

Baig, S., Khan, A. A., & Khan, A. A. (2021). A time series analysis of causality between tourist arrivals and climatic effects for nature-based tourism destinations: Evidence from Gilgit-Baltistan, Pakistan. *Environment, Development and Sustainability*, *23*(4), 5035–5057. doi:10.100710668-020-00803-0

Baig, S., Qasim, M., Xuemei, L., & Alam, K. M. (2020). Is the China-Pakistan economic corridor an opportunity or a threat for small and micro-entrepreneurs? Empirical evidence from Northern Pakistan. *Sustainability*, *12*(5), 1727. doi:10.3390u12051727

Baig, S., & Zehra, S. (2020). China-Pakistan economic corridor, governance, and tourism nexus: Evidence from Gilgit-Baltistan, Pakistan. *Current Issues in Tourism*, *23*(23), 2884–2889. doi:10.1080/13683500.2020.1763266

Baker, K., & Coulter, A. (2007). Terrorism and tourism: The vulnerability of beach vendors' livelihoods in Bali. *Journal of Sustainable Tourism*, *15*(3), 249–266. doi:10.2167/jost643.0

Bamberger, M. (1986). *The Role of Community Participation in Development Planning and Project Management.* EDI Policy Seminar Report No. 13. The World Bank.

Bandarin, F., Hosagrahar, J., & Sailer Albernaz, F. (2011). Why development needs culture. *Journal of Cultural Heritage Management and Sustainable Development*, *1*(1), 15–25. doi:10.1108/20441261111129906

Baron, R. M., & Kenny, D. A. (1986). The moderator-mediator variable distinction in social psychological research: Conceptual, strategic, and statistical considerations. *Journal of Personality and Social Psychology*, *51*(6), 1173–1182. doi:10.1037/0022-3514.51.6.1173 PMID:3806354

Bassey, F. O., & Egon, C. I. (2016). Nigerian tourism policy: A critical appraisal and re-appraisal. *Journal of Tourism. Hospitality and Sports.*, *20*(1), 26–29.

Baum, T. (2018). Sustainable human resource management as a driver in tourism policy and planning: A serious sin of omission? *Journal of Sustainable Tourism*, *26*(6), 873–889. doi:10.1080/09669582.2017.1423318

Baum, T., & Hagen, L. (1999). Responses to seasonality: The experiences of peripheral destinations. *International Journal of Tourism Research*, *1*(5), 299–312. doi:10.1002/(SICI)1522-1970(199909/10)1:5<299::AID-JTR198>3.0.CO;2-L

Bec, A., Moyle, B., Timms, K., Schaffer, V., Skavronskaya, L., & Little, C. (2019). Management of immersive heritage tourism experiences: A conceptual model. *Tourism Management*, *72*, 117–120. doi:10.1016/j.tourman.2018.10.033

Beeton, S. (2006). *Community development through tourism.* Landlinks Press. doi:10.1071/9780643093881

Bello, F. G., Lovelock, B., & Carr, N. (2018). Enhancing community participation in tourism planning associated with protected areas in developing countries: Lessons from Malawi. *Tourism and Hospitality Research, 18*(3), 309–320. doi:10.1177/1467358416647763

Benyei, P., Arreola, G., & Reyes-García, V. (2020). Storing and sharing: A review of indigenous and local knowledge conservation initiatives. *Ambio, 49*(1), 218–230. doi:10.100713280-019-01153-6 PMID:30859400

Besculides, A., Lee, M. E., & McCormick, P. J. (2002). Residents' perceptions of the cultural benefits of tourism. *Annals of Tourism Research, 29*(2), 303–319. doi:10.1016/S0160-7383(01)00066-4

Blackstock, K. (2005). A critical look at community-based tourism. *Community Development Journal: An International Forum, 40*(1), 39–49.

Bodin, O., & Crona, B. I. (2008). Management of natural resources at the community level: Exploring the role of social capital and leadership in a rural fishing community. *World Development, 36*(12), 2763–2779. doi:10.1016/j.worlddev.2007.12.002

Boluk, K. A., Cavaliere, C. T., & Higgins-Desbiolles, F. (2019). A critical framework for interrogating the United Nations Sustainable Development Goals 2030 Agenda in tourism. *Journal of Sustainable Tourism, 27*(7), 847–864. doi:10.1080/09669582.2019.1619748

Boniface, P. (2013). *Managing quality cultural tourism*. Routledge. doi:10.4324/9780203424230

Borowiecki, K. J., & Castiglione, C. (2014). Cultural participation and tourism flows: An empirical investigation of Italian provinces. *Tourism Economics, 20*(2), 241-262. doi:10.5367/te.2013.0278

Bowker, L. (2015). Computer-aided translation: Translator training. In S.-W. Chan (Ed.), *The Routledge encyclopedia of translation technology* (pp. 88–104). Routledge.

Bramwall, B. (2015). Theoretical Activity in Sustainable Research. *Annals of Tourism Research, 54*, 204–218. doi:10.1016/j.annals.2015.07.005

Bramwall, B., Higham, J., Lane, B., & Miller, G. (2017). Twenty-Five Years of Sustainable Tourism and the Journal of Sustainable Tourism. *Journal of Sustainable Tourism, 25*(1), 1–9. doi:10.1080/09669582.2017.1251689

Braun, V., & Clarke, V. (2006). Using thematic analysis in psychology. *Qualitative Research in Psychology, 3*(2), 77–101. doi:10.1191/1478088706qp063oa

Briones, Z. B. H., Yusay, R. M. S., & Valdez, S. (2017). Enhancing community-based tourism programs of Gawad Kalinga Enchanted Farm towards sustainable tourism development. *Journal of Economic Development, Management, IT. Finance and Marketing, 9*(1), 51–60.

Britton, S. G. (1982). The political economy of tourism in the Third World. *Annals of Tourism Research, 9*(3), 331–358. doi:10.1016/0160-7383(82)90018-4

Brohman, J. (1996). New directions in tourism for third world development. *Annals of Tourism Research, 23*(1), 48–70. doi:10.1016/0160-7383(95)00043-7

Brohman, J. (1996). *Popular Development Rethinking the Theory and Practice of Development*. Blackwell Publishers.

Brown, H. (2018). Rethinking agency and creativity: Translation, collaboration and gender in early modern Germany. *Translation Studies*, *11*(1), 84–102. doi:10.1080/14781700.2017.1300103

Brunet, M., & Aubry, M. (2016). The three dimensions of a governance framework for major public projects. *International Journal of Project Management*, *34*(8), 1596–1607.

Brunt, P., & Courtney, P. (1999). Host perceptions of socio-cultural impacts. *Annals of Tourism Research*, *26*(3), 493–515. doi:10.1016/S0160-7383(99)00003-1

Buanes, A., Jentoft, S., Maurstad, A., Søreng, S. U., & Karlsen, G. R. (2005). Stakeholder participation in Norwegian coastal zone planning. *Ocean and Coastal Management*, *48*(9), 658–669. doi:10.1016/j.ocecoaman.2005.05.005

Buberwa, R. F., & Msusa, A. (2019). Developing a mobile gis tour guide app for dar-es-salaam city, Tanzania. *International Journal of Scientific and Technology Research*, *8*(4), 237–242.

Burgos, A., & Mertens, F. (2017). Participatory management of community-based tourism: A network perspective. *Community Development (Columbus, Ohio)*, *48*(4), 546–565.

Burton, S., & Steane, P. (2004). *Surviving your thesis* (1st ed.). New York: Routledge.

Butler, R. (2018). Sustainable Tourism in Sensitive Environments: A Wolf in Sheeps's Clothing? *Sustainability*, *10*(1789), 1-11.

Butler, G. (2017). Fostering community empowerment and capacity building through tourism: Perspectives from Dullstroom, South Africa. *Journal of Tourism and Cultural Change*, *15*(3), 199–212. doi:10.1080/14766825.2015.1133631

Byrd, E. T., Bosley, H. E., & Dronberger, M. G. (2009). Comparison of stakeholder perceptions of tourism impacts in rural eastern North Carolina. *Tourism Management*, *30*(5), 693–703. doi:10.1016/j.tourman.2008.10.021

Calderon, M. M., & Nawir, A. A. (2006). *An evaluation of the feasibility and benefits of forest partnerships to develop tree plantations: case studies in the Philippines. Center for International Forestry Research*. CIFOR.

Camilleri, M. A. (2018). The Tourism Industry. *An Overview*, (October), 3–27. Advance online publication. doi:10.1007/978-3-319-49849-2_1

Campbell, L. M., Gray, N. J., & Meletis, Z. A. (2008). Political ecology perspectives on ecotourism to parks and protected areas. In K. S. Hanna, D. A. Clark, & D. S. Slocombe (Eds.), Transforming parks and protected areas: Policy and Governance in a Changing World (pp. 200–221). Academic Press.

Carr, A., & Melubo, K. (2018). *Developing indigenous tourism in the bomas: Critiquing issues from within the Maasai community in Tanzania*. Academic Press.

Ceballos-Lascurain, H. (1996). *Tourism, ecotourism, and protected areas*. IUCN. doi:10.2305/IUCN.CH.1996.7.en

Cellini, R., & Cuccia, T. (2013). Museum and monument attendance and tourism flow: A time series analysis approach. *Applied Economics*, *45*(24), 3473–3482. doi:10.1080/00036846.2012.716150

Chambers, D., & Buzinde, C. (2015). Tourism and decolonisation: Locating research and self. *Annals of Tourism Research*, *51*, 1–16. doi:10.1016/j.annals.2014.12.002

Chang, K. C. (2021). The affecting tourism development attitudes based on the social exchange theory and the social network theory. *Asia Pacific Journal of Tourism Research*, *26*(2), 167–182. doi:10.1080/10941665.2018.1540438

Chang, K. G., Chien, H., Cheng, H., & Chen, H. (2018). The Impacts of Tourism Development in Rural Indigenous Destinations: An Investigation of the Local Residents' Perception Using Choice Modeling. *Sustainability*, *10*(4766), 1–15. doi:10.3390u10124766

Chan, K. L. (2009). The consumption of museum service experiences: Benefits and value of museum experiences. *Journal of Hospitality Marketing & Management*, *18*(2-3), 173–196.

Charoenkongka, A., & Nomnian, S. (2019). Needs analysis of communicative English through community-based tourism: A case study of Koh Klang community in Krabi province. *Journal of International Students*, *9*(2), 211–232.

Cheer, J. M., & Lew, A. A. (Eds.). (2017). *Tourism, resilience and sustainability: Adapting to social, political and economic change*. Routledge. doi:10.4324/9781315464053

Chen, D. (2014).. . *Application Of Gis in Environmental Impact Assessment.*, *994*, 4855–4860. doi:10.4028/www.scientific.net/AMR.989-994.4855

Cherifi, B., Smith, A., Maitland, R., & Stevenson, N. (2014). Destination images of non-visitors. *Annals of Tourism Research*, *49*, 190–202. doi:10.1016/j.annals.2014.09.008

Christie, I., Fernandes, E., Messerli, H., & Twining-Ward, L. (2012). *Tourism in Africa: The World Bank harnessing tourism for growth and improved livelihoods*. https://www.worldbank.org/content/dam/Worldbank/document/Africa/Report/africa-tourism-report-2013-overview.pdf

Christie, I. T. (2002). Tourism, growth and poverty: Framework conditions for tourism in developing counties. *Tourism Review*, *57*(1/2), 35–41.

Chuenpagdee, R., & Jentoft, S. (2009). Governability assessment for fisheries and coastal systems: A reality check. *Human Ecology*, *37*(1), 109–120. doi:10.100710745-008-9212-3

Cobinnah, P. B., Black, R., & Thwaites, R. (2015). Ecotourism Implementation in the Kakum Conservation Area, Ghana: Administrative Framework and Local Community Experiences. *Journal of Ecotourism*, *14*(2-3), 223–242. doi:10.1080/14724049.2015.1051536

Cole, S. (2006). Information and empowerment: The keys to achieving sustainable tourism. *Journal of Sustainable Tourism*, *14*(6), 629–644. doi:10.2167/jost607.0

Cole, S. T., & Illum, S. F. (2006). Examining the mediating role of festival visitors' satisfaction in the relationship between service quality and behavioral intentions. *Journal of Vacation Marketing*, *12*(2), 160–173.

Comerio, N., & Strozzi, F. (2019). Tourism and its economic impact: A literature review using bibliometric tools. *Tourism Economics*, *25*(1), 109–131.

Coppock, T. (1994). *GIS and Natural Hazards: An overview from a GIS perspectives*. Academic Press.

Cordingley, A., & Manning, C. F. (Eds.). (2016). *Collaborative translation: From the Renaissance to the digital age*. Bloomsbury.

Cornell, D. (1997). Participatory development: An approach sensitive to class and gender. *Development in Practice*, *7*(3), 248–259. doi:10.1080/09614529754486 PMID:12348314

Craig, G. (2003). Editorial introduction. *Community Development Journal: An International Forum*, *38*(1), 1–5. doi:10.1093/cdj/38.1.1

Dalziel, P., Saunders, C., & Saunders, J. (2018). *Wellbeing economics: the capabilities approach to prosperity*. Springer Nature. doi:10.1007/978-3-319-93194-4

Dangi, T. B. (2018). Exploring the intersections of emotional solidarity and ethic of care: An analysis of their synergistic contributions to sustainable community tourism development. *Sustainability*, *10*(8), 2713. doi:10.3390u10082713

Dangi, T. B., & Jamal, T. (2016). An integrated approach to "sustainable community-based tourism". *Sustainability*, *8*(5), 475. doi:10.3390u8050475

David, L. (2011). ICT and digital records management in the Ugandan public service. *Records Management Journal*, *21*(2), 135–144. doi:10.1108/09565691111152062

Dei, G. J. S. (2000). Rethinking the role of indigenous knowledges in the academy. *International Journal of Inclusive Education*, *4*(2), 111–132. doi:10.1080/136031100284849

Dei, L. A. (2000). Community Participation in Tourism in Africa. In P. U. C. Dieke (Ed.), *Political Economy of Tourism in Africa* (pp. 285–298). Cognizant Communication Corporation.

Deutsch, M. (1975). Equity, equality, and need: What determines which value will be used as the basis of distributive justice? *The Journal of Social Issues*, *31*(3), 137–149. doi:10.1111/j.1540-4560.1975.tb01000.x

Di Minin, E., Leader-Williams, N., & Bradshaw, C. J. (2016). Banning trophy hunting will exacerbate biodiversity loss. *Trends in Ecology & Evolution*, *31*(2), 99–102.

Diamantis, D. (1999). The concept of ecotourism: Evolution and trends. *Current Issues in Tourism*, *2*(2–3), 93–122. doi:10.1080/13683509908667847

Dieke, P. U. (2000). Developing tourism in Africa: Issues for policy consideration. *Development Policy Management in Sub-Saharan Africa*, *7*(1), 19–26.

Dieke, P. U. C. (2003). Tourism in Africa's economic development: Policy implications. *Management Decision*, *41*(3), 287–295.

Dimoska, T., & Trimcev, B. (2012). Competitiveness strategies for supporting economic development of the touristic destination. *Procedia: Social and Behavioral Sciences*, *44*, 279–288.

Din, K. H. (1988). Reports: Social Cultural Impacts of Tourism. *Annals of Tourism Research*, *15*(4), 563–566. doi:10.1016/0160-7383(88)90054-0

Dixey, L. (2008). The unsustainability of community tourism donor projects: Lessons from Zambia. In A. Spenceley (Ed.), *Responsible Tourism: Critical Issues for Conservation and Development* (pp. 323–340). Earthscan.

Dodds, R., Alisha, A., & Kelly, G. (2016). Mobilising Knowledge: Determining key elements for success and pitfalls in developing Community Based Tourism. *Current Issues in Tourism*, *21*(13), 1547–1568. doi:10.1080/13683500.2016.1150257

Dolezal, C., & Novelli, M. (2020). Power in community-based tourism: Empowerment and partnership in Bali. *Journal of Sustainable Tourism*, 1–19. Advance online publication. doi:10.1080/09669582.2020.1838527

Dolezal, C., Trupp, A., & Bui, H. T. (Eds.). (2020). *Tourism and development in Southeast Asia*. Routledge. doi:10.4324/9780429264191

Driciru, M., Rwego, I. B., Asiimwe, B., Travis, D. A., Alvarez, J., VanderWaal, K., & Pelican, K. (2018). Spatio-temporal epidemiology of anthrax in Hippopotamus amphibious in Queen Elizabeth protected area, Uganda. *PLoS One*, *13*(11), 1–21. doi:10.1371/journal.pone.0206922 PMID:30485342

Du Gay, P. (1997). Organizing identity: making up people at work. *Production of Culture/Cultures of Production*, 285-344.

Du, J., Way, A., Qiu, Z., Wasla, A., & Schaler, R. (2015). Domain adaptation for social localisation-based SMT: A case study using Trommons platform. In S. O'Brien, & M. Simard (Eds.), *Proceedings of the 4th Workshop on Post-Editing Technology and Practice* (pp. 57–75). Retrieved October 18, 2020, from https://amtaweb.org/wpcontent/uploads/2015/10/MTSummitXV_WPTP4Proceedings.pdf

Dunn, C. E., Atkins, P. J., & Townsend, J. G. (1997). GIS for development: A contradiction in terms? *Area*, *29*(2), 151–159. doi:10.1111/j.1475-4762.1997.tb00017.x

Duvail, S., Hamerlynck, O., Nandi, R. X., Mwambeso, P., & Elibariki, R. (2006). Participatory Mapping for Local Management of Natural Resources in Villages of the Rufiji District, Tanzania. *The Electronic Journal on Information Systems in Developing Countries*, *25*(1), 1–6. doi:10.1002/j.1681-4835.2006.tb00167.x

Dwyer, L., Forsyth, P. & Dwyer, W. (2010). *Tourism economics and policy*. Channel View Publications.

Edgell, D. L., & Swanson, J. R. (2013). *Tourism, policy and planning: yesterday, today and tomorrow* (2nd ed.). Routledge.

Eggert, L. S., Eggert, J. A., & Woodruff, D. S. (2003). Estimating population sizes for elusive animals: The forest elephants of Kakum National Park, Ghana. *Molecular Ecology*, *12*(6), 1389–1402. doi:10.1046/j.1365-294X.2003.01822.x PMID:12755869

Ellis, S., & Sheridan, L. (2015). The Role of Resident Perceptions in Achieving Effective Community Based Tourism For Least Developed Countries. *International Journal of Tourism and Hospitality Research*, *26*(2), 244–257.

Emeka, N., Ejim, E. P., & Amaka, O. (2015). Effect of strategy formulation on organizational performance: A study of Innoson manufacturing company Ltd Emene, Enugu. *Word Journal of Management and Behavioral Studies*, *3*(1), 9–20.

Esfehani, M. H., & Walters, T. (2018). Lost in translation? Cross-language thematic analysis in tourism and hospitality research. *International Journal of Contemporary Hospitality Management*, *30*(11), 3158–3174. doi:10.1108/IJCHM-10-2017-0701

Eslami, A., & Azad, I. (2016). *Implementation of GIS in Natural Resources*. Academic Press.

ESRI. (2010, Oct.). GIS for Wildlife Management. *GIS Best Practices*, 40. https://www.esri.com/library/bestpractices/wildlife-management.pdf

Estellés-Arolas, E., & González-Ladrón-De-Guevara, F. (2012). Towards an integrated crowdsourcing definition. *Journal of Information Science*, *38*(2), 189–200. doi:10.1177/0165551512437638

Esu, B. B. (2015). Transforming the Nigeria tourism industry through tourism entrepreneurial development. *African Journal of Business Management Review*, *9*(15), 569–580.

Eze-Uzomaka, P. I. (2006). Tourism & community development in issues in tourism planning and development. Afro-Orbis Publishing Co.

Fagbolu, A. O. (2021). Hospitality and tourism management education versus employment in Nigeria. *Global Scientific Journals*, *9*(5), 131-139. https://www.globalscientificjournal.com/journal_volume9_issue5_May_2021_edition.ht ml

Fagbolu, A. O., & Fallon, J. (2021). Strategies exploration for academic achievement improvement through organization development intervention (ODI): A study of Kwara state university. *ABAC ODI Journal*, *8*(1), 161–179.

Fallon, J., & Fagbolu, A. O. (2021). Developing possible strategies for academic achievement improvement of hospitality and tourism management students in Nigeria: A study of Kwara state university. *ABAC Journal*, *41*(2), 177–203. http://www.assumptionjournal.au.edu/index.php/abacjournal/issue/current

Fang, W. T. (2020). Cultural tourism. In W. T. Fang (Ed.), *Tourism in Emerging Economies* (pp. 75–101). Springer. doi:10.1007/978-981-15-2463-9_4

Fennell, D. A. (2006). *Tourism Ethics* (Vol. 30). Channel View Publications. doi:10.21832/9781845410360

Fiagbomeh, F. F., & Burger-Arndt, R. (2015). Prioritisation of Strategies for protected area management with local people using the hybrid SWOT-AHP analyses: The case of Kakum Conservation Area, Ghana. *Management Science Letters*, *5*(5), 457–470. doi:10.5267/j.msl.2015.3.008

Fletcher, J., Fyall, A., Gilbert, D., & Wanhill, S. (2018). *Tourism Principles and Practice* (6th ed.). Pearson Education Ltd.

Franklin, A., & Crang, M. (2001). The trouble with tourism and travel theory? *Tourist Studies*, *1*(1), 5–22. doi:10.1177/146879760100100101

Fredline, E., & Faulkner, B. (2000). Host community reactions: A cluster analysis. *Annals of Tourism Research*, *27*(3), 763–784. doi:10.1016/S0160-7383(99)00103-6

Fusté-Forné, F. (2019). Seasonality in food tourism: wild foods in peripheral areas. *Tourism Geographies*. doi:10.1080/14616688.2018.1558453

G¨Ossling, S. (2009). Carbon neutral destinations: A conceptual analysis. *Journal of Sustainable Tourism*, *17*(1), 17–37.

Gan, J. (2020). Uncovering the Environmental and Social Conflicts Behind Residents' Perception of CBT: A Case of Perak, Malaysia. *Tourism Planning & Development*, *17*(6), 674–692. doi:10.1080/21568316.2020.1749121

Genç, R. (2012). Subjective aspects of tourists' quality-of-life (QOL). In *Handbook of Tourism and Quality-of-life Research* (pp. 149–167). Springer. doi:10.1007/978-94-007-2288-0_9

Geol, K. (2014). Understanding Community and Community Development: Defining the Concept of Community. *ResearchGate*. file:C:/Users/Ama/Downloads/1.UnderstandingCommunityandCommunityDevelopment.pdf

Getz, D. (2008). Event tourism: Definition, evolution, and research. *Tourism Management*, *29*(3), 403–428.

Ghana News Agency. (2009, July 7). *Chief Appeals for Royalties from Kakum Park*. Modern Ghana. https://www.modernghana.com/news/225929/chief-appeals-for-royalties-from-kakum-park.html

Ghana News Agency. (2018, January 30). Tourism at Kakum National Park Drops. *Modern Ghana*. https://www.modernghana.com/news/831799/tourism-at-kakum-national-park-drops.html

Giampiccoli, A., & Mtapuri, O. (2012). Community-based tourism: An exploration of the concept (s) from a political perspective. *Tourism Review International*, *16*(1), 29–43. doi:10.3727/154427212X13431568321500

Giampiccoli, A., & Mtapuri, O. (2020). Towards a coalescence of the community-based tourism and 'Albergo Difusso' tourism models for Sustainable Local Economic Development. *African Journal of Hospitality, Tourism and Leisure*, *9*(1), 1–20.

Giampiccoli, A., & Saayman, M. (2018). Community-based tourism development model and community participation. *African Journal of Hospitality, Tourism and Leisure*, *7*(4), 1–27.

Gibson, C. (2008). Locating geographies of tourism. *Progress in Human Geography*, *32*(3), 407–422. doi:10.1177/0309132507086877

Gibson, C. (2009). Geographies of tourism: Critical research on capitalism and local livelihoods. *Progress in Human Geography*, *33*(4), 527–534. doi:10.1177/0309132508099797

Gibson, C. (2010). Geographies of tourism:(un) ethical encounters. *Progress in Human Geography*, *34*(4), 521–527. doi:10.1177/0309132509348688

Gibson, C. (2019). Critical tourism studies: New directions for volatile times. *Tourism Geographies*, 1–19. doi:10.1080/14616688.2019.1647453

Goeldner, C. R., & Ritchie, J. R. B. (2009). *Tourism principles, practices, philosophies* (11th ed.). John Wiley & Sons, Inc.

Gohori, O., & van der Merwe, P. (2021). Tourism and Community Empowerment: The Perspectives of Local People in Manicaland Province, Zimbabwe. *Tourism Planning & Development*, 1–19. doi:10.1080/21568316.2021.1873838

Gölgeci, I., Arslan, A., Khan, Z., & Kontkanen, M. (2021). Foreign firm operations and skills development of local employees in violence-hit countries. *Technological Forecasting and Social Change*, *162*, 120376. doi:10.1016/j.techfore.2020.120376

Goodwin, H. (2006). Community-based Tourism: Failing to Deliver? Id1 Insights 62. Brighton: Institute of Development Studies.

Goodwin, H., & Santilli, R. (2009). Community-based tourism: A success. *ICRT Occasional Paper, 11*(1), 37.

GraphicOnline. (2019, November 7). *Poor Roads Hamper Ecotourism Promotion.* Daily Graphic. https://www.graphic.com.gh/news/general-news/ghana-news-poor-roads-hamper-ecotourism-promotion.html

Gray, B. (1985). Conditions facilitating interorganizational domain. *Human Relations*, *38*(10), 911–936. doi:10.1177/001872678503801001

Gregersen, H. M., & Contreras, A. (1992). *Economic Assessment of Forestry Project Impacts* (Vol. 106). Food and Agriculture Organization.

Grimwood, Caton, & Muldoon. (2018). Tourism and Wellness: Travel for the Good of All? Lexington Books.

Guo, Y., Jiang, J., & Li, S. (2019). A sustainable tourism policy research review. *Sustainability, 11*(3187), 1–16.

Gursoy, D., & Rutherford, D. G. (2004). Host attitudes toward tourism: An improved structural model. *Annals of Tourism Research, 31*(3), 495–516. doi:10.1016/j.annals.2003.08.008

Gwang-Chol, C. (2006). Strategic planning in education: Some concepts and steps. Paris: Division of Educational Policies and Strategies, United Nations Educational, Scientific and Cultural Organization (UNESCO).

Hall, C. M. (2005). *Tourism: Rethinking the Social Science of Obility*. Prentice-Hall.

Hall, C. M. (2008). *Tourism Planning: Policies, Processes and Relationships*. Pearson Education.

Hall, S. (1996). Who needs identity? *Questions of Cultural Identity, 16*(2), 1–17.

Haraway, D. J. (2016). *Staying with the Trouble: Making Kin in the Chthulucene*. Duke University Press. doi:10.1215/9780822373780

Hardy, A., Beeton, R., & Pearson, L. (2002). Sustainable tourism:An overview of the concept and its position in relation to conceptualisations of tourism. *Journal of Sustainable Tourism, 10*(6), 475–496. doi:10.1080/09669580208667183

Harrison, D., & Schipani, S. (2007). Lao tourism and poverty alleviation: Community-based tourism and the private sector. *Current Issues in Tourism, 10*(2-3), 194–230. doi:10.2167/cit310.0

Harris, R. W. (2009). Tourism in Bario, Sarawak, Malaysia: A case study of pro-poor community-based tourism integrated into community development. *Asia Pacific Journal of Tourism Research, 14*(2), 125–135. doi:10.1080/10941660902847179

Hartley, A. (2009). Technology and translation. In J. Munday (Ed.), *The Routledge companion to translation studies* (pp. 106–127). Routledge.

Harwood, S. (2010). Planning for community based tourism in a remote location. *Sustainability, 2*(7), 1909–1923. doi:10.3390u2071909

Heffernan, A. (2020). *Community-Based Natural Resource Management and Global Climate Change in Namibia*. Academic Press.

Herricks, E., & Fisher, J. (2003). Application of GIS Technologies to Wildlife Management at DFW Airport. *2003 Bird Strike*. http://digitalcommons.unl.edu/cgi/viewcontent.cgi?article=1005&context=birdstrike2003

Higginbottom, K., & Tribe, A. (2004). Contributions of wildlife tourism to conservation. *Wildlife tourism: Impacts, management and planning*, 99-123.

Higgins-Desbiolles, F. (2006). More than an "industry": The forgotten power of tourism as a social force. *Tourism Management, 27*(6), 1192–1208. doi:10.1016/j.tourman.2005.05.020

Higgins-Desbiolles, F. (2008). Justice tourism and alternative globalisation. *Journal of Sustainable Tourism*, *16*(3), 345–364. doi:10.1080/09669580802154132

Higgins-Desbiolles, F. (2018). Sustainable tourism: Sustaining tourism or something more? *Tourism Management Perspectives*, *25*, 157–160. doi:10.1016/j.tmp.2017.11.017

Holden, E., Linnerud, K., & Banister, D. (2014). Sustainable development: Our common future revisited. *Global Environmental Change*, *26*, 130–139. doi:10.1016/j.gloenvcha.2014.04.006

Hollinshead, K. (1999). Surveillance of the worlds of tourism: Foucault and the eye-of-power. *Tourism Management*, *20*(1), 7–23. doi:10.1016/S0261-5177(98)00090-9

Hollinshead, K. (2007). 'Worldmaking' and the transformation of place and culture: The enlargement of Meethan's analysis of tourism and global change. In I. Ateljevic, N. Morgan, & A. Pritchard (Eds.), *The Critical Turn in Tourism Studies: Innovative Research Methodologies* (pp. 165–193). doi:10.1016/B978-0-08-045098-8.50016-7

Homans, G. C. (1961). *Social behavior*. Harcourt, Brace, and World.

Hooks, B. (2003). *Teaching Community: A Pedagogy of Hope* (Vol. 36). Psychology Press.

Howe, J. (2006, July 6). The rise of crowdsourcing. *Wired*. Retrieved October 18, 2020, from https://www.wired.com/2006/06/crowds

Huang, J. (2017). The dining experience of Beijing roast duck: A comparative study of Chinese and English online consumer reviews. *International Journal of Hospitality Management*, *66*, 117–129. doi:10.1016/j.ijhm.2017.07.003

Hultsman, J. (1995). Just tourism: An ethical framework. *Annals of Tourism Research*, *22*(3), 553–567. doi:10.1016/0160-7383(95)00011-T

Hung, K., Sirakaya-Turk, E., & Ingram, L. J. (2011). Testing the efficacy of an integrative model for community participation. *Journal of Travel Research*, *50*(3), 276–288. doi:10.1177/0047287510362781

Ibanescu, B., Stoleriu, O. M., Munteanu, A., & Oatu, C. (2018). The Impact of Tourism on Sustainable Development of Rural Areas: Evidence from Romania. *Sustainability*, *10*(3529), 1–19. doi:10.3390u10103529

Igbojekwe, P. A. (2011). Globalization: Challenges on the development of hospitality and tourism in Nigeria. *HATMAN Journal of Hospitality and Tourism*, *1*(1), 42–50.

Ijasan, K. C., & Izobo, M. O. (2012). Assessing community engagement in tourism planning and development in Nigeria: A case study of Arinta Waterfall Tourist Resort, Ipole Iloro Ekiti State. *Transnational Journal of Science and Technology*, *2*(4), 11–20.

Inpeng, S., & Nomnian, S. (2020). The use of Facebook in a TEFL program based on the TPACK model. *LEARN Journal: Language Education and Acquisition Research Network Journal*, *13*(2), 369–393.

Inphoo, P., & Nomnian, S. (2019). Dramatizing a Thai northeastern folklore to lessen high school students' communication anxiety. *PASAA*, *57*(1), 33–66.

International Union for Conservation of Nature and Natural Resources. (2010). *Parks and Reserves of Ghana: Management Effectiveness Assessment of Protected Areas*. https://portals.iucn.org/library/sites/library/files/documents/2010-073.pdf

Irvine, W., & Anderson, A. R. (2004). Small tourist firms in rural areas: Agility, vulnerability and survival in the face of crisis. *International Journal of Entrepreneurial Behaviour & Research*, *10*(4), 229–246. doi:10.1108/13552550410544204

Islam, M. S. (2015). Study on factors influencing tourism: Way forward for sustainable tourism in Bangladesh. *Journal of Tourism. Hospitality and Sports*, *6*(1), 1–13.

Jackson, P. (1993). *Towards a Cultural Politics of Consumption*. NP.

Jamal, T. & Menzel, C. (2009). Good actions in tourism. *Philosophical Issues in Tourism*, 227-243.

Jamal, T. (2019). *Justice and Ethics in Tourism*. Routledge. doi:10.4324/9781315162942

Jamal, T., & Camargo, B. A. (2014). Sustainable tourism, justice and an ethic of care: Toward the Just Destination. *Journal of Sustainable Tourism*, *22*(1), 11–30. doi:10.1080/09669582.2013.786084

James, P. (2006). *Globalism, nationalism, tribalism: Bringing theory back in towards a theory of abstract community*. Sage Publications.

Jannah, E. U., Muryani, C., & Rindarjono, M. G. (2018). Community-Based Tourism Development at Gajah Mungkur Wonogiri Tourist Attraction. *IOP Conference Series. Earth and Environmental Science*, *145*, 798–807. doi:10.1088/1755-1315/145/1/012017

Jennings, G. (2010). *Tourism Research*. John Wiley & Sons.

Jentoft, S. (2007). Limits of governability: Institutional implications for fisheries and coastal governance. *Marine Policy*, *31*(4), 360–370. doi:10.1016/j.marpol.2006.11.003

Jhaiyanuntana, A., & Nomnian, S. (2020). Intercultural communication challenges and strategies for the Thai undergraduate hotel interns. *PASAA*, *59*(1), 204–235.

Jiménez-Crespo, M. A. (2017). *Crowdsourcing and online collaborative translations: Expanding the limits of translation studies*. John Benjamins. doi:10.1075/btl.131

Joanna. (2020a, November 10). All about revision tasks. *Translators without Borders*. https://community.translatorswb.org/t/all-about-revision-tasks/25147

Joanna. (2020b, November 10). What are chunks? *Translators without Borders*. https://community.translatorswb.org/t/what-are-chunks/25146

Jones, S. (2005). Community-based ecotourism: The significance of social capital. *Annals of Tourism Research*, *32*(2), 303–324. doi:10.1016/j.annals.2004.06.007

Joppe, M. (1996). Sustainable community tourism development revisited. *Tourism Management*, *17*(7), 475–479. doi:10.1016/S0261-5177(96)00065-9

Journey, D. (2021). *Promlok...Best Shade of Khao Luang 2D1N.* http://journey-d.com/trip/phrom-lok-nakhon-si-2d1n?lang=en

Jun, W. (2019). The Application of cross-cultural awareness in English translation of tourist attractions. *International Conference on Humanities, Cultures, Arts and Design (ICHCAD 2019).* https://www.webofproceedings.org/proceedings_series/ART2L/ICHCAD%202019/AD051.pdf

Kabuusu, R. M., Amuno, J. B., Maseruka, Y., & Macpherson, C. (2018). Incidence and Risk Factors of Wildlife-Associated Human Injuries in Queen Elizabeth National Park, Uganda: A Retrospective Cohort Study. *Journal of Zoonotic Diseases and Public Health*, *2*, 1.

Kant, I. (1998). *Kant: Religion within the boundaries of mere reason: And other writings.* Cambridge University Press. doi:10.1017/CBO9780511809637

Kattiyapornpong, U., & Chuntamara, C. (2020). Assessing sustainable community-based tourism development in Thailand. In S. K. Walia (Ed.), *The Routledge Handbook of Community-Based Tourism Management* (pp. 397–414). Routledge. doi:10.4324/9780429274664-37

Keen, M., & Mahanty, S. (2005). Collaborative Learning: Bridging Scales and Interests. In M. Keen, V. A. Brown, & R. Dyball (Eds.), *Social Learning in Environmental Management: Towards a Sustainable Future* (pp. 1–21). EARTSCAN.

Kellert, S. R., Mehta, J. N., Ebbin, S. A., & Lichtenfeld, L. L. (2000). Community natural resource management: Promise, rhetoric and reality. *Society & Natural Resources*, *13*(8), 705–713. doi:10.1080/089419200750035575

Kenny, D. A., Kashy, D. A., & Bolger, N. (1998). Data analysis in social psychology. The handbook of social psychology, 1(4), 233-265

Khan, A., Khalid, M., & Ali, S. (2018). *Cultural Tourism under CPEC: A Case Study of Peshawar Valley, KPK.* CPEC. Available online at http://cpec.gov.pk/brain/public/uploads/documents/Magazine2-2.pdf

Kilahama, F. (2008). Impact of increased charcoal consumption to forests and woodlands in Tanzania. *Forest Ecology and Management*, *308*, 45–55.

Kim, K. (2002). *The effects of tourism impacts upon quality of life of residents in the community* (Doctoral dissertation, Virginia Tech). Available online at https://vtechworks.lib.vt.edu/bitstream/handle/10919/29954/Title_and_Text.pdf?sequence=1

Kincaid, J. (1988). *A small place*. Farrar Straus Giroux.

Kinyondo, A., & Pelizzo, R. (2020). Socioeconomic Impact of Tourism: The Case of Tanzania. *African Journal of Economic Review*, *8*(2), 239–257.

Kiper, T. (2013). Role of Ecotourism in Sustainable Tourism Development. *IntechOpen*, *31*, 773–802.

Kironde, J. L. (2009). Improving land sector governance in Africa: the case of Tanzania. Proceedings of "Land governance in support of the MDGs: Responding to New Challenges.

Kiss, A. (2004). Is community-based ecotourism a good use of biodiversity conservation funds? *Trends in Ecology & Evolution*, *19*(5), 232–237. doi:10.1016/j.tree.2004.03.010 PMID:16701261

Koc, E. (2020). *Cross-cultural aspects of tourism and hospitality: A services marketing and management perspective*. Routledge. doi:10.4324/9781003018193

Kolios, S., Vorobev, A. V., Vorobeva, G., & Stylios, C. (2017). *GIS and environmental monitoring. In Applications in the marine, atmospheric and geomagnetic fields*. Springer International Publishing AG.

Kontogeorgopoulous, N. (2005). Community-based ecotourism in Phuket and Ao Phangnga, Thailand: Partial victories and bittersweet remedies. *Journal of Sustainable Tourism*, *13*(1), 4–23. doi:10.1080/17501220508668470

Kooiman, J., & Bavinck, M. (2005). The governance perspective. *Fish for Life: Interactive Governance for Fisheries*, *3*, 11. doi:10.1017/9789048505326.002

Kooiman, J., Bavinck, M., Chuenpagdee, R., Mahon, R., & Pullin, R. (2008). Interactive governance and governability: An introduction. *The Journal of Transdisciplinary Environmental Studies*, *7*(1), 1–11.

Kozak, M., & Baloglu, S. (2011). *Managing and marketing tourist destinations: Strategies to gain a competitive edge* (1st ed.). Routledge.

Kraft, R. J. (1996). Service learning: An introduction to its theory, practice, and effects. *Education and Urban Society*, *28*(2), 131–159. doi:10.1177/0013124596028002001

Kruczek, Z., Kruczek, M., & Szromek, A. R. (2018). Possibilities of using the tourism area life cycle model to understand and provide sustainable solutions for tourism development in the Antarctic region. *Sustainability*, *10*(89), 1–16.

Kumar, N., Yamaç, S. S., & Velmurugan, A. (2015). Applications of remote sensing and GIS in natural resource management. *Journal of the Andaman Science Association*, *20*(1), 1–6.

Kwangseh, B. E. (2014). *Community based tourism (CBT) planning –An analysis of opportunities and barriers: A case study of Cameroon* (Doctoral Dissertation). Eastern Mediterranean University, Gazimağusa, Cyprus.

Kwanya, T. (2019). Leveraging tourism in Kenya through indigenous knowledge. Broadening Uganda's tourism product base through indigenous cultural resource utilization. In Positive tourism in Africa (pp. 106-116). Routledge. doi:10.4324/9780429428685-10

Kyara, V. C., Rahman, M. M., & Khanam, R. (2021). Tourism expansion and economic growth in Tanzania: A causality analysis. *Heliyon, 7*(5), e06966. doi:10.1016/j.heliyon.2021.e06966 PMID:34027171

Lachapelle, P. R., McCool, S. F., & Patterson, M. E. (2003). Barriers to effective natural resource planning in a 'Messy' world. *Society & Natural Resources, 16*(6), 473–490. doi:10.1080/08941920309151

Lansing, P., & De Vries, P. (2007). Sustainable tourism: Ethical alternative or marketing ploy? *Journal of Business Ethics, 72*(1), 77–85.

Latip, N. A., Rasoolimanesh, S. M., Jaafar, M., Marzuki, A., & Umar, M. U. (2018). Indigenous participation in conservation and tourism development: A case of native people of Sabah, Malaysia. *International Journal of Tourism Research, 20*(3), 400–409. doi:10.1002/jtr.2191

Lee, S., & Jamal, T. (2008). Environmental justice and environmental equity in tourism: Missing links to sustainability. *Journal of Ecotourism, 7*(1), 44–67. doi:10.2167/joe191.0

Lee, T. H., & Jan, F. H. (2019). Can community-based tourism contribute to sustainable development? Evidence from residents' perceptions of the sustainability. *Tourism Management, 70*, 368–380. doi:10.1016/j.tourman.2018.09.003

Lewis, R. C., & Chambers, R. E. (2000). *Marketing leadership in hospitality: Foundations and practices* (3rd ed.). John Wiley & Sons, Inc.

Lim, C. C., & Cooper, C. (2009). Beyond sustainability: Optimizing island tourism development. *International Journal of Tourism Research, 11*(1), 89–103. doi:10.1002/jtr.688

Lin, Z., Chen, Y., & Filieri, R. (2017). The Role of Resident's Perceived Tourism Impacts and Life Satisfaction. *Tourism Management, 61*, 436–442. doi:10.1016/j.tourman.2017.02.013

Li, W. (2006). Community decision-making participation in development. *Annals of Tourism Research, 33*(1), 132–143. doi:10.1016/j.annals.2005.07.003

Long, E. (Ed.). (1997). *From sociology to cultural studies: New perspectives*. Wiley-Blackwell.

Lovelock, B., & Lovelock, K. (2013). *The ethics of tourism: Critical and applied perspectives*. Routledge. doi:10.4324/9780203854532

Lo, Y.-C., & Janta, P. (2020). Resident's perspective on developing community-based tourism – A qualitative study of Muen Ngoen Kong Community, Chiang Mai, Thailand. *Frontiers in Psychology, 11*, 1493. doi:10.3389/fpsyg.2020.01493 PMID:32848976

Lu, J., & Nepal, S. (2009). Sustainable Tourism Research: An analysis of Papers Published in the Journal of Tourism. *Journal of Tourism, 17*(1), 5–16.

Lundberg, C., & Fredman, P. (2012). Success factors and constraints among nature-based tourism entrepreneurs. *Current Issues in Tourism, 15*(7), 649–671. doi:10.1080/13683500.2011.630458

Lwoga, M. N. B. (2013). Tourism development in Tanzania before and after Independence: Sustainability perspectives. *The Eastern African Journal of Hospitality, Leisure and Tourism, 1*(2), 1–26.

Lwoga, N. B. (2017). Local collaboration network in management of built heritage. *Journal of Cultural Heritage Management and Sustainable Development, 7*(3), 226–243. doi:10.1108/JCHMSD-05-2016-0034

Lyttleton, C., & Allcock, A. (2002). *Tourism as a tool for development: UNESCO-Lao National Tourism Authority Nam Ha Ecotourism Project.* External Review July 6–18, 2002. Vientiane: NTAL and UNESCO.

MacKay, K. J., & Campbell, J. M. (2012). Innovative community tourism in Uganda: International tourists, national networks, and local livelihoods. *Innovations in Sustainable Tourism (TEOROS),* 59-64.

Mackay, K. J., & Campbell, J. M. (2012). Innovative community tourism in Uganda. *Téoros - Journal of Tourism Research, 31*, 3.

Mak, K. L., Cheung, L. T. O., & Hui, D. L. H. (2017). Community Participation in the decision making process for sustainable tourism development in rural areas of Hong Kong, China. *Sustainability, 9*(6975), 1–13. doi:10.3390u9101695

Malik, A. R. (2018). The China–Pakistan Economic Corridor (CPEC): a game changer for Pakistan's economy. In B. R. Deepak (Ed.), *China's Global Rebalancing and the New Silk Road* (pp. 69–83). Springer. doi:10.1007/978-981-10-5972-8_7

Mambo, P. W., & Makunga, J. E. (2017). Application of Remote Sensing and Gis for Assessing Land Cover Resources Variability in the the Selous Game Reserves, Tanzania. *European Journal of Technology, 1*(2), 74. doi:10.47672/ejt.227

Mansvelt, J. (2005). Geographies of consumption. *Sage (Atlanta, Ga.).*

Manuel. (2019, October 1). Glossaries in Kató TM. *Translators without Borders.* https://community.translatorswb.org/t/glossaries-in-kato-tm/10955

Manyara, G., & Jones, E. (2007). Community-based tourism enterprises development in Kenya: An exploration of their potential as avenues of poverty reduction. *Journal of Sustainable Tourism, 15*(6), 628–644. doi:10.2167/jost723.0

Marczak, M. (2018). Translation Pedagogy in the Digital Age. How digital technologies have been altering translator education. *Angles. New Perspectives on the Anglophone World,* (7). doi:10.4000/angles.895

Marx, S. (2011). Community-based & pro-poor tourism. *International Trade Forum,* 22.

Marzuki, A. (2008). Decision making and community participation: A case study of the tourism industry in Langkawi. *Turizam: međunarodni znanstveno-stručni časopis, 56*(3), 227-241.

Marzuki, A., Hay, I., & James, J. (2012). Public participation shortcomings in tourism planning: The case of the Langkawi Islands, Malaysia. *Journal of Sustainable Tourism*, *20*(4), 585–602. doi:10.1080/09669582.2011.638384

Massey, D. (2005). Backfire at the Border: Why Enforcement without Legalization. *Trade Policy Analysis, 29.*

MateCat. (n.d.). *How to use glossary terms while translating*. https://site.matecat.com/support/translating-projects/use-glossary-terms-translating

Mathew, P. V., & Sreejesh, S. (2017). Impact of responsible tourism on destination sustainability and quality of life of community in tourism destinations. *Journal of Hospitality and Tourism Management*, *31*, 83–89. doi:10.1016/j.jhtm.2016.10.001

Mayaka, M., Croy, W. G., & Cox, J. W. (2019). A dimensional approach to community-based tourism: Recognising and differentiating form and context. *Annals of Tourism Research*, *74*(1), 177–190.

Mbaiwa, J. E. (2007). Local community attitudes towards wildlife conservation and community based natural resources management in Ngamiland District, Botswana. In B. Schuster & O. T. Thakadu (Eds.), *Natural Resources Management and People* (pp. 19–26). IUCN CBNRM Support Programme.

Mcconchie, J., & Mckinnon, J. (2002). Using GIS to Produce Community-Based. *Asean Biodiversity*, 27–34.

McIntosh, R. W., & Goeldner, C. R. (1986). *Tourism: Principles, practices and philosophies* (5th ed.). John Wiley & Sons.

Mearns, I. (2003). Community-based Tourism: The key to empowering the Sankuyo community in Botswana. *Africa Insight*, *33*(1), 29–32.

Menbere, I. P., & Menbere, T. P. (2017). Opportunities and challenges for community-based ecotourism development: A case study in Dinsho and Goba Woredas, Southeast Ethiopia. *International Journal of Ecology*, *4*(1), 5–16.

Menkiti, I. A. (1984). Person and community in African traditional thought. *African Philosophy: An Introduction*, *3*, 171–182.

Mensah, I. (2016). Effects of socio-demographic characteristics and perceived benefits of tourism on community participation in tourism in the Mesomagor Area of the Kakum National Park, Ghana. *Athens Journal of Tourism*, *3*(3), 211–230. doi:10.30958/ajt.3-3-3

Mensah, I. (2017). Benefits and Challenges of Community Based Ecotourism in Park-Fringe Communities: The Case of Mesomagor of Kakum National Park, Ghana. *Tourism Review International*, *21*(1), 81–98.

Mensah, I. (2017). Benefits and Challenges of Community-Based Ecotourism In Park-Fringe Communities: The Case of Mesomagor of Kakum National Park, Ghana. *Tourism Review International, 21*(1), 81–98. doi:10.3727/154427217X14866652018947

Mensah, I., & Adofo, E. (2013). Community Participation in Ecotourism: The Case of Bobiri Forest Reserve and Butterfly Sanctuary in Ashanti Region of Ghana. *American Journal of Tourism Management, 2*(1), 34–42.

Mgonja, J. T., Sirima, A., Backman, K. F., & Backman, S. J. (2015). Cultural community-based tourism in Tanzania: Lessons learned and way forward. *Development Southern Africa, 32*(3), 377–391. doi:10.1080/0376835X.2015.1010710

Michalos, A. C. (2008). Education, happiness, and wellbeing. *Social Indicators Research, 87*(3), 347–366. doi:10.100711205-007-9144-0

Miller, D. (2017). Fair trade: What does it mean and why does it matter? *Journal of Moral Philosophy, 14*(3), 249–269. doi:10.1163/17455243-46810053

Ministry of Lands and Natural Resources. (2012). *Ghana Forest and Wildlife Policy*. https://www.documents.clientearth.org/wp-content/uploads/library/2012-01-01-ghana-forest-and-wildlife-policy-2012-ext-en.pdf

Ministry of Tourism, Arts and Culture. (2019). *Medium Term Expenditure Framework for 2019-2022*. Accra: Republic of Ghana.

Ministry of Tourism. (2012). *Arts and Culture*. Ghana National Tourism Development Plan.

Mitchell, J., & Ashley, C. (2010). *Tourism and poverty reduction: Pathways to prosperity*. Earthscan.

Mkono, M. (2012). A netnographic examination of constructive authenticity in Victoria Falls tourist (restaurant) experiences. *International Journal of Hospitality Management, 31*(2), 387–394. doi:10.1016/j.ijhm.2011.06.013

Modern Ghana. (2009, July 9). *Formula for Sharing Proceeds from Kakum to be revisited*. Ghana News Agency. https://www.modernghana.com/news/226414/formula-for-sharing-proceeds-from-kakum-park-to-be-revisited.html

Molm, L. D. (1991). Social exchange: Satisfaction in power-dependence relations. *American Sociological Review, 56*(4), 475–493. doi:10.2307/2096269

Montefrio, M. J. F., & Sin, H. L. (2019). Elite governance of agritourism in the Philippines. *Journal of Sustainable Tourism, 27*(9), 1338–1354. doi:10.1080/09669582.2019.1621327

Moreno, A., & Becken, S. (2009). A climate change vulnerability assessment methodology for coastal tourism. *Journal of Sustainable Tourism, 17*(4), 473–488. doi:10.1080/09669580802651681

Moscardo, G. (Ed.). (2008). Building community capacity for tourism development. CAB International.

Moswete, N., & Lacey, G. (2015). "Women cannot lead": Empowering women through cultural tourism in Botswana. *Journal of Sustainable Tourism, 23*(4), 600–617. doi:10.1080/09669582.2014.986488

Moutinho, L. (2000). *Strategic management in tourism*. CABI Publishing.

Mtapuri, O., & Giampiccoli, A. (2016). Towards a comprehensive model of community-based tourism development. *The South African Geographical Journal, 98*(1), 154–168.

MTWA (Ministry of Tourism Wildlife and Antiquities). (2018). *Uganda Tourism Sector Annual Performance Report, Financial Year 2017/2018*. MTWA.

Mugizi, F., Ayorekire, J., & Obua, J. (2017). Factors that Influence Local Community Participation in Tourism in Murchison Falls Conservation Area. *Journal of Environmental Science & Engineering, A6*, 209–223.

Muir, C., & Sense, S. (2005). *The status of marine turtles in the United Republic of Tanzania, East Africa. Sea sense*. Tanzania Turtle & Dugong Conservation Programme.

Murphy, J. W. (2014). *Community-based interventions: Philosophy and action*. Springer. doi:10.1007/978-1-4899-8020-5

Murphy, P. E. (1985). *Tourism: A community approach*. Methuen.

Murray, G., Agyare, A., Dearden, P., & Rollins, R. (2019). Devolution, coordination, and community-based natural resource management in Ghana's community resource management areas. *African Geographical Review, 38*(4), 296–309. doi:10.1080/19376812.2018.1426022

Mwape, A. (2010). *Geographical Information Systems and Natural Resource Management in Zambia*. http://mro.massey.ac.nz/bitstream/handle/10179/1657/02_whole.pdf?sequence=1

Nanang, D., & Owusu, E. H. (2010). Estimating the Economic Value of recreation at the Kakum National Park, Ghana. *Nova Science Publishers, 12*, 247-267. https://www.researchgate.net/publication/284031218_Estimating_The_Economic_Value_of_Recreation_at_the_Kakum_National_Park_Ghana

Napu, N. (2019). Tourism promotional materials: Translation problems and implications on the text's effectiveness for tourism promotion. *The 1st International Conference on Education, Social Sciences and Humanities (ICESSHum 2019)*. 10.2991/icesshum-19.2019.71

Napu, N. (2018). Does commissioning process matter? Understanding translation quality with a case of tourism promotional texts. *Journal of Language and Linguistic Studies, 14*(4), 352–363.

National Bureau of Statistics (NBS). (2013). *2012 Population and housing census distribution by administrative areas*. Ministry of Finance.

National Environment Statute. (1995). Government of Uganda.

National Forestry and Tree Planting Act. (2003). Government of Uganda.

Neef, A. (2019). *Tourism, Land Grabs and Displacement: A Study with Particular Emphasis on the Global South*. Academic Press.

Nelson, F. (2003). *Community-based tourism in Northern Tanzania: Increasing opportunities, escalating conflicts and an uncertain future*. Rozenberg Publishers.

Nepal, S. K. (2020). Adventure travel and tourism after COVID-19–business as usual or opportunity to reset? *Tourism Geographies*, *22*(3), 646–650. doi:10.1080/14616688.2020.1760926

Nguyen, D. M. (2007). *Building a GIS database for ecotourism development in Ba Vi District*. Ha Tay Province.

Nickols, F. (2016). *Strategy: Definitions and meanings*. Distance Consulting LLC. Retrieved from http://www.nickols.us/strategy_definitions.pdf

Nigar, N. (2018). Ecotourism for Sustainable Development in Gilgit-Baltistan. *Strategic Studies*, *38*(3), 72–85.

Nitikasetsoontorn, S. (2015). The success factors of community-based tourism in Thailand. *NIDA Development Journal*, *5*(2), 24–58.

Nkwanyana, S. (2018). Conserving cultural heritage and the usage of indigenous knowledge: A case study of the Zululand District Municipality, Kwazulu-Natal. *African Journal of Hospitality, Tourism and Leisure*, *7*(2), 1–12.

Nomnian, S., Trupp, A., Niyomthong, W., Tangcharoensathaporn, P., & Charoenkonka, A. (2020). Language and community-based tourism: Use, needs, dependency, and limitations. *Austrian Journal of South-East Asian Studies*, *13*(1), 57–79. doi:10.14764/10.ASEAS-0029

Nord, C. (2018). *Translation as a purposeful activity: Functionalist approaches explained* (2nd ed.). St Jerome.

NPA. (2020). *National Planning Authority – Third National Development Plan (NDPIII) 2020/2021 – 2024/2025*. NPA.

Nuchdang, Y. (2019). Thailand development according to Pracharat's policy along with people's participation in Chachoengsao province. *NRRU Community Research Journal*, *13*(1), 13–25.

Nunkooa, R. (2016). Toward a More Comprehensive Use of Social Exchange Theory to Study Residents' Attitudes to Tourism. *Procedia Economics and Finance*, *39*, 588–596. doi:10.1016/S2212-5671(16)30303-3

Nurhidayati, S. E. (2015). The Evaluation Study of Implementation as Community-Based Tourism (CBT) in Supporting Sustainable Agritourism. *Society. Cultural Politics*, *28*(1), 1–10.

Nussbaum, M. C. (2017). Capabilities, entitlements, rights: Supplementation and critique. In *Justice and the Capabilities Approach* (pp. 173–187). Routledge. doi:10.4324/9781315251240-4

Nyaupane, G. P., & Chhetri, N. (2009). Vulnerability to climate change of nature-based tourism in the Nepalese Himalayas. *Tourism Geographies*, *11*(1), 95–119. doi:10.1080/14616680802643359

O'Brien, S. (2011). Collaborative translation. In Y. Gambier & L. Van Doorslaer (Eds.), Handbook of Translation Studies Volume 2 (pp. 17-20). John Benjamins. doi:10.1075/hts.2.col1

O'Hagan, M. (2009). Evolution of user-generated translation: Fansubs, translation hacking and crowdsourcing. *The Journal of Internationalization and Localization*, *1*(1), 94–121. doi:10.1075/jial.1.04hag

Ofori, B. Y., Owusu, E., & Attuquayefio, D. K. (2014). Ecological Status of the Mount Afadjato-Agumatsa Range in Ghana After a Decade of Local Community Management. *African Journal of Ecology*, *53*(1), 116–120. doi:10.1111/aje.12165

Oisebe, P. R. (2012). *GIS and natural resource management*. GIS Lounge.

Okazaki, E. (2008). A community-based tourism model: Its conception and use. *Journal of Sustainable Tourism*, *16*(5), 511–529. doi:10.1080/09669580802159594

Okoth-Ogendo, H. W. O., & Tumushabe, G. W. (1999). Governing the environment: political change and natural resources management in Eastern and Southern Africa (No. 10). Acts Press.

Olsen, D. H. (2011). Towards a religious view of tourism: Negotiating faith perspectives on tourism. *Tourism, Culture & Communication*, *11*(1), 17–30. doi:10.3727/109830411X13049571092633

Omoruyi Osagie, L. (2016). Significance of community Participation in Rural Development. *International Journal of Agriculture and Earth Science*, *2*, 44–53.

Onchwati, J., Sommerville, H., & Brockway, N. (2010). Sustainable Tourism Development in the Masai Mara National Reserve, Kenya, East Africa. *WIT Transaction on Ecology and the Environment*, *139*, 319–330. doi:10.2495/ST100281

Oreto, W. D. M., Lemieux, A. M., & Nobles, M. R. N. (2016). 'It's in my blood now': The satisfaction of rangers working in Queen Elizabeth National Park, Uganda Article. *Oryx*, 1–9.

Ota, T., Lonn, P., & Mizoue, N. (2021). Contribution of Community-Based Ecotourism to Forest Conservation and Local Livelihoods. In T. Yahara (Ed.), *Decision Science for Future Earth*. Springer. doi:10.1007/978-981-15-8632-3_9

Owusu, E. H. (2001). *Community Based Conservation in Ghana: The Potential of the Afadjato and Agumatsa Range for Ecotourism* [Unpublished PhD Thesis]. University of Kent at Canterbury.

Owusu, E. A. (2008). Ecotourism as a Conservation Tool – A Case of Afadjato-Agumatsa Convertion Area, Ghana. *Journal of Science and Technology (Ghana)*, *28*(3), 166–177.

Owusu, E. H., Ekpe, E. K., & Asamoah, A. A. (2005). Large Mammal Fauna of Afadjato and Agumatsa range in Ghana: An Important Bird Area. *West African Journal of Applied Ecology*, *8*(1), 18–28.

Özel, C. H., & Kozak, N. (2017). An exploratory study of resident perceptions toward the tourism industry in Cappadocia: A Social Exchange Theory approach. *Asia Pacific Journal of Tourism Research*, *22*(3), 284–300. doi:10.1080/10941665.2016.1236826

Pacione, M. (2009). *Urban Georgraphy: A Global Perspective* (3rd ed.). Routledge. doi:10.4324/9780203881927

Pailler, S., Naidoo, R., Burgess, N. D., Freeman, O. E., & Fisher, B. (2015). Impacts of community-based natural resource management on wealth, food security and child health in Tanzania. *PLoS One, 10*(7), e0133252. doi:10.1371/journal.pone.0133252 PMID:26186210

Pandey, R. N., Chettri, P., Kunwar, R. R., & Ghimire, G. (1995). *A case study on the effects of tourism on culture and the environment.* UNESCO Principal Regional Office for Asia and the Pacific. Available online at https://resourcecentre.savethechildren.net/sites/default/files/documents/2996.pdf

Pandit, B. H., Albano, A., & Kumar, C. (2009). Community-based forest enterprises in Nepal: An analysis of their role in increasing income benefits to the poor. *Small-scale Forestry, 8*(4), 447–462. doi:10.100711842-009-9094-2

Pangani District Council. (2003). *Investment Profile.* Unpublished.

Pangani District Council. (2003). *Land Annual Report.* Unpublished.

Pangani District Council. (2018). *Annual Report.* Unpublished.

Pant, D., Thapa, S., Singh, A., Bhattarai, M., & Molden, D. J. (2005). *Integrated management of water, forest and land resources in Nepal: opportunities for improved livelihood* (No. 618-2016-41137). Academic Press.

Pavasiriporn, R., & Juntaronanont, P. (2017). A study on Chinese services related to tourism in Bangkok inner Rattanakosin Island. *Journal of Chinese Language and Culture, Huachiew Chalermprakiet University, 4*(1), 119-128. https://so02.tci-thaijo.org/index.php/clcjn/article/view/194066

Pearce, D. G. (1992). *Tourist organizations.* Longman Scientific & Technical.

Pearce, P. L. (1994). Tourist-resident impact: Examples and emerging solutions. In W. F. Theobald (Ed.), *Global Tourism: The Next Decade* (pp. 103–123). Butterworth-Heinemann.

Pechlaner, H., & Sauerwein, E. (2002). Strategy implementation in the Alpine tourism Industry: International perspectives on travel and tourism development. *International Journal of Contemporary Hospitality Management, 14*(4), 157–168.

Pena-Regueiro, J., Sebastiá-Frasquet, M. T., Estornell, J., & Aguilar-Maldonado, J. A. (2020). Sentinel-2 Application to the Surface Characterization of Small Water Bodies in Wetlands. *Water (Basel), 12*(5), 1487. doi:10.3390/w12051487

Perdue, R. R., Tyrrell, T., & Uysal, M. (2010). Understanding the value of tourism: conceptual divergence. In D. Pearce & R. Butler (Eds.), *Tourism research: A 20:20 vision* (pp. 123–134). Goodfellow Publishers.

Phomnimitkul, K., Putjorn, T., & Vechpong, T. (2020). The culture management of the second-tier provinces to support local experience-based tourism under the LINK project. *Journal of Liberal Arts, Prince of Songkla University, 12*(1), 200-226. https://so03.tci-thaijo.org/index.php/journal-la/article/view/244540/165633

Phumpho, R., & Nomnian, S. (2019). Challenges of Thai businesspeople using English in ASEAN. *Kasetsart Journal of Social Sciences, 40*(3), 743–750.

Pillay, M., & Rogerson, C. (2013). Agriculture-tourism linkages and pro-poor impacts: The accommodation sector of urban coastal KwaZulu-Natal, South Africa. *Applied Geography (Sevenoaks, England), 36*(1), 49–58.

Popp, J. N., Priadka, P., & Kozmik, C. (2019). The rise of moose co-management and integration of Indigenous knowledge. *Human Dimensions of Wildlife, 24*(2), 159–167. doi:10.1080/10871209.2019.1545953

Pozzoni, B., & Kumar, N. (2005). *A review of the literature on participatory approaches to local development for an evaluation of the effectiveness of World Bank support for community-based and driven development approaches*. Academic Press.

Prayag, G., Cohen, S. A., & Yan, H. (2015). Potential Chinese travellers to western Europe: Segmenting motivations and service expectations. *Current Issues in Tourism, 18*(8), 725–743. doi:10.1080/13683500.2013.868413

Preacher, K. J., & Hayes, A. F. (2008). Asymptotic and resampling strategies for assessing and comparing indirect effects in multiple mediator models. *Behavior Research Methods, 40*(3), 879–891. doi:10.3758/BRM.40.3.879 PMID:18697684

Pretty, J. (1995). Participatory Learning for Sustainable Agriculture. *World Development, 23*(8), 1247–1263. doi:10.1016/0305-750X(95)00046-F

Pretty, J. (1995). The many interpretations of participation. *Focus (San Francisco, Calif.), 16*(4), 4–5.

Pritchard, A., Morgan, N., & Ateljevic, I. (2011). Hopeful tourism: A new transformative perspective. *Annals of Tourism Research, 38*(3), 941–963. doi:10.1016/j.annals.2011.01.004

Pym, A. (2011). What technology does to translating. *Translation and Interpreting, 3*(1), 1-9. http://www.trans-int.org/index.php/transint/article/view/121/81

Pym, A. (2013). Translation skill-sets in a machine translation age. *Meta, 58*(3), 487–503. doi:10.7202/1025047ar

Qin, X., Shen, H., Ye, S., & Lingqiang Zhou, L. (2021). Revisiting residents' support for tourism development: The role of tolerance. *Journal of Hospitality and Tourism Management, 47*, 114–123. doi:10.1016/j.jhtm.2021.02.010

Quan, J., Oudwater, N., Pender, J., & Martin, A. (2001). GIS and Participatory Approaches in Natural Resources Research. *Socio-Economic Methodologies for Natural Resources Research. Best Practice Guidelines.*, 1–36.

Rahman, M. (2010). Application of GIS in Ecotourism Development: a case study in sundarbans. Academic Press.

Ratten, V., Costa, C., & Bogers, M. (2019). Artisan, cultural and tourism entrepreneurship. *International Journal of Entrepreneurial Behaviour & Research*, *25*(4), 582–591. doi:10.1108/IJEBR-05-2018-0319

Ratz, T., & Michalko, G. (2011). The contribution of tourism to well-being and welfare: The case of Hungary. *International Journal of Sustainable Development*, *14*(3-4), 332–346. doi:10.1504/IJSD.2011.041968

Redpath, S. M., Young, J., Evely, A., Adams, W. M., Sutherland, W. J., Whitehouse, A., Amar, A., Lambert, R. A., Linnell, J. D. C., Watt, A., & Gutierrez, R. J. (2013). Understanding and managing conservation conflicts. *Trends in Ecology & Evolution*, *28*(2), 100–109. doi:10.1016/j.tree.2012.08.021 PMID:23040462

Ribeiro, M. A., Pinto, P., Silva, J. A., & Woosnam, K. M. (2017). Residents' attitudes and the adoption of pro-tourism behaviours: The case of developing island countries. *Tourism Management*, *61*, 523–537. doi:10.1016/j.tourman.2017.03.004

Richards, G. (2007). *Cultural tourism: Global and local perspectives*. Psychology Press.

Richards, G. (2018). Cultural tourism: A review of recent research and trends. *Journal of Hospitality and Tourism Management*, *36*, 12–21. doi:10.1016/j.jhtm.2018.03.005

Richards, G. (Ed.). (2007). *Cultural Tourism: Global and Local Perspectives*. Haworth Press.

Richards, G., & Richards, G. B. (Eds.). (1996). *Cultural tourism in Europe*. Cab International.

Ridder, H. (2017). The Theory Contribution of Case Study Research. *Business Research*, *10*, 281-305. https://link.springer.com/article/10.1007/s40685-017-0045-z

Ringo, J. E. (2019). The Usage of Indigenous Knowledge in Cultural Tourism Destinations. *Tengeru Community Development Journal*, *6*(1), 86–98.

Ritchie, W. B. (2004). Chaos, crises and disaster: A strategic approach to crisis management in the tourism industry. *Tourism Management*, *25*(6), 669–683.

Robinson, M., & Picard, D. (2006). *Tourism, Culture and Sustainable Development*. Programme "Culture, tourism, development", Division of cultural policies and intercultural dialogue, Culture Sector, UNESCO. Available online at https://unesdoc.unesco.org/ark:/48223/pf0000147578

Ross, S., & Wall, G. (1999). Ecotourism: Towards congruence between theory and practice. *Tourism Management*, *20*(1), 123–132. doi:10.1016/S0261-5177(98)00098-3

Ross, S., & Wall, G. (1999). Evaluating ecotourism: The case of North Sulawesi, Indonesia. *Tourism Management*, *20*(6), 673–682. doi:10.1016/S0261-5177(99)00040-0

Ruhanen, L., & Whitford, M. (2019). Cultural heritage and Indigenous tourism. *Journal of Heritage Tourism*, *19*(3), 179–191. doi:10.1080/1743873X.2019.1581788

Russell, M., & Scheyvens, R. (2012). Tourism and poverty alleviation in Fiji: Comparing the impacts of small-scale and large-scale tourism enterprises. *Journal of Sustainable Tourism*, *20*(3), 417–436.

Ryan, J., & Silvanto, S. (2011). A brand for all the nations: The development of the World Heritage Brand in emerging markets. *Marketing Intelligence & Planning*, *29*(3), 305–318. doi:10.1108/02634501111129266

Saarinen, J., & Manwa, H. (2008). Tourism as a socio-cultural encounter: Host-guest relations in tourism development in Botswana. *Botswana Notes and Records*, *39*, 43–53.

Saayman, M., & Giampiccoli, A. (2016). Community-based and pro-poor tourism: Initial assessment of their relation to community development. *European Journal of Tourism Research*, *12*, 145–190.

Salazar, N. B. (2012). Community-based cultural tourism: Issues, threats and opportunities. *Journal of Sustainable Tourism*, *20*(1), 9–22. doi:10.1080/09669582.2011.596279

Sandel, M. J. (1998). Religious liberty: Freedom of choice or freedom of conscience. *Secularism and Its Critics*, 73-93.

Sarod, T., & Kraisame, S. (2019). Urban multilingualism: A case study of the relationship between linguistic landscape and business in North Nana. *Journal of Language and Culture*, *38*(1), 24–40. https://so03.tci-thaijo.org/index.php/JLC/article/view/241369/163900

Satria, A., Matsuda, Y., & Sano, M. (2006). Questioning community based coral reef management systems: Case study of Awig-Awig in Gili Indah, Indonesia. *Environment, Development and Sustainability*, *8*(1), 99–118. doi:10.100710668-005-0909-9

Scheyvens, J. (2009). Empowerment. In R. Kitchin & N. Thrift (Eds.), *International Encyclopaedia of Human Geography* (pp. 464–470). Elsevier. doi:10.1016/B978-008044910-4.00092-4

Scheyvens, R. (1999). *The potential for ecotourism to facilitate the empowerment of local communities in Southern Africa: A summary report using selected case studies*. Massey University.

Scheyvens, R. (2002). *Tourism for development: Empowering communities*. Prentice Hall.

Schiffman, L. G., & Kanuk, L. L. (2000). *Consumer behavior* (7th ed.). Prentice Hall/Pearson Education.

Schmidt, P. R. (2017). *Community-based heritage in Africa: Unveiling local research and development initiatives*. Routledge. doi:10.4324/9781315270050

Scholtz, M., & Slabbert, E. (2017). Tourism is not just about the Money: A Comparison of Three South African Communities. *African Journal of Hospitality, Tourism and Leisure*, *6*(3).

Schwass, R. D. (2011). World Conservation Strategy of International Union for the Conservation of Nature and Natural Resources (IUCN). In D. V. J. Bell & Y. A. Cheung (Eds.), *Introduction to Sustainable Development* (pp. 147–171). Eolss Publishers.

Seaton, A. V., & Bennett, M. M. (1996). *Marketing tourism products: Concepts, issues, cases* (1st ed.). International Thompson Business Press.

Sebastiá-Frasquet, M. T., Altur, V., & Sanchis, J. A. (2014). Wetland planning: Current problems and environmental management proposals at supra-municipal scale (Spanish Mediterranean Coast). *Water (Basel)*, *6*(3), 620–641. doi:10.3390/w6030620

Sebele, L. S. (2010). Community-based tourism ventures, benefits and challenges: Khama rhino sanctuary trust, central district, Botswana. *Tourism Management*, *31*(1), 136–146. doi:10.1016/j.tourman.2009.01.005

Sène-Harper, A., & Séye, M. (2019). Community-based tourism around national parks in Senegal: The Implications of colonial legacies in current Management policies. *Tourism Planning & Development*, *16*(2), 217–234.

Shani, A., & Uriely, N. (2012). VFR tourism: The host experience. *Annals of Tourism Research*, *39*(1), 421–440. doi:10.1016/j.annals.2011.07.003

Sharifian, F. (2011). *Cultural conceptualisations and language: Theoretical framework and applications*. John Benjamins. doi:10.1075/clscc.1

Sharma, K. K. (2004). *Tourism and socio-cultural development*. Sarup and Sons.

Shorten, A., & Smith, J. (2017). Mixed Methods Research: Expanding the Evidence Base. *Evidence-Based Nursing*, *20*(3), 74–75. doi:10.1136/eb-2017-102699 PMID:28615184

Shrout, P. E., & Bolger, N. (2002). Mediation in experimental and nonexperimental studies: New procedures and recommendations. *Psychological Methods*, *7*(4), 422–445. doi:10.1037/1082-989X.7.4.422 PMID:12530702

Siakwah, P. (2018). Tourism Geography and Spatial Distribution of Tourism Sites in Ghana. *African Journal of Hospitality, Tourism and Leisure*, *7*(1), 1–19.

Silverman, D. (1993). *Interpreting Qualitative Data: Methods for Analysing Talk, Text and Interaction*. Sage.

Simmons, D. G. (1994). Community participation in tourism planning. *Tourism Management*, *15*(2), 98–108. doi:10.1016/0261-5177(94)90003-5

Simpson, K. (2001). Strategic Planning and Community Involvement in Contributors to Sustainable Tourism Development. *Current Issues in Tourism*, *4*(1), 3–41. doi:10.1080/13683500108667880

Simpson, M. C. (2008). Community benefit tourism initiatives—A conceptual oxymoron? *Tourism Management*, *29*(1), 1–18. doi:10.1016/j.tourman.2007.06.005

Sindiga, I. (1995). Wildlife based tourism in Kenya: Land use conflicts and government compensation policies over protected areas. *Journal of Tourism Studies*, *6*(2), 45.

Sipkes, J. (2002). Use of GIS and remote sensing in East Africa. *GIM International*, *16*(3), 6–9.

Skibitska, O. (2015). The language of tourism: Translating terms in tourist texts. *Translation Journal, 18*(4). https://www.translationjournal.net/October-2015/the-language-of-tourism-translating-terms-in-tourist-texts.html

Smith, M. K. (2009). *Issues in cultural tourism studies*. Routledge. doi:10.4324/9780203869857

Smith, V. L. (1996). The four Hs of tribal tourism: Acoma—a pueblo case study. *Progress in Tourism and Hospitality Research, 2*(3-4), 295–306. doi:10.1002/pth.6070020310

Snyder, C. R., Harris, C., Anderson, J. R., Holleran, S. A., Irving, L. M., Sigmon, S. T., ... Harney, P. (1991). The will and the ways: Development and validation of an individual-differences measure of hope. *Journal of Personality and Social Psychology, 60*(4), 570–585.

Snyder, H. (2019). Literature review as a research methodology: An overview and guidelines. *Journal of Business Research, 104*, 333–339. doi:10.1016/j.jbusres.2019.07.039

Sonti, S. (2015). Application of Geographic Information System (GIS) in Forest Management. *Journal of Geography & Natural Disasters, 5*(3). Advance online publication. doi:10.4172/2167-0587.1000145

Soontayatron, S. (2013). Thai interpretation of socio-cultural impacts of tourism development in beach resort. *American Journal of Tourism Management, 2*(2), 29–35.

Soutsas, K., Tsantopoulos, G., Arabatzis, G., & Christopoulou, O. (2006). Characteristics of tourism development in mountainous regions using categorical regression: The case of Metsovo (Greece). *International Journal of Sustainable Development and Planning, 1*(1), 32–45. doi:10.2495/SDP-V1-N1-32-45

Spenceley, A., & Meyer, D. (2012). Tourism and poverty reduction: Theory and practice in less economically developed countries. *Journal of Sustainable Tourism, 20*(3), 297–317. doi:10.1080/09669582.2012.668909

Sproule, K. W. (1996). Community-based ecotourism development: Identifying partners in the process. In E. Malek-Zadeh (Eds.), The Ecotourism Equation: Measuring the Impacts (pp. 233–250). New Haven, CT: Yale University.

Stavros, J. M., & Hinrichs, G. (2009). The thin book of SOAR: Building strengths-based strategy. The Thin Book Publishing Co.

Steven, D., & Jennifer, T. (2002). *Challenges and barriers to community participation in policy development.* Retrieved from http://www.ruralnovascotia.ca/documents/policy/challenges %20 and%20barriers.pdf

Steyn, J. N., & Spencer, J. P. (2013). Are sustainable tourism policies and strategies working in Tanzania?: Tourism and hospitality. *African Journal for Physical Health Education, Recreation and Dance, 19*(2), 299–312.

Stone, M. T., & Nyaupane, G. P. (2016). Protected areas, tourism and community livelihoods linkages: A comprehensive analysis approach. *Journal of Sustainable Tourism, 24*(5), 673–693. doi:10.1080/09669582.2015.1072207

Stone, M. T., & Rogerson, C. M. (2011). Community-based natural resource management and tourism: Nata bird sanctuary, Botswana. *Tourism Review International, 15*(1-2), 159–169. doi: 10.3727/154427211X13139345020570

Store, R., & Kangas, J. (2001). Integrating spatial multi-criteria evaluation and expert knowledge for GIS-based habitat suitability modelling. *Landscape and Urban Planning, 55*(2), 79–93. doi:10.1016/S0169-2046(01)00120-7

Stronza, A. (2007). The economic promise of ecotourism for conservation. *Journal of Ecotourism, 6*(3), 170–190. doi:10.2167/joe177.0

Stukas, A. A., & Dunlap, M. R. (2002). Community involvement: Theoretical approaches and educational initiatives. *The Journal of Social Issues, 58*(3), 411–427. doi:10.1111/1540-4560.00268

Suansri, P. (2003). *Community based tourism handbook.* Responsible Ecological Social Tour-REST. Retrieved from https://www.mekongtourism.org/wp-content/uploads/REST-CBT-Handbook-2003.pdf

Suebwongsuwan, W., & Nomnian, S. (2020). Thai hotel undergraduate interns' awareness and attitudes towards English as a lingua franca. *Indonesian Journal of Applied Linguistics, 9*(3), 704–716. doi:10.17509/ijal.v9i3.23221

Suhr, D. D. (2005). Principal component analysis vs. exploratory factor analysis (paper 203-30). In *Proceedings of the thirtieth annual SAS® users group international conference* (Vol. 203, p. 30). Available online at https://support.sas.com/resources/papers/proceedings/proceedings/sugi30/203-30.pdf

Sulaiman, M. Z. (2016). The misunderstood concept of translation in tourism promotion. *Translation & Interpreting, 8*(1), 53–68. doi:10.12807/ti.108201.2016.a04

Sulaiman, M. Z., & Wilson, R. (2019). *Translation and tourism: Strategies for effective cross-cultural promotion.* Springer. doi:10.1007/978-981-13-6343-6

Sullivan, S. (2003). Protest, conflict and litigation: dissent or libel in resistance to a conservancy in north-west Namibia. In E. Berglund & D. Anderson (Eds.), *Ethnographies of Conservation: Environmentalism and the Distribution of Privilege* (pp. 69–86). Bergahn Press.

Su, M. M., Long, Y., Wall, G., & Jin, M. (2016). Tourist–community interactions in ethnic tourism: Tuva villages, Kanas Scenic Area, China. *Journal of Tourism and Cultural Change, 14*(1), 1–26. doi:10.1080/14766825.2014.976228

Sumari, N. S., Shao, Z., & Kira, E. (2018). Challenges and opportunities for the advancement of GIS education in Tanzania. *Journal of Education and Practice, 8*(28), 67–75.

SYSTRAN. (n.d.). *What is machine translation? Rule-based machine translation vs. statistical machine translation*. https://www.systransoft.com/systran/translation-technology/what-is-machine-translation

Tanjua, A. (2003). Innovative human resource practices and organizational commitment: An empirical investigation. *International Journal of Human Resource Management, 14*(2), 175–197.

Tao, T. C. H., & Wall, G. (2009). Tourism as a Sustainable Livelihood Strategy. *Tourism Management, 30*(2), 90–98. doi:10.1016/j.tourman.2008.03.009

Tashakkori, A. (2007). Exploring the Nature of Research Questions in Mixed Research Methods. *Journal of Mixed Research Methods, 1*, 207–211. doi:10.1177/1558689807302814

Tawfik, G. M., Dila, K. A. S., Mohamed, M. Y. F., Tam, D. N. H., Kien, N. D., Ahmed, A. M., & Huy, N. T. (2019). A step by step guide for conducting a systematic review and meta-analysis with simulation data. *Tropical Medicine and Health, 47*(1), 1–9. doi:10.118641182-019-0165-6 PMID:31388330

Taylor, R. (2009). Community based natural resource management in Zimbabwe: The experience of CAMPFIRE. *Biodiversity and Conservation, 18*(10), 2563–2583. doi:10.100710531-009-9612-8

Terencia, N. M. (2018). Community based tourism and development in third world countries: the case of the Bamileke region of Cameroon: World academy of science, engineering and technology. *International Journal of Social and Tourism Sciences, 12*(1), 26–30.

The Mountain Institute. (2000). *Community based tourism for conservation and development: A resource kit*. Retrieved from http://www.mountain.org

The Republic of Uganda. (2011). *Office of the Auditor General: Value for money audit report on the management of wildlife conservation by the Uganda Wildlife Authority*. http://www.oag.go.ug/wp-content/uploads/2016/07/uwa-final-report-for-signature.pdf

The Republic of Uganda. (2014). *The Uganda Wildlife Policy: Ministry of Tourism, Wildlife and antiquities*. National Policy on Conservation and Sustainable Development of Wildlife Resources. http://extwprlegs1.fao.org/docs/pdf/uga174672.pdf

The Republic of Uganda. (2019). *The Uganda Wildlife Act*. file:///C:/Users/JM505%20Computers/Downloads/Uganda%20Wildlife%20Act,%202019.pdf

The World Bank in Uganda. (2021). *Economic Overview*. https://www.worldbank.org/en/country/uganda/overview

The World Bank. (2020). *Community-Driven Development*. The World Bank. https://www.worldbank.org/en/topic/communitydrivendevelopment

Theerapappisit, P. (2009). Pro-poor ethnic tourism in the Mekong: A study of three approaches in northern Thailand. *Asia Pacific Journal of Tourism Research, 14*(2), 20–31.

Thomas, F., Barya, G., & Katongole, C. (2011). *Opportunity study: Inclusive tourism programme: Uganda inclusive tourism*. File:///C:/Users/Jm505%20computers/Downloads/Itc_Opportunity_Study_Uganda.Pdf. Retrieved On 19/08/2021

TIES (The International Ecotourism Society). (n.d.). *Definition and principles: Ecotourism*. http://www.ecotourism.org/webmodules/webarticlesnet/templates/eco_template.aspx?articleid=95&zoneid=2

Timothy, D. J. (1999). Participatory planning a view of tourism in Indonesia. *Annals of Tourism Research*, *26*(2), 371–391. doi:10.1016/S0160-7383(98)00104-2

Timothy, D. J. (Ed.). (2017). *Managing heritage and cultural tourism resources: Critical essays, Volume one*. Routledge.

Toharudin, T., Suprijadi, J., Caraka, R. E., Pontoh, R. S., Chen, R. C., Lee, Y., & Pardamean, B. (2021). Social Vulnerability and How It Matters: A Bibliometric Analysis. *International Journal of Criminology and Sociology*, *10*, 610–619. doi:10.6000/1929-4409.2021.10.71

Tolkach, D., & King, B. (2015). Strengthening community-based tourism in a new resource-based island nation: Why and how? *Tourism Management*, *48*, 386–398. doi:10.1016/j.tourman.2014.12.013

Tonya, E. M., & Matilda, L. (2019, April). Unexplored Tourist Sites in Tanzania: Experience of Mbeya and Songwe Regions. In *International Conference on the Future of Tourism (ICFT)* (pp. 1-17). OUT.

Tony, B., & Etienne, N. (2002). Tourism as a local development strategy in South Africa. *The Geographical Journal*, *168*(3), 235–247.

Tosun, C. (1998). Roots of Unsustainable Tourism Development at the Local Level: The Case of Ürgüp in Turkey. *Tourism Management*, *19*(6), 595–610. doi:10.1016/S0261-5177(98)00068-5

Tosun, C. (1999). Towards a typology of community participation in the tourism development process. *Anatolia*, *10*(2), 113–134. doi:10.1080/13032917.1999.9686975

Tosun, C. (2000). Limits to community participation in the tourism development process in developing countries. *Tourism Management*, *21*(6), 613–633. doi:10.1016/S0261-5177(00)00009-1

Tosun, C. (2006). Expected nature of community participation in tourism development. *Tourism Management*, *27*(3), 493–504. doi:10.1016/j.tourman.2004.12.004

Tosun, C., & Timothy, D. J. (2001). Shortcomings in planning approaches to tourism development in developing countries: The case of Turkey. *International Journal of Contemporary Hospitality Management*, *13*(7), 352–359. doi:10.1108/09596110110403910

Tourism Authority of Thailand. (2011). *Nakhon Si Thammarat*. Retrieved from https://www.turismotailandes.com/wpcontent/uploads/2015/07/Nakhon%20Si%20Thammarat.pdf

Tourism Development International. (2006*). Nigeria tourism development master plan: Institutional capacity strengthening to the tourism sector in Nigeria.* Retrieved from http://www.ibadektyma.com/tourism_masterplan.pdf

Translators without Borders. (2017, June 15). *Translators without Borders and The Rosetta Foundation are merging*. https://translatorswithoutborders.org/translators-without-borders-and-the-rosetta-foundation-are-merging

Translators without Borders. (2020). *TWB Kató*. https://kato.translatorswb.org

Tribe, J. (1997). *Corporate strategy for tourism*. International Thomson Business Press.

Tribe, J. (2007). Critical tourism: Rules and resistance. In I. Ateljevic, N. Morgan, & A. Pritchard (Eds.), *The Critical Turn in Tourism Studies: Innovative Research Methodologies* (pp. 51–62). Routledge. doi:10.1016/B978-0-08-045098-8.50007-6

Tsegaw, W. E., & Teressa, D. K. (2017). Tour guiding quality assurance mechanisms and respective tourist satisfaction: Evidence from South Ethiopia. *Journal of Tourism & Hospitality (Los Angeles, Calif.)*, *6*(6), 1–9. doi:10.4172/2167-0269.1000317

Tucker, H. (2011). Critiquing pro-poor tourism – by Scheyvens Regina. *Asia Pacific Viewpoint*, *52*(3), 379–380.

Turk, T., & Gumusay, M. U. (2004, July). GIS design and application for tourism. In *XXth ISPRS Congress* (pp. 12-23). Academic Press.

Tweheyo, M., Hill, C. M., & Obua, J. (2005). Patterns of crop raiding by primates around the Budongo Forest Reserve, Uganda. *Wildlife Biology*, *11*(3), 237–247. doi:10.2981/0909-6396(2005)11[237:POCRBP]2.0.CO;2

Twerefou, D. K., & Ababio, D. K. A. (2012). An Economic Valuation of Kakum National Park: An Individual Travel Cost Approach. *African Journal of Environmental Science and Technology*, *6*(4), 199–207.

Uganda Vision Resource Centre. (2021). *Community tourism*. Accessed from website https://v2o40rc.wordpress.com/tour-and-volunteer-program/community-tourism/

Uganda Wildlife Act. (2019). Government of Uganda.

Uganda Wildlife Policy. (2014). Government of Uganda.

Uganda Wildlife Statute. (1996). Government of Uganda.

Ullah, I., & Kim, D. (2020). A model of collaborative governance for community-based trophy-hunting programs in developing countries. *Perspectives in Ecology and Conservation*, *18*(1), 145–160.

UNESCO. (2019). *Local and Indigenous Knowledge Systems (LINKS)*. https://en.unesco.org/links

United Nations – UNISDR, UNDP, & ISDR. (2009). *Making Disaster Risk Reduction Gender Sensitive: Policy and Practical Guidelines*. United Nations. Retrieved from https://www.unisdr.org/files/9922_MakingDisasterRiskReductionGenderSe.pdf

United Nations Department of Economic and Social Affairs (UNDESA). (2017). *The 17 Goals: History*. Retrieved from https://sdgs.un.org/goals

United Nations General Assembly. (1987). *Report of the world commission on environment and development: Our common future*. Oslo, Norway: United Nations General Assembly, Development and International Co-operation *Environment*.

United Nations World Tourism Organisation. (2013). *Sustainable Tourism for Development Guidebook: Enhancing Capabilities for Sustainable Tourism for Developing Countries*. https://www.e-unwto.org/doi/pdf/10.18111/9789284415496

United Nations World Tourism Organization (UNWTO). (2002). *Tourism and Poverty Alleviation*. UNWTO.

United Republic of Tanzania (URT). (1997). *National Land Policy*. Ministry of Land and Human Settlement Development.

United Republic of Tanzania (URT). (1999). *Land Act*. Government Printers.

United Republic of Tanzania (URT). (1999). *National Tourism Policy*. Ministry of Natural Resources and Tourism.

United Republic of Tanzania (URT). (2001). *Overview of the Tourism Sector 2001*(May). URT.

United Republic of Tanzania (URT). (2007). *Community Based Forest Management Guidelines; for the establishment of village land forest reserves and community forest reserves. Ministry of Natural Resources and Tourism*. Forest and Beekeeping Division.

United Republic of Tanzania (URT). (2007). *The Wildlife Policy of Tanzania*. Government Printer.

UNWTO. (2005). *City Tourism & Culture: The European Experience*. World Tourism Organization and European Travel Commission. Available online at_http://81.47.175.201/stodomingo/attachments/article/122/CityTourismCulture.pdf

Usadolo, S. E., & Caldwel, M. (2016). A stakeholder approach to community participation in a rural development project. *SAGE Open*, *6*(1), 2158244016638132. doi:10.1177/2158244016638132

USAID. (2020). *Making Wildlife Management Areas Deliver for Conservation and Communities*. USAID.

UWA. (2012). *Guidelines for Revenue Sharing between wildlife protected areas and adjacent local governments and communities*. Uganda Wildlife Authority.

UWA. (2021). *Queen Elizabeth National Park at a Glance*. https://www.ugandawildlife.org/explore-our-parks/parks-by-name-a-z/queen-elizabeth-national-park

Van Bibber, M. (1997). *It takes a community: A resource manual for community-based prevention of Fetal Alcohol Syndrome and Fetal Alcohol Effects*. Aboriginal Nurses Association of Canada.

Vanhove, N. (2002). Tourism policy - between competitiveness and sustainability: The case of Bruges. *Tourism Review*, *57*(3), 34–40.

Vanijdee, A. (2015). Translating Thai cultural information into language learning activities: the case of English for local museum personnel. *ASEAN Journal of Education*, *1*(1), 53-68. https://so01.tci-thaijo.org/index.php/AJE/article/view/180800

Veal, A. J. (2011). *Research methods for leisure & tourism* (4th ed.). Prentice Hall/Pearson Education.

Verka, J., & Angelina, N. (2008). The application of GIS and its components in tourism. *Yugoslav Journal of Operations Research*, *2*(18), 261–272.

Virtanen, P. (2003). Local management of global values: Community-based wildlife management in Zimbabwe and Zambia. *Society & Natural Resources*, *16*(3), 179–190. doi:10.1080/08941920309164

Vodouhê, F. G., Coulibaly, O., Adégbidi, A., & Sinsin, B. (2010). Community perception of biodiversity conservation within protected areas in Benin. *Forest Policy and Economics*, *12*(7), 505–512. doi:10.1016/j.forpol.2010.06.008

Wallace, G. N., & Pierce, S. M. (1996). An evaluation of ecotourism in Amazonas, Brazil. *Annals of Tourism Research*, *23*(4), 843–873. doi:10.1016/0160-7383(96)00009-6

Wang, B., & Ping, Y. (2020). Perceptions of machine translation and computer-aided translation by professionals and the general public: A survey study based on articles in professional journals and in the media. *International Journal of Translation, Interpretation, and Applied Linguistics*, *2*(2), 1–14. doi:10.4018/IJTIAL.20200701.oa1

Wang, M., Jiang, J., Xu, S., & Guo, Y. (2021). Community Participation and Residents' Support for Tourism Development in Ancient Villages: The Mediating Role of Perceptions of Conflicts in Tourism Community. *Sustainability*, *13*(2455), 1–15.

Wearing, S. L., Wearing, M., & McDonald, M. (2010). Understanding local power and interactional processes in sustainable tourism: Exploring village-tour operator relations on the Kokoda Track, Papua New Guinea. *Journal of Sustainable Tourism*, *18*(1), 61–76. doi:10.1080/09669580903071995

Weaver, D. (2010). Community-based tourism as strategic dead-end. *Tourism Recreation Research*, *35*(2), 206–208. doi:10.1080/02508281.2010.11081635

Wen, S., Cai, X., & Li, J. (2021). Pro-poor Tourism and Local Practices: An Empirical Study of an Enormous County in China. *SAGE Open*, 1–11. https://journals.sagepub.com/doi/pdf/10.1177/21582440211022740

Wester, P., Mishra, A., Mukherji, A., & Shrestha, A. B. (2019). *The Hindu Kush Himalaya assessment: Mountains, climate change, sustainability and people*. Cham: Springer Nature.

Western, D. (1994). Ecosystem Conservation and Rural Development: the Case of Amboseli. In Natural Cormections: Perspectives in Community-based Conservation. Academic Press.

Wiafe, E. D., & Agyei, D. (2013). Spices Richness, Diversity and Distribution of Amphibians Along Elevation Gradient on Mountain Afadjato Ghana. *Eurasian Journal of Forest Science, 1*(2), 68–76.

Wiafe, E. D., & Arku, F. S. (2014). The Role of Women in Community-Based Resource Management in the Afadjato Mountain Ecosystem, Ghana. *Applied Research Journal, 1*(1), 1–11.

Wilfred, P. (2010). Towards sustainable wildlife management areas in Tanzania. *Tropical Conservation Science, 3*(1), 103–116. doi:10.1177/194008291000300102

Wood, M. E. (2017). *Sustainable tourism on a finite planet: Environmental, business and policy solutions*. Taylor & Francis. doi:10.4324/9781315439808

World Bank. (2000). *Attacking Poverty: World Development Report 2000/01*. World Bank.

World Bank. (2018). *The World Bank Annual Report 2018*. The World Bank.

World Commission on Environment and Development. Special Working Session. (1987). World commission on environment and development. *Our Common Future, 17*(1), 1-91.

World Tourism Organization (UNWTO). (2016). *Tourism Highlights*. Calle Capitán Haya. https://www.e-unwto.org/doi/book/10.18111/9789284418145

World Tourism Organization and United Nations Development Programme. (2017). *Tourism and the Sustainable Development Goals – Journey to 2030*. Retrieved from https://www.e-unwto.org/doi/epdf/10.18111/9789284419401

World Travel & Tourism Council (2021). *Ghana 2021 Annual Research Key Highlights, Economic Impact Report*. Wttc.org.

Wright, S. C., & Boese, G. D. (2015). Meritocracy and tokenism. In International Encyclopedia of the Social and Behavioural Sciences. doi:10.1016/B978-0-08-097086-8.24074-9

Wu, M. Y., & Pearce, P. L. (2014). Tourism blogging motivations: Why do Chinese tourists create little 'lonely planets'? *Journal of Travel Research, 55*(4), 537–549. doi:10.1177/0047287514553057

WWF (World Wildlife Fund). (2001). *Guidelines for Community-Based Ecotourism Development*. WWF International.

Xu, C. (2019). Cross-culture in tourism English translation. *Education Research Frontier, 9*(3), 106–109.

Yanes, A., Zielinski, S., Diaz Cano, M., & Kim, S. I. (2019). Community-based tourism in developing countries: A framework for policy evaluation. *Sustainability, 11*(9), 2506. doi:10.3390u11092506

Yang, J., Ryan, C., & Zhang, L. (2013). Social Conflict in Communities Impacted by Tourism. *Tourism Management, 35*, 82–93. doi:10.1016/j.tourman.2012.06.002

Yang, L., & Wall, G. (2009). Ethnic Tourism: A framework and an application. *Tourism Management, 30*(4), 559–570. doi:10.1016/j.tourman.2008.09.008

Yang, X., & Hung, K. (2014). Poverty alleviation via tourism cooperatives in china: The story of Yuhu. *International Journal of Contemporary Hospitality Management, 26*(6), 879–906.

Yeatman, L., Du Toit, M., & Andreasen, L. (2003). *Developing approaches for improved natural woodland resource use and management: Lessons learnt in the communal areas of Bushbuckridge (Part 2)*. DWAF.

Yeboah, S. (1998). Small Mammal Diversity in the Kakum National Park in Ghana. *Journal of Science, 38*, 25–32.

Young, M., & Markham, F. (2020). Tourism, capital, and the commodification of place. *Progress in Human Geography, 44*(2), 276–296. doi:10.1177/0309132519826679

Yurdugul, H. (2008). Minimum sample size for Cronbach's coefficient alpha: a Monte-Carlo study. *Hacettepe Üniversitesi eğitim fakültesi dergisi*. Available online at http://www.efdergi.hacettepe.edu.tr/yonetim/icerik/makaleler/571-published.pdf

Zadel, Z., & Bogdan, S. (2013). Economic impact of cultural tourism. *UTMS Journal of Economics (Skopje), 4*(3), 355–366.

Zamani-Farahani, H., & Musa, G. (2012). The relationship between Islamic religiosity and residents' perceptions of socio-cultural impacts of tourism in Iran: Case studies of Sare'in and Masooleh. *Tourism Management, 33*(4), 802–814. doi:10.1016/j.tourman.2011.09.003

Zapata, M. J., Hall, C. M., Lindo, P., & Vanderschaeghe, M. (2011). Can community-based tourism contribute to development and poverty alleviation? Lessons from Nicaragua. *Current Issues in Tourism, 14*(8), 725–749. doi:10.1080/13683500.2011.559200

Zdenek, R. O., & Walsh, D. (2017). The Background and History of Community Development Organizations. In Navigating Community Development (pp. 21-52). Palgrave Macmillan. doi:10.1057/978-1-137-47701-9_3

Zeppel, H. (2006). *Indigenous ecotourism: Sustainable development and management* (Vol. 3). Cabi. doi:10.1079/9781845931247.0000

Zhang, H., Cho, T., Wang, H., & Ge, Q. (2018). The influence of cross-cultural awareness and tourist experience on authenticity, tourist satisfaction and acculturation in World Cultural Heritage Sites of Korea. *Sustainability, 10*(4), 927. doi:10.3390u10040927

Zieba, M. (2016). Tourism Flows and the Demand for Regional and City Theatres in Austria. *Journal of Cultural Economics, 40*(2), 191–221. doi:10.100710824-015-9250-9

Zulu, L. C. (2012). Neoliberalization, decentralization and community-based natural resources management in Malawi: The first sixteen years and looking ahead. *Progress in Development Studies*, *12*(2-3), 193–212. doi:10.1177/146499341101200307

Related References

To continue our tradition of advancing information science and technology research, we have compiled a list of recommended IGI Global readings. These references will provide additional information and guidance to further enrich your knowledge and assist you with your own research and future publications.

Abtahi, M. S., Behboudi, L., & Hasanabad, H. M. (2017). Factors Affecting Internet Advertising Adoption in Ad Agencies. *International Journal of Innovation in the Digital Economy*, *8*(4), 18–29. doi:10.4018/IJIDE.2017100102

Agrawal, S. (2017). The Impact of Emerging Technologies and Social Media on Different Business(es): Marketing and Management. In O. Rishi & A. Sharma (Eds.), *Maximizing Business Performance and Efficiency Through Intelligent Systems* (pp. 37–49). Hershey, PA: IGI Global. doi:10.4018/978-1-5225-2234-8.ch002

Alnoukari, M., Razouk, R., & Hanano, A. (2016). BSC-SI: A Framework for Integrating Strategic Intelligence in Corporate Strategic Management. *International Journal of Social and Organizational Dynamics in IT*, *5*(2), 1–14. doi:10.4018/IJSODIT.2016070101

Alnoukari, M., Razouk, R., & Hanano, A. (2016). BSC-SI, A Framework for Integrating Strategic Intelligence in Corporate Strategic Management. *International Journal of Strategic Information Technology and Applications*, *7*(1), 32–44. doi:10.4018/IJSITA.2016010103

Altındağ, E. (2016). Current Approaches in Change Management. In A. Goksoy (Ed.), *Organizational Change Management Strategies in Modern Business* (pp. 24–51). Hershey, PA: IGI Global. doi:10.4018/978-1-4666-9533-7.ch002

Alvarez-Dionisi, L. E., Turner, R., & Mittra, M. (2016). Global Project Management Trends. *International Journal of Information Technology Project Management, 7*(3), 54–73. doi:10.4018/IJITPM.2016070104

Anantharaman, R. N., Rajeswari, K. S., Angusamy, A., & Kuppusamy, J. (2017). Role of Self-Efficacy and Collective Efficacy as Moderators of Occupational Stress Among Software Development Professionals. *International Journal of Human Capital and Information Technology Professionals, 8*(2), 45–58. doi:10.4018/IJHCITP.2017040103

Aninze, F., El-Gohary, H., & Hussain, J. (2018). The Role of Microfinance to Empower Women: The Case of Developing Countries. *International Journal of Customer Relationship Marketing and Management, 9*(1), 54–78. doi:10.4018/IJCRMM.2018010104

Arsenijević, O. M., Orčić, D., & Kastratović, E. (2017). Development of an Optimization Tool for Intangibles in SMEs: A Case Study from Serbia with a Pilot Research in the Prestige by Milka Company. In M. Vemić (Ed.), *Optimal Management Strategies in Small and Medium Enterprises* (pp. 320–347). Hershey, PA: IGI Global. doi:10.4018/978-1-5225-1949-2.ch015

Aryanto, V. D., Wismantoro, Y., & Widyatmoko, K. (2018). Implementing Eco-Innovation by Utilizing the Internet to Enhance Firm's Marketing Performance: Study of Green Batik Small and Medium Enterprises in Indonesia. *International Journal of E-Business Research, 14*(1), 21–36. doi:10.4018/IJEBR.2018010102

Atiku, S. O., & Fields, Z. (2017). Multicultural Orientations for 21st Century Global Leadership. In N. Baporikar (Ed.), *Management Education for Global Leadership* (pp. 28–51). Hershey, PA: IGI Global. doi:10.4018/978-1-5225-1013-0.ch002

Atiku, S. O., & Fields, Z. (2018). Organisational Learning Dimensions and Talent Retention Strategies for the Service Industries. In N. Baporikar (Ed.), *Global Practices in Knowledge Management for Societal and Organizational Development* (pp. 358–381). Hershey, PA: IGI Global. doi:10.4018/978-1-5225-3009-1.ch017

Ávila, L., & Teixeira, L. (2018). The Main Concepts Behind the Dematerialization of Business Processes. In M. Khosrow-Pour, D.B.A. (Ed.), Encyclopedia of Information Science and Technology, Fourth Edition (pp. 888-898). Hershey, PA: IGI Global. doi:10.4018/978-1-5225-2255-3.ch076

Bartens, Y., Chunpir, H. I., Schulte, F., & Voß, S. (2017). Business/IT Alignment in Two-Sided Markets: A COBIT 5 Analysis for Media Streaming Business Models. In S. De Haes & W. Van Grembergen (Eds.), *Strategic IT Governance and Alignment in Business Settings* (pp. 82–111). Hershey, PA: IGI Global. doi:10.4018/978-1-5225-0861-8.ch004

Bashayreh, A. M. (2018). Organizational Culture and Organizational Performance. In W. Lee & F. Sabetzadeh (Eds.), *Contemporary Knowledge and Systems Science* (pp. 50–69). Hershey, PA: IGI Global. doi:10.4018/978-1-5225-5655-8.ch003

Bedford, D. A. (2018). Sustainable Knowledge Management Strategies: Aligning Business Capabilities and Knowledge Management Goals. In N. Baporikar (Ed.), *Global Practices in Knowledge Management for Societal and Organizational Development* (pp. 46–73). Hershey, PA: IGI Global. doi:10.4018/978-1-5225-3009-1.ch003

Benmoussa, F., Nakara, W. A., & Jaouen, A. (2016). The Use of Social Media by SMEs in the Tourism Industry. In I. Lee (Ed.), *Encyclopedia of E-Commerce Development, Implementation, and Management* (pp. 2159–2170). Hershey, PA: IGI Global. doi:10.4018/978-1-4666-9787-4.ch155

Berger, R. (2016). Indigenous Management and Bottom of Pyramid Countries: The Role of National Institutions. In U. Aung & P. Ordoñez de Pablos (Eds.), *Managerial Strategies and Practice in the Asian Business Sector* (pp. 107–123). Hershey, PA: IGI Global. doi:10.4018/978-1-4666-9758-4.ch007

Bharwani, S., & Musunuri, D. (2018). Reflection as a Process From Theory to Practice. In M. Khosrow-Pour, D.B.A. (Ed.), Encyclopedia of Information Science and Technology, Fourth Edition (pp. 1529-1539). Hershey, PA: IGI Global. doi:10.4018/978-1-5225-2255-3.ch132

Bhatt, G. D., Wang, Z., & Rodger, J. A. (2017). Information Systems Capabilities and Their Effects on Competitive Advantages: A Study of Chinese Companies. *Information Resources Management Journal*, *30*(3), 41–57. doi:10.4018/IRMJ.2017070103

Bhushan, M., & Yadav, A. (2017). Concept of Cloud Computing in ESB. In R. Bhadoria, N. Chaudhari, G. Tomar, & S. Singh (Eds.), *Exploring Enterprise Service Bus in the Service-Oriented Architecture Paradigm* (pp. 116–127). Hershey, PA: IGI Global. doi:10.4018/978-1-5225-2157-0.ch008

Bhushan, S. (2017). System Dynamics Base-Model of Humanitarian Supply Chain (HSCM) in Disaster Prone Eco-Communities of India: A Discussion on Simulation and Scenario Results. *International Journal of System Dynamics Applications*, *6*(3), 20–37. doi:10.4018/IJSDA.2017070102

Biswas, A., & De, A. K. (2017). On Development of a Fuzzy Stochastic Programming Model with Its Application to Business Management. In S. Trivedi, S. Dey, A. Kumar, & T. Panda (Eds.), *Handbook of Research on Advanced Data Mining Techniques and Applications for Business Intelligence* (pp. 353–378). Hershey, PA: IGI Global. doi:10.4018/978-1-5225-2031-3.ch021

Bücker, J., & Ernste, K. (2018). Use of Brand Heroes in Strategic Reputation Management: The Case of Bacardi, Adidas, and Daimler. In A. Erdemir (Ed.), *Reputation Management Techniques in Public Relations* (pp. 126–150). Hershey, PA: IGI Global. doi:10.4018/978-1-5225-3619-2.ch007

Bureš, V. (2018). Industry 4.0 From the Systems Engineering Perspective: Alternative Holistic Framework Development. In R. Brunet-Thornton & F. Martinez (Eds.), *Analyzing the Impacts of Industry 4.0 in Modern Business Environments* (pp. 199–223). Hershey, PA: IGI Global. doi:10.4018/978-1-5225-3468-6.ch011

Buzady, Z. (2017). Resolving the Magic Cube of Effective Case Teaching: Benchmarking Case Teaching Practices in Emerging Markets – Insights from the Central European University Business School, Hungary. In D. Latusek (Ed.), *Case Studies as a Teaching Tool in Management Education* (pp. 79–103). Hershey, PA: IGI Global. doi:10.4018/978-1-5225-0770-3.ch005

Campatelli, G., Richter, A., & Stocker, A. (2016). Participative Knowledge Management to Empower Manufacturing Workers. *International Journal of Knowledge Management, 12*(4), 37–50. doi:10.4018/IJKM.2016100103

Căpusneanu, S., & Topor, D. I. (2018). Business Ethics and Cost Management in SMEs: Theories of Business Ethics and Cost Management Ethos. In I. Oncioiu (Ed.), *Ethics and Decision-Making for Sustainable Business Practices* (pp. 109–127). Hershey, PA: IGI Global. doi:10.4018/978-1-5225-3773-1.ch007

Carneiro, A. (2016). Maturity in Health Organization Information Systems: Metrics and Privacy Perspectives. *International Journal of Privacy and Health Information Management, 4*(2), 1–18. doi:10.4018/IJPHIM.2016070101

Chan, R. L., Mo, P. L., & Moon, K. K. (2018). Strategic and Tactical Measures in Managing Enterprise Risks: A Study of the Textile and Apparel Industry. In K. Strang, M. Korstanje, & N. Vajjhala (Eds.), *Research, Practices, and Innovations in Global Risk and Contingency Management* (pp. 1–19). Hershey, PA: IGI Global. doi:10.4018/978-1-5225-4754-9.ch001

Chandan, H. C. (2016). Motivations and Challenges of Female Entrepreneurship in Developed and Developing Economies. In N. Baporikar (Ed.), *Handbook of Research on Entrepreneurship in the Contemporary Knowledge-Based Global Economy* (pp. 260–286). Hershey, PA: IGI Global. doi:10.4018/978-1-4666-8798-1.ch012

Charlier, S. D., Burke-Smalley, L. A., & Fisher, S. L. (2018). Undergraduate Programs in the U.S: A Contextual and Content-Based Analysis. In J. Mendy (Ed.), *Teaching Human Resources and Organizational Behavior at the College Level* (pp. 26–57). Hershey, PA: IGI Global. doi:10.4018/978-1-5225-2820-3.ch002

Chaudhuri, S. (2016). Application of Web-Based Geographical Information System (GIS) in E-Business. In U. Panwar, R. Kumar, & N. Ray (Eds.), *Handbook of Research on Promotional Strategies and Consumer Influence in the Service Sector* (pp. 389–405). Hershey, PA: IGI Global. doi:10.4018/978-1-5225-0143-5.ch023

Choudhuri, P. S. (2016). An Empirical Study on the Quality of Services Offered by the Private Life Insurers in Burdwan. In U. Panwar, R. Kumar, & N. Ray (Eds.), *Handbook of Research on Promotional Strategies and Consumer Influence in the Service Sector* (pp. 31–55). Hershey, PA: IGI Global. doi:10.4018/978-1-5225-0143-5.ch002

Dahlberg, T., Kivijärvi, H., & Saarinen, T. (2017). IT Investment Consistency and Other Factors Influencing the Success of IT Performance. In S. De Haes & W. Van Grembergen (Eds.), *Strategic IT Governance and Alignment in Business Settings* (pp. 176–208). Hershey, PA: IGI Global. doi:10.4018/978-1-5225-0861-8.ch007

Damnjanović, A. M. (2017). Knowledge Management Optimization through IT and E-Business Utilization: A Qualitative Study on Serbian SMEs. In M. Vemić (Ed.), *Optimal Management Strategies in Small and Medium Enterprises* (pp. 249–267). Hershey, PA: IGI Global. doi:10.4018/978-1-5225-1949-2.ch012

Daneshpour, H. (2017). Integrating Sustainable Development into Project Portfolio Management through Application of Open Innovation. In M. Vemić (Ed.), *Optimal Management Strategies in Small and Medium Enterprises* (pp. 370–387). Hershey, PA: IGI Global. doi:10.4018/978-1-5225-1949-2.ch017

Daniel, A. D., & Reis de Castro, V. (2018). Entrepreneurship Education: How to Measure the Impact on Nascent Entrepreneurs. In A. Carrizo Moreira, J. Guilherme Leitão Dantas, & F. Manuel Valente (Eds.), *Nascent Entrepreneurship and Successful New Venture Creation* (pp. 85–110). Hershey, PA: IGI Global. doi:10.4018/978-1-5225-2936-1.ch004

David, F., van der Sijde, P., & van den Besselaar, P. (2016). Enterpreneurial Incentives, Obstacles, and Management in University-Business Co-Operation: The Case of Indonesia. In J. Saiz-Álvarez (Ed.), *Handbook of Research on Social Entrepreneurship and Solidarity Economics* (pp. 499–518). Hershey, PA: IGI Global. doi:10.4018/978-1-5225-0097-1.ch024

David, R., Swami, B. N., & Tangirala, S. (2018). Ethics Impact on Knowledge Management in Organizational Development: A Case Study. In N. Baporikar (Ed.), *Global Practices in Knowledge Management for Societal and Organizational Development* (pp. 19–45). Hershey, PA: IGI Global. doi:10.4018/978-1-5225-3009-1.ch002

Delias, P., & Lakiotaki, K. (2018). Discovering Process Horizontal Boundaries to Facilitate Process Comprehension. *International Journal of Operations Research and Information Systems*, *9*(2), 1–31. doi:10.4018/IJORIS.2018040101

Denholm, J., & Lee-Davies, L. (2018). Success Factors for Games in Business and Project Management. In *Enhancing Education and Training Initiatives Through Serious Games* (pp. 34–68). Hershey, PA: IGI Global. doi:10.4018/978-1-5225-3689-5.ch002

Deshpande, M. (2017). Best Practices in Management Institutions for Global Leadership: Policy Aspects. In N. Baporikar (Ed.), *Management Education for Global Leadership* (pp. 1–27). Hershey, PA: IGI Global. doi:10.4018/978-1-5225-1013-0.ch001

Deshpande, M. (2018). Policy Perspectives for SMEs Knowledge Management. In N. Baporikar (Ed.), *Knowledge Integration Strategies for Entrepreneurship and Sustainability* (pp. 23–46). Hershey, PA: IGI Global. doi:10.4018/978-1-5225-5115-7.ch002

Dezdar, S. (2017). ERP Implementation Projects in Asian Countries: A Comparative Study on Iran and China. *International Journal of Information Technology Project Management*, *8*(3), 52–68. doi:10.4018/IJITPM.2017070104

Domingos, D., Martinho, R., & Varajão, J. (2016). Controlled Flexibility in Healthcare Processes: A BPMN-Extension Approach. In M. Cruz-Cunha, I. Miranda, R. Martinho, & R. Rijo (Eds.), *Encyclopedia of E-Health and Telemedicine* (pp. 521–535). Hershey, PA: IGI Global. doi:10.4018/978-1-4666-9978-6.ch040

Domingos, D., Respício, A., & Martinho, R. (2017). Reliability of IoT-Aware BPMN Healthcare Processes. In C. Reis & M. Maximiano (Eds.), *Internet of Things and Advanced Application in Healthcare* (pp. 214–248). Hershey, PA: IGI Global. doi:10.4018/978-1-5225-1820-4.ch008

Dosumu, O., Hussain, J., & El-Gohary, H. (2017). An Exploratory Study of the Impact of Government Policies on the Development of Small and Medium Enterprises in Developing Countries: The Case of Nigeria. *International Journal of Customer Relationship Marketing and Management*, *8*(4), 51–62. doi:10.4018/IJCRMM.2017100104

Durst, S., Bruns, G., & Edvardsson, I. R. (2017). Retaining Knowledge in Smaller Building and Construction Firms. *International Journal of Knowledge and Systems Science*, *8*(3), 1–12. doi:10.4018/IJKSS.2017070101

Edvardsson, I. R., & Durst, S. (2017). Outsourcing, Knowledge, and Learning: A Critical Review. *International Journal of Knowledge-Based Organizations*, *7*(2), 13–26. doi:10.4018/IJKBO.2017040102

Edwards, J. S. (2018). Integrating Knowledge Management and Business Processes. In M. Khosrow-Pour, D.B.A. (Ed.), Encyclopedia of Information Science and Technology, Fourth Edition (pp. 5046-5055). Hershey, PA: IGI Global. doi:10.4018/978-1-5225-2255-3.ch437

Ejiogu, A. O. (2018). Economics of Farm Management. In *Agricultural Finance and Opportunities for Investment and Expansion* (pp. 56–72). Hershey, PA: IGI Global. doi:10.4018/978-1-5225-3059-6.ch003

Ekanem, I., & Abiade, G. E. (2018). Factors Influencing the Use of E-Commerce by Small Enterprises in Nigeria. *International Journal of ICT Research in Africa and the Middle East*, *7*(1), 37–53. doi:10.4018/IJICTRAME.2018010103

Ekanem, I., & Alrossais, L. A. (2017). Succession Challenges Facing Family Businesses in Saudi Arabia. In P. Zgheib (Ed.), *Entrepreneurship and Business Innovation in the Middle East* (pp. 122–146). Hershey, PA: IGI Global. doi:10.4018/978-1-5225-2066-5.ch007

El Faquih, L., & Fredj, M. (2017). Ontology-Based Framework for Quality in Configurable Process Models. *Journal of Electronic Commerce in Organizations*, *15*(2), 48–60. doi:10.4018/JECO.2017040104

El-Gohary, H., & El-Gohary, Z. (2016). An Attempt to Explore Electronic Marketing Adoption and Implementation Aspects in Developing Countries: The Case of Egypt. *International Journal of Customer Relationship Marketing and Management*, *7*(4), 1–26. doi:10.4018/IJCRMM.2016100101

Entico, G. J. (2016). Knowledge Management and the Medical Health Librarians: A Perception Study. In J. Yap, M. Perez, M. Ayson, & G. Entico (Eds.), *Special Library Administration, Standardization and Technological Integration* (pp. 52–77). Hershey, PA: IGI Global. doi:10.4018/978-1-4666-9542-9.ch003

Faisal, M. N., & Talib, F. (2017). Building Ambidextrous Supply Chains in SMEs: How to Tackle the Barriers? *International Journal of Information Systems and Supply Chain Management, 10*(4), 80–100. doi:10.4018/IJISSCM.2017100105

Fernandes, T. M., Gomes, J., & Romão, M. (2017). Investments in E-Government: A Benefit Management Case Study. *International Journal of Electronic Government Research, 13*(3), 1–17. doi:10.4018/IJEGR.2017070101

Fouda, F. A. (2016). A Suggested Curriculum in Career Education to Develop Business Secondary Schools Students' Career Knowledge Management Domains and Professional Thinking. *International Journal of Technology Diffusion, 7*(2), 42–62. doi:10.4018/IJTD.2016040103

Gallardo-Vázquez, D., & Pajuelo-Moreno, M. L. (2016). How Spanish Universities are Promoting Entrepreneurship through Your Own Lines of Teaching and Research? In L. Carvalho (Ed.), *Handbook of Research on Entrepreneurial Success and its Impact on Regional Development* (pp. 431–454). Hershey, PA: IGI Global. doi:10.4018/978-1-4666-9567-2.ch019

Gao, S. S., Oreal, S., & Zhang, J. (2018). Contemporary Financial Risk Management Perceptions and Practices of Small-Sized Chinese Businesses. In I. Management Association (Ed.), Global Business Expansion: Concepts, Methodologies, Tools, and Applications (pp. 917-931). Hershey, PA: IGI Global. doi:10.4018/978-1-5225-5481-3.ch041

Garg, R., & Berning, S. C. (2017). Indigenous Chinese Management Philosophies: Key Concepts and Relevance for Modern Chinese Firms. In B. Christiansen & G. Koc (Eds.), *Transcontinental Strategies for Industrial Development and Economic Growth* (pp. 43–57). Hershey, PA: IGI Global. doi:10.4018/978-1-5225-2160-0.ch003

Gencer, Y. G. (2017). Supply Chain Management in Retailing Business. In U. Akkucuk (Ed.), *Ethics and Sustainability in Global Supply Chain Management* (pp. 197–210). Hershey, PA: IGI Global. doi:10.4018/978-1-5225-2036-8.ch011

Giacosa, E. (2016). Innovation in Luxury Fashion Businesses as a Means for the Regional Development. In L. Carvalho (Ed.), *Handbook of Research on Entrepreneurial Success and its Impact on Regional Development* (pp. 206–222). Hershey, PA: IGI Global. doi:10.4018/978-1-4666-9567-2.ch010

Giacosa, E. (2018). The Increasing of the Regional Development Thanks to the Luxury Business Innovation. In L. Carvalho (Ed.), *Handbook of Research on Entrepreneurial Ecosystems and Social Dynamics in a Globalized World* (pp. 260–273). Hershey, PA: IGI Global. doi:10.4018/978-1-5225-3525-6.ch011

Gianni, M., & Gotzamani, K. (2016). Integrated Management Systems and Information Management Systems: Common Threads. In P. Papajorgji, F. Pinet, A. Guimarães, & J. Papathanasiou (Eds.), *Automated Enterprise Systems for Maximizing Business Performance* (pp. 195–214). Hershey, PA: IGI Global. doi:10.4018/978-1-4666-8841-4.ch011

Gianni, M., Gotzamani, K., & Linden, I. (2016). How a BI-wise Responsible Integrated Management System May Support Food Traceability. *International Journal of Decision Support System Technology*, *8*(2), 1–17. doi:10.4018/IJDSST.2016040101

Glykas, M., & George, J. (2017). Quality and Process Management Systems in the UAE Maritime Industry. *International Journal of Productivity Management and Assessment Technologies*, *5*(1), 20–39. doi:10.4018/IJPMAT.2017010102

Glykas, M., Valiris, G., Kokkinaki, A., & Koutsoukou, Z. (2018). Banking Business Process Management Implementation. *International Journal of Productivity Management and Assessment Technologies*, *6*(1), 50–69. doi:10.4018/IJPMAT.2018010104

Gomes, J., & Romão, M. (2017). The Balanced Scorecard: Keeping Updated and Aligned with Today's Business Trends. *International Journal of Productivity Management and Assessment Technologies*, *5*(2), 1–15. doi:10.4018/IJPMAT.2017070101

Gomes, J., & Romão, M. (2017). Aligning Information Systems and Technology with Benefit Management and Balanced Scorecard. In S. De Haes & W. Van Grembergen (Eds.), *Strategic IT Governance and Alignment in Business Settings* (pp. 112–131). Hershey, PA: IGI Global. doi:10.4018/978-1-5225-0861-8.ch005

Grefen, P., & Turetken, O. (2017). Advanced Business Process Management in Networked E-Business Scenarios. *International Journal of E-Business Research*, *13*(4), 70–104. doi:10.4018/IJEBR.2017100105

Haider, A., & Saetang, S. (2017). Strategic IT Alignment in Service Sector. In S. Rozenes & Y. Cohen (Eds.), *Handbook of Research on Strategic Alliances and Value Co-Creation in the Service Industry* (pp. 231–258). Hershey, PA: IGI Global. doi:10.4018/978-1-5225-2084-9.ch012

Haider, A., & Tang, S. S. (2016). Maximising Value Through IT and Business Alignment: A Case of IT Governance Institutionalisation at a Thai Bank. *International Journal of Technology Diffusion*, *7*(3), 33–58. doi:10.4018/IJTD.2016070104

Hajilari, A. B., Ghadaksaz, M., & Fasghandis, G. S. (2017). Assessing Organizational Readiness for Implementing ERP System Using Fuzzy Expert System Approach. *International Journal of Enterprise Information Systems*, *13*(1), 67–85. doi:10.4018/IJEIS.2017010105

Haldorai, A., Ramu, A., & Murugan, S. (2018). Social Aware Cognitive Radio Networks: Effectiveness of Social Networks as a Strategic Tool for Organizational Business Management. In H. Bansal, G. Shrivastava, G. Nguyen, & L. Stanciu (Eds.), *Social Network Analytics for Contemporary Business Organizations* (pp. 188–202). Hershey, PA: IGI Global. doi:10.4018/978-1-5225-5097-6.ch010

Hall, O. P. Jr. (2017). Social Media Driven Management Education. *International Journal of Knowledge-Based Organizations*, *7*(2), 43–59. doi:10.4018/IJKBO.2017040104

Hanifah, H., Halim, H. A., Ahmad, N. H., & Vafaei-Zadeh, A. (2017). Innovation Culture as a Mediator Between Specific Human Capital and Innovation Performance Among Bumiputera SMEs in Malaysia. In N. Ahmad, T. Ramayah, H. Halim, & S. Rahman (Eds.), *Handbook of Research on Small and Medium Enterprises in Developing Countries* (pp. 261–279). Hershey, PA: IGI Global. doi:10.4018/978-1-5225-2165-5.ch012

Hartlieb, S., & Silvius, G. (2017). Handling Uncertainty in Project Management and Business Development: Similarities and Differences. In Y. Raydugin (Ed.), *Handbook of Research on Leveraging Risk and Uncertainties for Effective Project Management* (pp. 337–362). Hershey, PA: IGI Global. doi:10.4018/978-1-5225-1790-0.ch016

Hass, K. B. (2017). Living on the Edge: Managing Project Complexity. In Y. Raydugin (Ed.), *Handbook of Research on Leveraging Risk and Uncertainties for Effective Project Management* (pp. 177–201). Hershey, PA: IGI Global. doi:10.4018/978-1-5225-1790-0.ch009

Hassan, A., & Privitera, D. S. (2016). Google AdSense as a Mobile Technology in Education. In J. Holland (Ed.), *Wearable Technology and Mobile Innovations for Next-Generation Education* (pp. 200–223). Hershey, PA: IGI Global. doi:10.4018/978-1-5225-0069-8.ch011

Hassan, A., & Rahimi, R. (2016). Consuming "Innovation" in Tourism: Augmented Reality as an Innovation Tool in Digital Tourism Marketing. In N. Pappas & I. Bregoli (Eds.), *Global Dynamics in Travel, Tourism, and Hospitality* (pp. 130–147). Hershey, PA: IGI Global. doi:10.4018/978-1-5225-0201-2.ch008

Hawking, P., & Carmine Sellitto, C. (2017). Developing an Effective Strategy for Organizational Business Intelligence. In M. Tavana (Ed.), *Enterprise Information Systems and the Digitalization of Business Functions* (pp. 222–237). Hershey, PA: IGI Global. doi:10.4018/978-1-5225-2382-6.ch010

Hawking, P., & Sellitto, C. (2017). A Fast-Moving Consumer Goods Company and Business Intelligence Strategy Development. *International Journal of Enterprise Information Systems, 13*(2), 22–33. doi:10.4018/IJEIS.2017040102

Hawking, P., & Sellitto, C. (2017). Business Intelligence Strategy: Two Case Studies. *International Journal of Business Intelligence Research*, *8*(2), 17–30. doi:10.4018/IJBIR.2017070102

Haynes, J. D., Arockiasamy, S., Al Rashdi, M., & Al Rashdi, S. (2016). Business and E Business Strategies for Coopetition and Thematic Management as a Sustained Basis for Ethics and Social Responsibility in Emerging Markets. In M. Al-Shammari & H. Masri (Eds.), *Ethical and Social Perspectives on Global Business Interaction in Emerging Markets* (pp. 25–39). Hershey, PA: IGI Global. doi:10.4018/978-1-4666-9864-2.ch002

Hee, W. J., Jalleh, G., Lai, H., & Lin, C. (2017). E-Commerce and IT Projects: Evaluation and Management Issues in Australian and Taiwanese Hospitals. *International Journal of Public Health Management and Ethics*, *2*(1), 69–90. doi:10.4018/IJPHME.2017010104

Hernandez, A. A. (2018). Exploring the Factors to Green IT Adoption of SMEs in the Philippines. *Journal of Cases on Information Technology*, *20*(2), 49–66. doi:10.4018/JCIT.2018040104

Hernandez, A. A., & Ona, S. E. (2016). Green IT Adoption: Lessons from the Philippines Business Process Outsourcing Industry. *International Journal of Social Ecology and Sustainable Development*, *7*(1), 1–34. doi:10.4018/IJSESD.2016010101

Hollman, A., Bickford, S., & Hollman, T. (2017). Cyber InSecurity: A Post-Mortem Attempt to Assess Cyber Problems from IT and Business Management Perspectives. *Journal of Cases on Information Technology*, *19*(3), 42–70. doi:10.4018/JCIT.2017070104

Igbinakhase, I. (2017). Responsible and Sustainable Management Practices in Developing and Developed Business Environments. In Z. Fields (Ed.), *Collective Creativity for Responsible and Sustainable Business Practice* (pp. 180–207). Hershey, PA: IGI Global. doi:10.4018/978-1-5225-1823-5.ch010

Ilahi, L., Ghannouchi, S. A., & Martinho, R. (2016). A Business Process Management Approach to Home Healthcare Processes: On the Gap between Intention and Reality. In M. Cruz-Cunha, I. Miranda, R. Martinho, & R. Rijo (Eds.), *Encyclopedia of E-Health and Telemedicine* (pp. 439–457). Hershey, PA: IGI Global. doi:10.4018/978-1-4666-9978-6.ch035

Iwata, J. J., & Hoskins, R. G. (2017). Managing Indigenous Knowledge in Tanzania: A Business Perspective. In P. Jain & N. Mnjama (Eds.), *Managing Knowledge Resources and Records in Modern Organizations* (pp. 198–214). Hershey, PA: IGI Global. doi:10.4018/978-1-5225-1965-2.ch012

Jabeen, F., Ahmad, S. Z., & Alkaabi, S. (2016). The Internationalization Decision-Making of United Arab Emirates Family Businesses. In N. Zakaria, A. Abdul-Talib, & N. Osman (Eds.), *Handbook of Research on Impacts of International Business and Political Affairs on the Global Economy* (pp. 1–22). Hershey, PA: IGI Global. doi:10.4018/978-1-4666-9806-2.ch001

Jain, P. (2017). Ethical and Legal Issues in Knowledge Management Life-Cycle in Business. In P. Jain & N. Mnjama (Eds.), *Managing Knowledge Resources and Records in Modern Organizations* (pp. 82–101). Hershey, PA: IGI Global. doi:10.4018/978-1-5225-1965-2.ch006

Jamali, D., Abdallah, H., & Matar, F. (2016). Opportunities and Challenges for CSR Mainstreaming in Business Schools. *International Journal of Technology and Educational Marketing*, *6*(2), 1–29. doi:10.4018/IJTEM.2016070101

James, S., & Hauli, E. (2017). Holistic Management Education at Tanzanian Rural Development Planning Institute. In N. Baporikar (Ed.), *Management Education for Global Leadership* (pp. 112–136). Hershey, PA: IGI Global. doi:10.4018/978-1-5225-1013-0.ch006

Janošková, M., Csikósová, A., & Čulková, K. (2018). Measurement of Company Performance as Part of Its Strategic Management. In R. Leon (Ed.), *Managerial Strategies for Business Sustainability During Turbulent Times* (pp. 309–335). Hershey, PA: IGI Global. doi:10.4018/978-1-5225-2716-9.ch017

Jean-Vasile, A., & Alecu, A. (2017). Theoretical and Practical Approaches in Understanding the Influences of Cost-Productivity-Profit Trinomial in Contemporary Enterprises. In A. Jean Vasile & D. Nicolò (Eds.), *Sustainable Entrepreneurship and Investments in the Green Economy* (pp. 28–62). Hershey, PA: IGI Global. doi:10.4018/978-1-5225-2075-7.ch002

Jha, D. G. (2016). Preparing for Information Technology Driven Changes. In S. Tiwari & L. Nafees (Eds.), *Innovative Management Education Pedagogies for Preparing Next-Generation Leaders* (pp. 258–274). Hershey, PA: IGI Global. doi:10.4018/978-1-4666-9691-4.ch015

Joia, L. A., & Correia, J. C. (2018). CIO Competencies From the IT Professional Perspective: Insights From Brazil. *Journal of Global Information Management, 26*(2), 74–103. doi:10.4018/JGIM.2018040104

Juma, A., & Mzera, N. (2017). Knowledge Management and Records Management and Competitive Advantage in Business. In P. Jain & N. Mnjama (Eds.), *Managing Knowledge Resources and Records in Modern Organizations* (pp. 15–28). Hershey, PA: IGI Global. doi:10.4018/978-1-5225-1965-2.ch002

K., I., & A, V. (2018). Monitoring and Auditing in the Cloud. In K. Munir (Ed.), *Cloud Computing Technologies for Green Enterprises* (pp. 318-350). Hershey, PA: IGI Global. doi:10.4018/978-1-5225-3038-1.ch013

Kabra, G., Ghosh, V., & Ramesh, A. (2018). Enterprise Integrated Business Process Management and Business Intelligence Framework for Business Process Sustainability. In A. Paul, D. Bhattacharyya, & S. Anand (Eds.), *Green Initiatives for Business Sustainability and Value Creation* (pp. 228–238). Hershey, PA: IGI Global. doi:10.4018/978-1-5225-2662-9.ch010

Kaoud, M. (2017). Investigation of Customer Knowledge Management: A Case Study Research. *International Journal of Service Science, Management, Engineering, and Technology, 8*(2), 12–22. doi:10.4018/IJSSMET.2017040102

Kara, M. E., & Fırat, S. Ü. (2016). Sustainability, Risk, and Business Intelligence in Supply Chains. In M. Erdoğdu, T. Arun, & I. Ahmad (Eds.), *Handbook of Research on Green Economic Development Initiatives and Strategies* (pp. 501–538). Hershey, PA: IGI Global. doi:10.4018/978-1-5225-0440-5.ch022

Katuu, S. (2018). A Comparative Assessment of Enterprise Content Management Maturity Models. In N. Gwangwava & M. Mutingi (Eds.), *E-Manufacturing and E-Service Strategies in Contemporary Organizations* (pp. 93–118). Hershey, PA: IGI Global. doi:10.4018/978-1-5225-3628-4.ch005

Khan, M. A. (2016). MNEs Management Strategies in Developing Countries: Establishing the Context. In M. Khan (Ed.), *Multinational Enterprise Management Strategies in Developing Countries* (pp. 1–33). Hershey, PA: IGI Global. doi:10.4018/978-1-5225-0276-0.ch001

Khan, M. A. (2016). Operational Approaches in Organizational Structure: A Case for MNEs in Developing Countries. In M. Khan (Ed.), *Multinational Enterprise Management Strategies in Developing Countries* (pp. 129–151). Hershey, PA: IGI Global. doi:10.4018/978-1-5225-0276-0.ch007

Kinnunen, S., Ylä-Kujala, A., Marttonen-Arola, S., Kärri, T., & Baglee, D. (2018). Internet of Things in Asset Management: Insights from Industrial Professionals and Academia. *International Journal of Service Science, Management, Engineering, and Technology*, *9*(2), 104–119. doi:10.4018/IJSSMET.2018040105

Klein, A. Z., Sabino de Freitas, A., Machado, L., Freitas, J. C. Jr, Graziola, P. G. Jr, & Schlemmer, E. (2017). Virtual Worlds Applications for Management Education. In L. Tomei (Ed.), *Exploring the New Era of Technology-Infused Education* (pp. 279–299). Hershey, PA: IGI Global. doi:10.4018/978-1-5225-1709-2.ch017

Kożuch, B., & Jabłoński, A. (2017). Adopting the Concept of Business Models in Public Management. In M. Lewandowski & B. Kożuch (Eds.), *Public Sector Entrepreneurship and the Integration of Innovative Business Models* (pp. 10–46). Hershey, PA: IGI Global. doi:10.4018/978-1-5225-2215-7.ch002

Kumar, J., Adhikary, A., & Jha, A. (2017). Small Active Investors' Perceptions and Preferences Towards Tax Saving Mutual Fund Schemes in Eastern India: An Empirical Note. *International Journal of Asian Business and Information Management*, *8*(2), 35–45. doi:10.4018/IJABIM.2017040103

Lassoued, Y., Bouzguenda, L., & Mahmoud, T. (2016). Context-Aware Business Process Versions Management. *International Journal of e-Collaboration*, *12*(3), 7–33. doi:10.4018/IJeC.2016070102

Lavassani, K. M., & Movahedi, B. (2017). Applications Driven Information Systems: Beyond Networks toward Business Ecosystems. *International Journal of Innovation in the Digital Economy*, *8*(1), 61–75. doi:10.4018/IJIDE.2017010104

Lazzareschi, V. H., & Brito, M. S. (2017). Strategic Information Management: Proposal of Business Project Model. In G. Jamil, A. Soares, & C. Pessoa (Eds.), *Handbook of Research on Information Management for Effective Logistics and Supply Chains* (pp. 59–88). Hershey, PA: IGI Global. doi:10.4018/978-1-5225-0973-8.ch004

Lederer, M., Kurz, M., & Lazarov, P. (2017). Usage and Suitability of Methods for Strategic Business Process Initiatives: A Multi Case Study Research. *International Journal of Productivity Management and Assessment Technologies*, *5*(1), 40–51. doi:10.4018/IJPMAT.2017010103

Lee, I. (2017). A Social Enterprise Business Model and a Case Study of Pacific Community Ventures (PCV). In V. Potocan, M. Üngan, & Z. Nedelko (Eds.), *Handbook of Research on Managerial Solutions in Non-Profit Organizations* (pp. 182–204). Hershey, PA: IGI Global. doi:10.4018/978-1-5225-0731-4.ch009

Lee, L. J., & Leu, J. (2016). Exploring the Effectiveness of IT Application and Value Method in the Innovation Performance of Enterprise. *International Journal of Enterprise Information Systems*, *12*(2), 47–65. doi:10.4018/IJEIS.2016040104

Lee, Y. (2016). Alignment Effect of Entrepreneurial Orientation and Marketing Orientation on Firm Performance. *International Journal of Customer Relationship Marketing and Management*, *7*(4), 58–69. doi:10.4018/IJCRMM.2016100104

Leon, L. A., Seal, K. C., Przasnyski, Z. H., & Wiedenman, I. (2017). Skills and Competencies Required for Jobs in Business Analytics: A Content Analysis of Job Advertisements Using Text Mining. *International Journal of Business Intelligence Research*, *8*(1), 1–25. doi:10.4018/IJBIR.2017010101

Leu, J., Lee, L. J., & Krischke, A. (2016). Value Engineering-Based Method for Implementing the ISO14001 System in the Green Supply Chains. *International Journal of Strategic Decision Sciences*, *7*(4), 1–20. doi:10.4018/IJSDS.2016100101

Levy, C. L., & Elias, N. I. (2017). SOHO Users' Perceptions of Reliability and Continuity of Cloud-Based Services. In M. Moore (Ed.), *Cybersecurity Breaches and Issues Surrounding Online Threat Protection* (pp. 248–287). Hershey, PA: IGI Global. doi:10.4018/978-1-5225-1941-6.ch011

Levy, M. (2018). Change Management Serving Knowledge Management and Organizational Development: Reflections and Review. In N. Baporikar (Ed.), *Global Practices in Knowledge Management for Societal and Organizational Development* (pp. 256–270). Hershey, PA: IGI Global. doi:10.4018/978-1-5225-3009-1.ch012

Lewandowski, M. (2017). Public Organizations and Business Model Innovation: The Role of Public Service Design. In M. Lewandowski & B. Kożuch (Eds.), *Public Sector Entrepreneurship and the Integration of Innovative Business Models* (pp. 47–72). Hershey, PA: IGI Global. doi:10.4018/978-1-5225-2215-7.ch003

Lhannaoui, H., Kabbaj, M. I., & Bakkoury, Z. (2017). A Survey of Risk-Aware Business Process Modelling. *International Journal of Risk and Contingency Management*, *6*(3), 14–26. doi:10.4018/IJRCM.2017070102

Li, J., Sun, W., Jiang, W., Yang, H., & Zhang, L. (2017). How the Nature of Exogenous Shocks and Crises Impact Company Performance?: The Effects of Industry Characteristics. *International Journal of Risk and Contingency Management*, *6*(4), 40–55. doi:10.4018/IJRCM.2017100103

Lu, C., & Liu, S. (2016). Cultural Tourism O2O Business Model Innovation-A Case Study of CTrip. *Journal of Electronic Commerce in Organizations*, *14*(2), 16–31. doi:10.4018/JECO.2016040102

Machen, B., Hosseini, M. R., Wood, A., & Bakhshi, J. (2016). An Investigation into using SAP-PS as a Multidimensional Project Control System (MPCS). *International Journal of Enterprise Information Systems*, *12*(2), 66–81. doi:10.4018/IJEIS.2016040105

Malega, P. (2017). Small and Medium Enterprises in the Slovak Republic: Status and Competitiveness of SMEs in the Global Markets and Possibilities of Optimization. In M. Vemić (Ed.), *Optimal Management Strategies in Small and Medium Enterprises* (pp. 102–124). Hershey, PA: IGI Global. doi:10.4018/978-1-5225-1949-2.ch006

Malewska, K. M. (2017). Intuition in Decision-Making on the Example of a Non-Profit Organization. In V. Potocan, M. Üngan, & Z. Nedelko (Eds.), *Handbook of Research on Managerial Solutions in Non-Profit Organizations* (pp. 378–399). Hershey, PA: IGI Global. doi:10.4018/978-1-5225-0731-4.ch018

Maroofi, F. (2017). Entrepreneurial Orientation and Organizational Learning Ability Analysis for Innovation and Firm Performance. In N. Baporikar (Ed.), *Innovation and Shifting Perspectives in Management Education* (pp. 144–165). Hershey, PA: IGI Global. doi:10.4018/978-1-5225-1019-2.ch007

Martins, P. V., & Zacarias, M. (2017). A Web-based Tool for Business Process Improvement. *International Journal of Web Portals*, *9*(2), 68–84. doi:10.4018/IJWP.2017070104

Matthies, B., & Coners, A. (2017). Exploring the Conceptual Nature of e-Business Projects. *Journal of Electronic Commerce in Organizations*, *15*(3), 33–63. doi:10.4018/JECO.2017070103

McKee, J. (2018). Architecture as a Tool to Solve Business Planning Problems. In M. Khosrow-Pour, D.B.A. (Ed.), Encyclopedia of Information Science and Technology, Fourth Edition (pp. 573-586). Hershey, PA: IGI Global. doi:10.4018/978-1-5225-2255-3.ch050

McMurray, A. J., Cross, J., & Caponecchia, C. (2018). The Risk Management Profession in Australia: Business Continuity Plan Practices. In N. Bajgoric (Ed.), *Always-On Enterprise Information Systems for Modern Organizations* (pp. 112–129). Hershey, PA: IGI Global. doi:10.4018/978-1-5225-3704-5.ch006

Meddah, I. H., & Belkadi, K. (2018). Mining Patterns Using Business Process Management. In R. Hamou (Ed.), *Handbook of Research on Biomimicry in Information Retrieval and Knowledge Management* (pp. 78–89). Hershey, PA: IGI Global. doi:10.4018/978-1-5225-3004-6.ch005

Mendes, L. (2017). TQM and Knowledge Management: An Integrated Approach Towards Tacit Knowledge Management. In D. Jaziri-Bouagina & G. Jamil (Eds.), *Handbook of Research on Tacit Knowledge Management for Organizational Success* (pp. 236–263). Hershey, PA: IGI Global. doi:10.4018/978-1-5225-2394-9.ch009

Mnjama, N. M. (2017). Preservation of Recorded Information in Public and Private Sector Organizations. In P. Jain & N. Mnjama (Eds.), *Managing Knowledge Resources and Records in Modern Organizations* (pp. 149–167). Hershey, PA: IGI Global. doi:10.4018/978-1-5225-1965-2.ch009

Mokoqama, M., & Fields, Z. (2017). Principles of Responsible Management Education (PRME): Call for Responsible Management Education. In Z. Fields (Ed.), *Collective Creativity for Responsible and Sustainable Business Practice* (pp. 229–241). Hershey, PA: IGI Global. doi:10.4018/978-1-5225-1823-5.ch012

Muniapan, B. (2017). Philosophy and Management: The Relevance of Vedanta in Management. In P. Ordóñez de Pablos (Ed.), *Managerial Strategies and Solutions for Business Success in Asia* (pp. 124–139). Hershey, PA: IGI Global. doi:10.4018/978-1-5225-1886-0.ch007

Muniapan, B., Gregory, M. L., & Ling, L. A. (2016). Marketing Education in Sarawak: Looking at It from the Employers' Viewpoint. In B. Smith & A. Porath (Eds.), *Global Perspectives on Contemporary Marketing Education* (pp. 112–130). Hershey, PA: IGI Global. doi:10.4018/978-1-4666-9784-3.ch008

Murad, S. E., & Dowaji, S. (2017). Using Value-Based Approach for Managing Cloud-Based Services. In A. Turuk, B. Sahoo, & S. Addya (Eds.), *Resource Management and Efficiency in Cloud Computing Environments* (pp. 33–60). Hershey, PA: IGI Global. doi:10.4018/978-1-5225-1721-4.ch002

Mutahar, A. M., Daud, N. M., Thurasamy, R., Isaac, O., & Abdulsalam, R. (2018). The Mediating of Perceived Usefulness and Perceived Ease of Use: The Case of Mobile Banking in Yemen. *International Journal of Technology Diffusion*, *9*(2), 21–40. doi:10.4018/IJTD.2018040102

Naidoo, V. (2017). E-Learning and Management Education at African Universities. In N. Baporikar (Ed.), *Management Education for Global Leadership* (pp. 181–201). Hershey, PA: IGI Global. doi:10.4018/978-1-5225-1013-0.ch009

Naidoo, V., & Igbinakhase, I. (2018). Opportunities and Challenges of Knowledge Retention in SMEs. In N. Baporikar (Ed.), *Knowledge Integration Strategies for Entrepreneurship and Sustainability* (pp. 70–94). Hershey, PA: IGI Global. doi:10.4018/978-1-5225-5115-7.ch004

Nayak, S., & Prabhu, N. (2017). Paradigm Shift in Management Education: Need for a Cross Functional Perspective. In N. Baporikar (Ed.), *Management Education for Global Leadership* (pp. 241–255). Hershey, PA: IGI Global. doi:10.4018/978-1-5225-1013-0.ch012

Ndede-Amadi, A. A. (2016). Student Interest in the IS Specialization as Predictor of the Success Potential of New Information Systems Programmes within the Schools of Business in Kenyan Public Universities. *International Journal of Information Systems and Social Change*, *7*(2), 63–79. doi:10.4018/IJISSC.2016040104

Nedelko, Z., & Potocan, V. (2016). Management Practices for Processes Optimization: Case of Slovenia. In G. Alor-Hernández, C. Sánchez-Ramírez, & J. García-Alcaraz (Eds.), *Handbook of Research on Managerial Strategies for Achieving Optimal Performance in Industrial Processes* (pp. 545–561). Hershey, PA: IGI Global. doi:10.4018/978-1-5225-0130-5.ch025

Nedelko, Z., & Potocan, V. (2017). Management Solutions in Non-Profit Organizations: Case of Slovenia. In V. Potocan, M. Üngan, & Z. Nedelko (Eds.), *Handbook of Research on Managerial Solutions in Non-Profit Organizations* (pp. 1–22). Hershey, PA: IGI Global. doi:10.4018/978-1-5225-0731-4.ch001

Nedelko, Z., & Potocan, V. (2017). Priority of Management Tools Utilization among Managers: International Comparison. In V. Wang (Ed.), *Encyclopedia of Strategic Leadership and Management* (pp. 1083–1094). Hershey, PA: IGI Global. doi:10.4018/978-1-5225-1049-9.ch075

Nedelko, Z., Raudeliūnienė, J., & Črešnar, R. (2018). Knowledge Dynamics in Supply Chain Management. In N. Baporikar (Ed.), *Knowledge Integration Strategies for Entrepreneurship and Sustainability* (pp. 150–166). Hershey, PA: IGI Global. doi:10.4018/978-1-5225-5115-7.ch008

Nguyen, H. T., & Hipsher, S. A. (2018). Innovation and Creativity Used by Private Sector Firms in a Resources-Constrained Environment. In S. Hipsher (Ed.), *Examining the Private Sector's Role in Wealth Creation and Poverty Reduction* (pp. 219–238). Hershey, PA: IGI Global. doi:10.4018/978-1-5225-3117-3.ch010

Nycz, M., & Pólkowski, Z. (2016). Business Intelligence as a Modern IT Supporting Management of Local Government Units in Poland. *International Journal of Knowledge and Systems Science, 7*(4), 1–18. doi:10.4018/IJKSS.2016100101

Obaji, N. O., Senin, A. A., & Olugu, M. U. (2016). Supportive Government Policy as a Mechanism for Business Incubation Performance in Nigeria. *International Journal of Information Systems and Social Change, 7*(4), 52–66. doi:10.4018/IJISSC.2016100103

Obicci, P. A. (2017). Risk Sharing in a Partnership. In *Risk Management Strategies in Public-Private Partnerships* (pp. 115–152). Hershey, PA: IGI Global. doi:10.4018/978-1-5225-2503-5.ch004

Obidallah, W. J., & Raahemi, B. (2017). Managing Changes in Service Oriented Virtual Organizations: A Structural and Procedural Framework to Facilitate the Process of Change. *Journal of Electronic Commerce in Organizations, 15*(1), 59–83. doi:10.4018/JECO.2017010104

Ojasalo, J., & Ojasalo, K. (2016). Service Logic Business Model Canvas for Lean Development of SMEs and Start-Ups. In N. Baporikar (Ed.), *Handbook of Research on Entrepreneurship in the Contemporary Knowledge-Based Global Economy* (pp. 217–243). Hershey, PA: IGI Global. doi:10.4018/978-1-4666-8798-1.ch010

Ojo, O. (2017). Impact of Innovation on the Entrepreneurial Success in Selected Business Enterprises in South-West Nigeria. *International Journal of Innovation in the Digital Economy, 8*(2), 29–38. doi:10.4018/IJIDE.2017040103

Okdinawati, L., Simatupang, T. M., & Sunitiyoso, Y. (2017). Multi-Agent Reinforcement Learning for Value Co-Creation of Collaborative Transportation Management (CTM). *International Journal of Information Systems and Supply Chain Management, 10*(3), 84–95. doi:10.4018/IJISSCM.2017070105

Ortner, E., Mevius, M., Wiedmann, P., & Kurz, F. (2016). Design of Interactional Decision Support Applications for E-Participation in Smart Cities. *International Journal of Electronic Government Research, 12*(2), 18–38. doi:10.4018/IJEGR.2016040102

Pal, K. (2018). Building High Quality Big Data-Based Applications in Supply Chains. In A. Kumar & S. Saurav (Eds.), *Supply Chain Management Strategies and Risk Assessment in Retail Environments* (pp. 1–24). Hershey, PA: IGI Global. doi:10.4018/978-1-5225-3056-5.ch001

Palos-Sanchez, P. R., & Correia, M. B. (2018). Perspectives of the Adoption of Cloud Computing in the Tourism Sector. In J. Rodrigues, C. Ramos, P. Cardoso, & C. Henriques (Eds.), *Handbook of Research on Technological Developments for Cultural Heritage and eTourism Applications* (pp. 377–400). Hershey, PA: IGI Global. doi:10.4018/978-1-5225-2927-9.ch018

Parry, V. K., & Lind, M. L. (2016). Alignment of Business Strategy and Information Technology Considering Information Technology Governance, Project Portfolio Control, and Risk Management. *International Journal of Information Technology Project Management*, *7*(4), 21–37. doi:10.4018/IJITPM.2016100102

Pashkova, N., Trujillo-Barrera, A., Apostolakis, G., Van Dijk, G., Drakos, P. D., & Baourakis, G. (2016). Business Management Models of Microfinance Institutions (MFIs) in Africa: A Study into Their Enabling Environments. *International Journal of Food and Beverage Manufacturing and Business Models*, *1*(2), 63–82. doi:10.4018/IJFBMBM.2016070105

Patiño, B. E. (2017). New Generation Management by Convergence and Individual Identity: A Systemic and Human-Oriented Approach. In N. Baporikar (Ed.), *Innovation and Shifting Perspectives in Management Education* (pp. 119–143). Hershey, PA: IGI Global. doi:10.4018/978-1-5225-1019-2.ch006

Pawliczek, A., & Rössler, M. (2017). Knowledge of Management Tools and Systems in SMEs: Knowledge Transfer in Management. In A. Bencsik (Ed.), *Knowledge Management Initiatives and Strategies in Small and Medium Enterprises* (pp. 180–203). Hershey, PA: IGI Global. doi:10.4018/978-1-5225-1642-2.ch009

Pejic-Bach, M., Omazic, M. A., Aleksic, A., & Zoroja, J. (2018). Knowledge-Based Decision Making: A Multi-Case Analysis. In R. Leon (Ed.), *Managerial Strategies for Business Sustainability During Turbulent Times* (pp. 160–184). Hershey, PA: IGI Global. doi:10.4018/978-1-5225-2716-9.ch009

Perano, M., Hysa, X., & Calabrese, M. (2018). Strategic Planning, Cultural Context, and Business Continuity Management: Business Cases in the City of Shkoder. In A. Presenza & L. Sheehan (Eds.), *Geopolitics and Strategic Management in the Global Economy* (pp. 57–77). Hershey, PA: IGI Global. doi:10.4018/978-1-5225-2673-5.ch004

Pereira, R., Mira da Silva, M., & Lapão, L. V. (2017). IT Governance Maturity Patterns in Portuguese Healthcare. In S. De Haes & W. Van Grembergen (Eds.), *Strategic IT Governance and Alignment in Business Settings* (pp. 24–52). Hershey, PA: IGI Global. doi:10.4018/978-1-5225-0861-8.ch002

Perez-Uribe, R., & Ocampo-Guzman, D. (2016). Conflict within Colombian Family Owned SMEs: An Explosive Blend between Feelings and Business. In J. Saiz-Álvarez (Ed.), *Handbook of Research on Social Entrepreneurship and Solidarity Economics* (pp. 329–354). Hershey, PA: IGI Global. doi:10.4018/978-1-5225-0097-1.ch017

Pérez-Uribe, R. I., Torres, D. A., Jurado, S. P., & Prada, D. M. (2018). Cloud Tools for the Development of Project Management in SMEs. In R. Perez-Uribe, C. Salcedo-Perez, & D. Ocampo-Guzman (Eds.), *Handbook of Research on Intrapreneurship and Organizational Sustainability in SMEs* (pp. 95–120). Hershey, PA: IGI Global. doi:10.4018/978-1-5225-3543-0.ch005

Petrisor, I., & Cozmiuc, D. (2017). Global Supply Chain Management Organization at Siemens in the Advent of Industry 4.0. In L. Saglietto & C. Cezanne (Eds.), *Global Intermediation and Logistics Service Providers* (pp. 123–142). Hershey, PA: IGI Global. doi:10.4018/978-1-5225-2133-4.ch007

Pierce, J. M., Velliaris, D. M., & Edwards, J. (2017). A Living Case Study: A Journey Not a Destination. In N. Silton (Ed.), *Exploring the Benefits of Creativity in Education, Media, and the Arts* (pp. 158–178). Hershey, PA: IGI Global. doi:10.4018/978-1-5225-0504-4.ch008

Radosavljevic, M., & Andjelkovic, A. (2017). Multi-Criteria Decision Making Approach for Choosing Business Process for the Improvement: Upgrading of the Six Sigma Methodology. In J. Stanković, P. Delias, S. Marinković, & S. Rochhia (Eds.), *Tools and Techniques for Economic Decision Analysis* (pp. 225–247). Hershey, PA: IGI Global. doi:10.4018/978-1-5225-0959-2.ch011

Radovic, V. M. (2017). Corporate Sustainability and Responsibility and Disaster Risk Reduction: A Serbian Overview. In M. Camilleri (Ed.), *CSR 2.0 and the New Era of Corporate Citizenship* (pp. 147–164). Hershey, PA: IGI Global. doi:10.4018/978-1-5225-1842-6.ch008

Raghunath, K. M., Devi, S. L., & Patro, C. S. (2018). Impact of Risk Assessment Models on Risk Factors: A Holistic Outlook. In K. Strang, M. Korstanje, & N. Vajjhala (Eds.), *Research, Practices, and Innovations in Global Risk and Contingency Management* (pp. 134–153). Hershey, PA: IGI Global. doi:10.4018/978-1-5225-4754-9.ch008

Raman, A., & Goyal, D. P. (2017). Extending IMPLEMENT Framework for Enterprise Information Systems Implementation to Information System Innovation. In M. Tavana (Ed.), *Enterprise Information Systems and the Digitalization of Business Functions* (pp. 137–177). Hershey, PA: IGI Global. doi:10.4018/978-1-5225-2382-6.ch007

Rao, Y., & Zhang, Y. (2017). The Construction and Development of Academic Library Digital Special Subject Databases. In L. Ruan, Q. Zhu, & Y. Ye (Eds.), *Academic Library Development and Administration in China* (pp. 163–183). Hershey, PA: IGI Global. doi:10.4018/978-1-5225-0550-1.ch010

Ravasan, A. Z., Mohammadi, M. M., & Hamidi, H. (2018). An Investigation Into the Critical Success Factors of Implementing Information Technology Service Management Frameworks. In K. Jakobs (Ed.), *Corporate and Global Standardization Initiatives in Contemporary Society* (pp. 200–218). Hershey, PA: IGI Global. doi:10.4018/978-1-5225-5320-5.ch009

Renna, P., Izzo, C., & Romaniello, T. (2016). The Business Process Management Systems to Support Continuous Improvements. In W. Nuninger & J. Châtelet (Eds.), *Handbook of Research on Quality Assurance and Value Management in Higher Education* (pp. 237–256). Hershey, PA: IGI Global. doi:10.4018/978-1-5225-0024-7.ch009

Rezaie, S., Mirabedini, S. J., & Abtahi, A. (2018). Designing a Model for Implementation of Business Intelligence in the Banking Industry. *International Journal of Enterprise Information Systems, 14*(1), 77–103. doi:10.4018/IJEIS.2018010105

Riccò, R. (2016). Diversity Management: Bringing Equality, Equity, and Inclusion in the Workplace. In J. Prescott (Ed.), *Handbook of Research on Race, Gender, and the Fight for Equality* (pp. 335–359). Hershey, PA: IGI Global. doi:10.4018/978-1-5225-0047-6.ch015

Romano, L., Grimaldi, R., & Colasuonno, F. S. (2017). Demand Management as a Success Factor in Project Portfolio Management. In L. Romano (Ed.), *Project Portfolio Management Strategies for Effective Organizational Operations* (pp. 202–219). Hershey, PA: IGI Global. doi:10.4018/978-1-5225-2151-8.ch008

Rostek, K. B. (2016). Risk Management: Role and Importance in Business Organization. In D. Jakóbczak (Ed.), *Analyzing Risk through Probabilistic Modeling in Operations Research* (pp. 149–178). Hershey, PA: IGI Global. doi:10.4018/978-1-4666-9458-3.ch007

Rouhani, S., & Savoji, S. R. (2016). A Success Assessment Model for BI Tools Implementation: An Empirical Study of Banking Industry. *International Journal of Business Intelligence Research, 7*(1), 25–44. doi:10.4018/IJBIR.2016010103

Ruan, Z. (2016). A Corpus-Based Functional Analysis of Complex Nominal Groups in Written Business Discourse: The Case of "Business". *International Journal of Computer-Assisted Language Learning and Teaching*, *6*(2), 74–90. doi:10.4018/IJCALLT.2016040105

Ruhi, U. (2018). Towards an Interdisciplinary Socio-Technical Definition of Virtual Communities. In M. Khosrow-Pour, D.B.A. (Ed.), Encyclopedia of Information Science and Technology, Fourth Edition (pp. 4278-4295). Hershey, PA: IGI Global. doi:10.4018/978-1-5225-2255-3.ch371

Ryan, J., Doster, B., Daily, S., & Lewis, C. (2016). A Case Study Perspective for Balanced Perioperative Workflow Achievement through Data-Driven Process Improvement. *International Journal of Healthcare Information Systems and Informatics*, *11*(3), 19–41. doi:10.4018/IJHISI.2016070102

Safari, M. R., & Jiang, Q. (2018). The Theory and Practice of IT Governance Maturity and Strategies Alignment: Evidence From Banking Industry. *Journal of Global Information Management*, *26*(2), 127–146. doi:10.4018/JGIM.2018040106

Sahoo, J., Pati, B., & Mohanty, B. (2017). Knowledge Management as an Academic Discipline: An Assessment. In B. Gunjal (Ed.), *Managing Knowledge and Scholarly Assets in Academic Libraries* (pp. 99–126). Hershey, PA: IGI Global. doi:10.4018/978-1-5225-1741-2.ch005

Saini, D. (2017). Relevance of Teaching Values and Ethics in Management Education. In N. Baporikar (Ed.), *Management Education for Global Leadership* (pp. 90–111). Hershey, PA: IGI Global. doi:10.4018/978-1-5225-1013-0.ch005

Sambhanthan, A. (2017). Assessing and Benchmarking Sustainability in Organisations: An Integrated Conceptual Model. *International Journal of Systems and Service-Oriented Engineering*, *7*(4), 22–43. doi:10.4018/IJSSOE.2017100102

Sambhanthan, A., & Potdar, V. (2017). A Study of the Parameters Impacting Sustainability in Information Technology Organizations. *International Journal of Knowledge-Based Organizations*, *7*(3), 27–39. doi:10.4018/IJKBO.2017070103

Sánchez-Fernández, M. D., & Manríquez, M. R. (2018). The Entrepreneurial Spirit Based on Social Values: The Digital Generation. In P. Isaias & L. Carvalho (Eds.), *User Innovation and the Entrepreneurship Phenomenon in the Digital Economy* (pp. 173–193). Hershey, PA: IGI Global. doi:10.4018/978-1-5225-2826-5.ch009

Sanchez-Ruiz, L., & Blanco, B. (2017). Process Management for SMEs: Barriers, Enablers, and Benefits. In M. Vemić (Ed.), *Optimal Management Strategies in Small and Medium Enterprises* (pp. 293–319). Hershey, PA: IGI Global. doi:10.4018/978-1-5225-1949-2.ch014

Sanz, L. F., Gómez-Pérez, J., & Castillo-Martinez, A. (2018). Analysis of the European ICT Competence Frameworks. In V. Ahuja & S. Rathore (Eds.), *Multidisciplinary Perspectives on Human Capital and Information Technology Professionals* (pp. 225–245). Hershey, PA: IGI Global. doi:10.4018/978-1-5225-5297-0.ch012

Sarvepalli, A., & Godin, J. (2017). Business Process Management in the Classroom. *Journal of Cases on Information Technology*, *19*(2), 17–28. doi:10.4018/JCIT.2017040102

Satpathy, B., & Muniapan, B. (2016). Ancient Wisdom for Transformational Leadership and Its Insights from the Bhagavad-Gita. In U. Aung & P. Ordoñez de Pablos (Eds.), *Managerial Strategies and Practice in the Asian Business Sector* (pp. 1–10). Hershey, PA: IGI Global. doi:10.4018/978-1-4666-9758-4.ch001

Saygili, E. E., Ozturkoglu, Y., & Kocakulah, M. C. (2017). End Users' Perceptions of Critical Success Factors in ERP Applications. *International Journal of Enterprise Information Systems*, *13*(4), 58–75. doi:10.4018/IJEIS.2017100104

Saygili, E. E., & Saygili, A. T. (2017). Contemporary Issues in Enterprise Information Systems: A Critical Review of CSFs in ERP Implementations. In M. Tavana (Ed.), *Enterprise Information Systems and the Digitalization of Business Functions* (pp. 120–136). Hershey, PA: IGI Global. doi:10.4018/978-1-5225-2382-6.ch006

Seidenstricker, S., & Antonino, A. (2018). Business Model Innovation-Oriented Technology Management for Emergent Technologies. In M. Khosrow-Pour, D.B.A. (Ed.), Encyclopedia of Information Science and Technology, Fourth Edition (pp. 4560-4569). Hershey, PA: IGI Global. doi:10.4018/978-1-5225-2255-3.ch396

Senaratne, S., & Gunarathne, A. D. (2017). Excellence Perspective for Management Education from a Global Accountants' Hub in Asia. In N. Baporikar (Ed.), *Management Education for Global Leadership* (pp. 158–180). Hershey, PA: IGI Global. doi:10.4018/978-1-5225-1013-0.ch008

Sensuse, D. I., & Cahyaningsih, E. (2018). Knowledge Management Models: A Summative Review. *International Journal of Information Systems in the Service Sector*, *10*(1), 71–100. doi:10.4018/IJISSS.2018010105

Sensuse, D. I., Wibowo, W. C., & Cahyaningsih, E. (2016). Indonesian Government Knowledge Management Model: A Theoretical Model. *Information Resources Management Journal, 29*(1), 91–108. doi:10.4018/irmj.2016010106

Seth, M., Goyal, D., & Kiran, R. (2017). Diminution of Impediments in Implementation of Supply Chain Management Information System for Enhancing its Effectiveness in Indian Automobile Industry. *Journal of Global Information Management, 25*(3), 1–20. doi:10.4018/JGIM.2017070101

Seyal, A. H., & Rahman, M. N. (2017). Investigating Impact of Inter-Organizational Factors in Measuring ERP Systems Success: Bruneian Perspectives. In M. Tavana (Ed.), *Enterprise Information Systems and the Digitalization of Business Functions* (pp. 178–204). Hershey, PA: IGI Global. doi:10.4018/978-1-5225-2382-6.ch008

Shaikh, A. A., & Karjaluoto, H. (2016). On Some Misconceptions Concerning Digital Banking and Alternative Delivery Channels. *International Journal of E-Business Research, 12*(3), 1–16. doi:10.4018/IJEBR.2016070101

Shams, S. M. (2016). Stakeholder Relationship Management in Online Business and Competitive Value Propositions: Evidence from the Sports Industry. *International Journal of Online Marketing, 6*(2), 1–17. doi:10.4018/IJOM.2016040101

Shamsuzzoha, A. (2016). Management of Risk and Resilience within Collaborative Business Network. In R. Addo-Tenkorang, J. Kantola, P. Helo, & A. Shamsuzzoha (Eds.), *Supply Chain Strategies and the Engineer-to-Order Approach* (pp. 143–159). Hershey, PA: IGI Global. doi:10.4018/978-1-5225-0021-6.ch008

Shaqrah, A. A. (2018). Analyzing Business Intelligence Systems Based on 7s Model of McKinsey. *International Journal of Business Intelligence Research, 9*(1), 53–63. doi:10.4018/IJBIR.2018010104

Sharma, A. J. (2017). Enhancing Sustainability through Experiential Learning in Management Education. In N. Baporikar (Ed.), *Management Education for Global Leadership* (pp. 256–274). Hershey, PA: IGI Global. doi:10.4018/978-1-5225-1013-0.ch013

Shetty, K. P. (2017). Responsible Global Leadership: Ethical Challenges in Management Education. In N. Baporikar (Ed.), *Innovation and Shifting Perspectives in Management Education* (pp. 194–223). Hershey, PA: IGI Global. doi:10.4018/978-1-5225-1019-2.ch009

Sinthupundaja, J., & Kohda, Y. (2017). Effects of Corporate Social Responsibility and Creating Shared Value on Sustainability. *International Journal of Sustainable Entrepreneurship and Corporate Social Responsibility*, *2*(1), 27–38. doi:10.4018/IJSECSR.2017010103

Škarica, I., & Hrgović, A. V. (2018). Implementation of Total Quality Management Principles in Public Health Institutes in the Republic of Croatia. *International Journal of Productivity Management and Assessment Technologies*, *6*(1), 1–16. doi:10.4018/IJPMAT.2018010101

Smuts, H., Kotzé, P., Van der Merwe, A., & Loock, M. (2017). Framework for Managing Shared Knowledge in an Information Systems Outsourcing Context. *International Journal of Knowledge Management*, *13*(4), 1–30. doi:10.4018/IJKM.2017100101

Soares, E. R., & Zaidan, F. H. (2016). Information Architecture and Business Modeling in Modern Organizations of Information Technology: Professional Career Plan in Organizations IT. In G. Jamil, J. Poças Rascão, F. Ribeiro, & A. Malheiro da Silva (Eds.), *Handbook of Research on Information Architecture and Management in Modern Organizations* (pp. 439–457). Hershey, PA: IGI Global. doi:10.4018/978-1-4666-8637-3.ch020

Sousa, M. J., Cruz, R., Dias, I., & Caracol, C. (2017). Information Management Systems in the Supply Chain. In G. Jamil, A. Soares, & C. Pessoa (Eds.), *Handbook of Research on Information Management for Effective Logistics and Supply Chains* (pp. 469–485). Hershey, PA: IGI Global. doi:10.4018/978-1-5225-0973-8.ch025

Spremic, M., Turulja, L., & Bajgoric, N. (2018). Two Approaches in Assessing Business Continuity Management Attitudes in the Organizational Context. In N. Bajgoric (Ed.), *Always-On Enterprise Information Systems for Modern Organizations* (pp. 159–183). Hershey, PA: IGI Global. doi:10.4018/978-1-5225-3704-5.ch008

Steenkamp, A. L. (2018). Some Insights in Computer Science and Information Technology. In *Examining the Changing Role of Supervision in Doctoral Research Projects: Emerging Research and Opportunities* (pp. 113–133). Hershey, PA: IGI Global. doi:10.4018/978-1-5225-2610-0.ch005

Studdard, N., Dawson, M., Burton, S. L., Jackson, N., Leonard, B., Quisenberry, W., & Rahim, E. (2016). Nurturing Social Entrepreneurship and Building Social Entrepreneurial Self-Efficacy: Focusing on Primary and Secondary Schooling to Develop Future Social Entrepreneurs. In Z. Fields (Ed.), *Incorporating Business Models and Strategies into Social Entrepreneurship* (pp. 154–175). Hershey, PA: IGI Global. doi:10.4018/978-1-4666-8748-6.ch010

Sun, Z. (2016). A Framework for Developing Management Intelligent Systems. *International Journal of Systems and Service-Oriented Engineering*, *6*(1), 37–53. doi:10.4018/IJSSOE.2016010103

Swami, B., & Mphele, G. T. (2016). Problems Preventing Growth of Small Entrepreneurs: A Case Study of a Few Small Entrepreneurs in Botswana Sub-Urban Areas. In N. Baporikar (Ed.), *Handbook of Research on Entrepreneurship in the Contemporary Knowledge-Based Global Economy* (pp. 479–508). Hershey, PA: IGI Global. doi:10.4018/978-1-4666-8798-1.ch020

Tabach, A., & Croteau, A. (2017). Configurations of Information Technology Governance Practices and Business Unit Performance. *International Journal of IT/Business Alignment and Governance*, *8*(2), 1–27. doi:10.4018/IJITBAG.2017070101

Talaue, G. M., & Iqbal, T. (2017). Assessment of e-Business Mode of Selected Private Universities in the Philippines and Pakistan. *International Journal of Online Marketing*, *7*(4), 63–77. doi:10.4018/IJOM.2017100105

Tam, G. C. (2017). Project Manager Sustainability Competence. In *Managerial Strategies and Green Solutions for Project Sustainability* (pp. 178–207). Hershey, PA: IGI Global. doi:10.4018/978-1-5225-2371-0.ch008

Tambo, T. (2018). Fashion Retail Innovation: About Context, Antecedents, and Outcome in Technological Change Projects. In I. Management Association (Ed.), Fashion and Textiles: Breakthroughs in Research and Practice (pp. 233-260). Hershey, PA: IGI Global. doi:10.4018/978-1-5225-3432-7.ch010

Tambo, T., & Mikkelsen, O. E. (2016). Fashion Supply Chain Optimization: Linking Make-to-Order Purchasing and B2B E-Commerce. In S. Joshi & R. Joshi (Eds.), *Designing and Implementing Global Supply Chain Management* (pp. 1–21). Hershey, PA: IGI Global. doi:10.4018/978-1-4666-9720-1.ch001

Tandon, K. (2016). Innovative Andragogy: The Paradigm Shift to Heutagogy. In S. Tiwari & L. Nafees (Eds.), *Innovative Management Education Pedagogies for Preparing Next-Generation Leaders* (pp. 238–257). Hershey, PA: IGI Global. doi:10.4018/978-1-4666-9691-4.ch014

Tantau, A. D., & Frăţilă, L. C. (2018). Information and Management System for Renewable Energy Business. In *Entrepreneurship and Business Development in the Renewable Energy Sector* (pp. 200–244). Hershey, PA: IGI Global. doi:10.4018/978-1-5225-3625-3.ch006

Teixeira, N., Pardal, P. N., & Rafael, B. G. (2018). Internationalization, Financial Performance, and Organizational Challenges: A Success Case in Portugal. In L. Carvalho (Ed.), *Handbook of Research on Entrepreneurial Ecosystems and Social Dynamics in a Globalized World* (pp. 379–423). Hershey, PA: IGI Global. doi:10.4018/978-1-5225-3525-6.ch017

Trad, A., & Kalpić, D. (2016). The E-Business Transformation Framework for E-Commerce Architecture-Modeling Projects. In I. Lee (Ed.), *Encyclopedia of E-Commerce Development, Implementation, and Management* (pp. 733–753). Hershey, PA: IGI Global. doi:10.4018/978-1-4666-9787-4.ch052

Trad, A., & Kalpić, D. (2016). The E-Business Transformation Framework for E-Commerce Control and Monitoring Pattern. In I. Lee (Ed.), *Encyclopedia of E-Commerce Development, Implementation, and Management* (pp. 754–777). Hershey, PA: IGI Global. doi:10.4018/978-1-4666-9787-4.ch053

Trad, A., & Kalpić, D. (2018). The Business Transformation Framework, Agile Project and Change Management. In M. Khosrow-Pour, D.B.A. (Ed.), Encyclopedia of Information Science and Technology, Fourth Edition (pp. 620-635). Hershey, PA: IGI Global. doi:10.4018/978-1-5225-2255-3.ch054

Trad, A., & Kalpić, D. (2018). The Business Transformation and Enterprise Architecture Framework: The Financial Engineering E-Risk Management and E-Law Integration. In B. Sergi, F. Fidanoski, M. Ziolo, & V. Naumovski (Eds.), *Regaining Global Stability After the Financial Crisis* (pp. 46–65). Hershey, PA: IGI Global. doi:10.4018/978-1-5225-4026-7.ch003

Turulja, L., & Bajgoric, N. (2018). Business Continuity and Information Systems: A Systematic Literature Review. In N. Bajgoric (Ed.), *Always-On Enterprise Information Systems for Modern Organizations* (pp. 60–87). Hershey, PA: IGI Global. doi:10.4018/978-1-5225-3704-5.ch004

van Wessel, R. M., de Vries, H. J., & Ribbers, P. M. (2016). Business Benefits through Company IT Standardization. In K. Jakobs (Ed.), *Effective Standardization Management in Corporate Settings* (pp. 34–53). Hershey, PA: IGI Global. doi:10.4018/978-1-4666-9737-9.ch003

Vargas-Hernández, J. G. (2017). Professional Integrity in Business Management Education. In N. Baporikar (Ed.), *Management Education for Global Leadership* (pp. 70–89). Hershey, PA: IGI Global. doi:10.4018/978-1-5225-1013-0.ch004

Vasista, T. G., & AlAbdullatif, A. M. (2017). Role of Electronic Customer Relationship Management in Demand Chain Management: A Predictive Analytic Approach. *International Journal of Information Systems and Supply Chain Management, 10*(1), 53–67. doi:10.4018/IJISSCM.2017010104

Vergidis, K. (2016). Rediscovering Business Processes: Definitions, Patterns, and Modelling Approaches. In P. Papajorgji, F. Pinet, A. Guimarães, & J. Papathanasiou (Eds.), *Automated Enterprise Systems for Maximizing Business Performance* (pp. 97–122). Hershey, PA: IGI Global. doi:10.4018/978-1-4666-8841-4.ch007

Vieru, D., & Bourdeau, S. (2017). Survival in the Digital Era: A Digital Competence-Based Multi-Case Study in the Canadian SME Clothing Industry. *International Journal of Social and Organizational Dynamics in IT, 6*(1), 17–34. doi:10.4018/IJSODIT.2017010102

Vijayan, G., & Kamarulzaman, N. H. (2017). An Introduction to Sustainable Supply Chain Management and Business Implications. In M. Khan, M. Hussain, & M. Ajmal (Eds.), *Green Supply Chain Management for Sustainable Business Practice* (pp. 27–50). Hershey, PA: IGI Global. doi:10.4018/978-1-5225-0635-5.ch002

Vlachvei, A., & Notta, O. (2017). Firm Competitiveness: Theories, Evidence, and Measurement. In A. Vlachvei, O. Notta, K. Karantininis, & N. Tsounis (Eds.), *Factors Affecting Firm Competitiveness and Performance in the Modern Business World* (pp. 1–42). Hershey, PA: IGI Global. doi:10.4018/978-1-5225-0843-4.ch001

von Rosing, M., Fullington, N., & Walker, J. (2016). Using the Business Ontology and Enterprise Standards to Transform Three Leading Organizations. *International Journal of Conceptual Structures and Smart Applications, 4*(1), 71–99. doi:10.4018/IJCSSA.2016010104

von Rosing, M., & von Scheel, H. (2016). Using the Business Ontology to Develop Enterprise Standards. *International Journal of Conceptual Structures and Smart Applications, 4*(1), 48–70. doi:10.4018/IJCSSA.2016010103

Walczak, S. (2016). Artificial Neural Networks and other AI Applications for Business Management Decision Support. *International Journal of Sociotechnology and Knowledge Development, 8*(4), 1–20. doi:10.4018/IJSKD.2016100101

Wamba, S. F., Akter, S., Kang, H., Bhattacharya, M., & Upal, M. (2016). The Primer of Social Media Analytics. *Journal of Organizational and End User Computing, 28*(2), 1–12. doi:10.4018/JOEUC.2016040101

Wang, C., Schofield, M., Li, X., & Ou, X. (2017). Do Chinese Students in Public and Private Higher Education Institutes Perform at Different Level in One of the Leadership Skills: Critical Thinking?: An Exploratory Comparison. In V. Wang (Ed.), *Encyclopedia of Strategic Leadership and Management* (pp. 160–181). Hershey, PA: IGI Global. doi:10.4018/978-1-5225-1049-9.ch013

Wang, F., Raisinghani, M. S., Mora, M., & Wang, X. (2016). Strategic E-Business Management through a Balanced Scored Card Approach. In I. Lee (Ed.), *Encyclopedia of E-Commerce Development, Implementation, and Management* (pp. 361–386). Hershey, PA: IGI Global. doi:10.4018/978-1-4666-9787-4.ch027

Wang, J. (2017). Multi-Agent based Production Management Decision System Modelling for the Textile Enterprise. *Journal of Global Information Management, 25*(4), 1–15. doi:10.4018/JGIM.2017100101

Wiedemann, A., & Gewald, H. (2017). Examining Cross-Domain Alignment: The Correlation of Business Strategy, IT Management, and IT Business Value. *International Journal of IT/Business Alignment and Governance*, *8*(1), 17–31. doi:10.4018/IJITBAG.2017010102

Wolf, R., & Thiel, M. (2018). Advancing Global Business Ethics in China: Reducing Poverty Through Human and Social Welfare. In S. Hipsher (Ed.), *Examining the Private Sector's Role in Wealth Creation and Poverty Reduction* (pp. 67–84). Hershey, PA: IGI Global. doi:10.4018/978-1-5225-3117-3.ch004

Wu, J., Ding, F., Xu, M., Mo, Z., & Jin, A. (2016). Investigating the Determinants of Decision-Making on Adoption of Public Cloud Computing in E-government. *Journal of Global Information Management, 24*(3), 71–89. doi:10.4018/JGIM.2016070104

Xu, L., & de Vrieze, P. (2016). Building Situational Applications for Virtual Enterprises. In I. Lee (Ed.), *Encyclopedia of E-Commerce Development, Implementation, and Management* (pp. 715–724). Hershey, PA: IGI Global. doi:10.4018/978-1-4666-9787-4.ch050

Yablonsky, S. (2018). Innovation Platforms: Data and Analytics Platforms. In *Multi-Sided Platforms (MSPs) and Sharing Strategies in the Digital Economy: Emerging Research and Opportunities* (pp. 72–95). Hershey, PA: IGI Global. doi:10.4018/978-1-5225-5457-8.ch003

Yusoff, A., Ahmad, N. H., & Halim, H. A. (2017). Agropreneurship among Gen Y in Malaysia: The Role of Academic Institutions. In N. Ahmad, T. Ramayah, H. Halim, & S. Rahman (Eds.), *Handbook of Research on Small and Medium Enterprises in Developing Countries* (pp. 23–47). Hershey, PA: IGI Global. doi:10.4018/978-1-5225-2165-5.ch002

Zanin, F., Comuzzi, E., & Costantini, A. (2018). The Effect of Business Strategy and Stock Market Listing on the Use of Risk Assessment Tools. In *Management Control Systems in Complex Settings: Emerging Research and Opportunities* (pp. 145–168). Hershey, PA: IGI Global. doi:10.4018/978-1-5225-3987-2.ch007

Zgheib, P. W. (2017). Corporate Innovation and Intrapreneurship in the Middle East. In P. Zgheib (Ed.), *Entrepreneurship and Business Innovation in the Middle East* (pp. 37–56). Hershey, PA: IGI Global. doi:10.4018/978-1-5225-2066-5.ch003

About the Contributors

Ishmael Mensah is an Associate Professor of tourism and hospitality management at the University of Cape Coast and Ghana Director of the Confucius Institute at the University of Cape Coast. He holds a PhD Tourism degree from the same university and is a Certified Hospitality Educator (CHE) by the American Hotel and Lodging Association as well as a Member of the Institute of Hospitality (MIH). His research interests include environmental management in tourism and hospitality, service quality, destination marketing, community participation in tourism development, career development in tourism and events management.

Ewoenam Afenyo-Agbe is a Lecturer at the Department of Hospitality and Tourism Management of the University of Cape Coast, Ghana. She holds a PhD in Tourism Management from the University of Cape Coast, Ghana. Her teaching and research interests are in community tourism development, tourism entrepreneurship, gender and livelihood issues in tourism. She is also the founder of 'The Sustainable Tourism Initiatives' Hub – Ghana; a non-governmental organization aimed at supporting sustainable rural tourism initiatives.

* * *

Victor Mawutor Agbo is a PhD candidate in the Department of Recreation and Leisure Studies at the University of Waterloo, Ontario, Canada. He holds a Master of Environmental Studies (MES) in Geography from the Department of Geography and Environmental Management, University of Waterloo, and a Bachelor of Science (BSc.) in Tourism Management from the University of Cape Coast, Ghana. Victor's area of interest is within the intersection of community-based ecotourism development, human-nature relationships, tourism ethics and justice, and critical theories. His current work focusses on using critical theories to critique and challenge traditional forms of tourism that have had significant adverse impacts on communities and nature, and to advocate for just and sustainable forms of tourism development for communities that enable tourism.

Grace Ampaire is a student the Department of Forestry, Biodiversity and Tourism at Makerere University with research interests in community tourism development. She holds a Bachelors degree in Tourism Management from Makerere university and with interests to pursue graduate studies in community-based tourism and conservation planning and management.

Ahmad Arslan is currently working as an Associate Professor at Department of Marketing, Management and International Business, Oulu Business School, University of Oulu, Finland. He also holds the position of Honorary Chair in Business Management at the University of Aberdeen, Scotland, UK. Previously, he has worked in academia in different universities in the UK and Finland. His earlier research has been published in prestigious academic journals like British Journal of Management, International Business Review, International Marketing Review, International Journal of Organizational Analysis, Journal of Business Research, Scandinavian Journal of Management, Journal of Strategic Marketing, Journal of Knowledge Management, Journal of Organizational Change Management, Production Planning & Control, and Technological Forecasting and Social Change among others. Moreover, he has also contributed book chapters to several edited handbooks addressing different management related topics. Finally, he holds several editorial board memberships and is currently a Senior Editor of International Journal of Emerging Markets (Emerald).

Jim Ayorekire is a Senior Lecturer and researcher in the Department of Forestry, Biodiversity and Tourism at Makerere University with interest in tourism planning and policy analysis, tourism destination management and sustainable regional tourism development. He holds a PhD in Tourism and has over the years conducted studies interrogating policy, planning and institutional practices and how they influence sustainable tourism destination development. As a consulting researcher he has contributed to the formulation of several tourism plans, policies and institutional frameworks across Eastern Africa.

Saranjam Baig is currently working as an Assistant Professor at the College of Economics and Political Science, Sultan Qaboos University, Muscat, Oman. Previously, he has worked in academia in different universities in the US, Pakistan and Oman. His earlier research has been published in academic journals like Current Issues in Tourism, Networks and Spatial Economics, Sustainability among others. He also contributed a book chapter published by Oxford University Press.

Ama Amponsaa Birch Freeman received her MPhil in Tourism Management from the University of Cape Coast in 2015. She has over 15 years' experience in the maritime industry and headed the branch of a shorehandling company for 5 years. Currently, she is an Assistant Lecturer at the Takoradi Technical University in Ghana and has written three dissertations in the field of tourism. She is a Commonwealth Scholar who was awarded the scholarship to study M.Sc Tourism Management and Marketing at the University of Nottingham in 2004. In 2005 she become an associate member of the Chartered Institute of Marketing, UK which awarded her Chartered Marketer status in 2020 and has also been a Chartered Member of the Chartered Institute of Marketing, Ghana since 2012.

Olukemi Deborah Fagbolu has a PhD in Hospitality and Tourism Management from the Assumption University, Thailand. Currently, she is working as a Lecturer I in the department of Tourism Studies, National Open University of Nigeria.

Francis Mugizi is a Lecturer in the Department of Forestry Biodiversity and Tourism at Makerere University. He has research interest in community tourism enterprise development and livelihoods and tourism destination management. He holds a PhD in Environment and Natural Resources focusing on tourism as a driver to community development. He has participated in several engagements that facilitate tourism and community development at a destination and national levels.

Norbert Ngowi is a Natural Resources Manager by training. He has over 20 years of practice in the field of conservation, teaching, and research experience in Biodiversity, Ecosystem Services, Livelihoods, and Natural Resources Management. He received his training both at undergraduate and postgraduate levels from the University of Dar es Salaam, Addis Ababa University, and Griffith University in Australia. Currently, he works as a Lecturer at the Centre for Environment, Poverty & Sustainable Development of the Institute of Development Studies, Mzumbe University. Dr. Ngowi also serves as an investigator for the Wood Charcoal Wastes Project under VLIR-OUS (4SiTe) Programme. His research interests include Biodiversity Conservation, Ecosystem Services, Public Health, Livelihoods, Environment, and Natural Resources Management.

Singhanat Nomnian is an Associate Professor in an MA program (Language and Intercultural Communication, Language Teaching Major) at Research Institute for Languages and Cultures of Asia, Mahidol University, Thailand. He obtained his EdD (TESOL and Applied Linguistics) from the University of Leicester, UK. Awarded Endeavour Postdoctoral Fellowship from the Australian Government, he was a research fellow at the School of International Studies and Education at the University of Technology Sydney (UTS), Australia. He is currently the President of Thailand TESOL and the editor-in-chief of THAITESOL Journal.

Joseph Obua is Professor in the Department of Forestry, Biodiversity and Tourism at Makerere University with research interest in ecotourism and forest conservation. He teaches and supervises undergraduate and postgraduate research in ecotourism and related areas. He has published over 70 papers and book chapters on topics in tourism, ecotourism and forest conservation. He has participated in initiatives to establish a regional tourism council for IGAD. He is a member of the Evidence Leadership in Africa championing use of research evidence in decision and policy making. He is Fellow of the Uganda National Academy of Sciences and Fellow of the African Academy of Sciences.

Narongdej Phanthaphoommee is a lecturer in an MA program (Language and Intercultural Communication, Translation Major) at Research Institute for Languages and Cultures of Asia, Mahidol University. He obtained his PhD in Translation Studies from the University of Leeds, UK.

Arifa Shabbnum has been working with the Center for Research on China-Pakistan Economic Corridor as a research fellow. Currently, she is a Fulbright Fellow and pursuing a PhD in Economics from Colorado State University, Colorado USA.

Index

A

C

D

E

G

I

J

www.ingramcontent.com/pod-product-compliance
Lightning Source LLC
LaVergne TN
LVHW061547290926
849162LV00004B/931
* 9 7 8 1 7 9 9 8 7 3 3 5 8 *